Church, Faith and Culture in the Medieval West
General Editor: Brenda Bolton

About the series

The series *Church, Faith and Culture in the Medieval West* reflects the central concerns necessary for any in-depth study of the medieval Church – greater cultural awareness and interdisciplinarity. Including both monographs and edited collections, this new series draws on the most innovative work from established and younger scholars alike, offering a balance of interests, vertically though the period from *c.* 400 to *c.* 1500 or horizontally across Latin Christendom. Topics covered range from cultural history, the monastic life, relations between Church and State to law and ritual, palaeography and textual transmission. All authors, from a wide range of disciplinary backgrounds, share a commitment to innovation, analysis and historical accuracy.

About the volume

Gateway to the Heavenly City presents a penetrating analysis of the attitudes of Latin Christendom towards Jerusalem in the period from the First Crusade to the Muslim capture of the city in 1187. Sylvia Schein starts by exploring the changes in the Western image of Jerusalem, first as the goal of the crusade, then after its conquest. She examines the theories used to justify the conquest and rule of the Holy City and the attitudes of the papacy towards this new rival centre of sanctity.

Subsequent chapters describe the new character of Jerusalem's sanctity as the city of the Old and New Testaments, as the earthly gateway to the heavenly city, and in apocalyptic terms as the centre of the world and the place where the events of the end of the world would unroll. The reaction to the fall of crusader Jerusalem in 1187 is the subject of the final chapter. Based on a detailed examination of the source materials, from poetry and song to chronicles and charters, this book paints a clear picture of the place of the Earthly and the Heavenly Jerusalem in Latin Christendom.

Church, Faith and Culture in the Medieval West
General Editor: Brenda Bolton

Other titles in the series:

Gillian R. Knight
The Correspondence between Peter the Venerable and Bernard of Clairvaux
A Semantic and Structural Analysis

Michelle Still
The Abbot and the Rule
Religious Life at St Albans, 1290–1349

Edited by Brenda Bolton and Anne J. Duggan
Adrian IV The English Pope (1154–1159)
Studies and Texts

Richard Kay
The Council of Bourges, 1225
A Documentary History

Damian J. Smith
Innocent III and the Crown of Aragon
The Limits of Papal Authority

Anke Holdenried
The Sibyl and Her Scribes
Manuscripts and Interpretation of the Latin *Sibylla Tiburtina c.* 1050–1500

Edited by Kathleen G. Cushing and Richard F. Gyug
Ritual, Text and Law
Studies in Medieval Canon Law and Liturgy Presented to Roger E. Reynolds

Edited by Anne J. Duggan, Joan Greatrex and Brenda Bolton
Omnia disce – Medieval Studies in Memory of Leonard Boyle, O.P.

GATEWAY TO THE HEAVENLY CITY: CRUSADER JERUSALEM AND THE CATHOLIC WEST (1099–1187)

To Joshua Prawer, a temporary citizen of the Earthly Jerusalem

Gateway to the Heavenly City: Crusader Jerusalem and the Catholic West (1099–1187)

SYLVIA SCHEIN

Routledge
Taylor & Francis Group
LONDON AND NEW YORK

First published 2005 by Ashgate Publishing

Published 2016 by Routledge
2 Park Square, Milton Park, Abingdon, Oxon OX14 4RN
711 Third Avenue, New York, NY 10017, USA

Routledge is an imprint of the Taylor & Francis Group, an informa business

British Library Cataloguing in Publication Data
Schein, Sylvia
Gateway to the Heavenly City: Crusader Jerusalem and the Catholic West (1099–1187). – (Church, Faith and Culture in the Medieval West).
1. Jerusalem in Christianity—History of doctrines—Middle Ages, 600–1500. 2. Jerusalem in the Bible. 3. Church history—11th century. 4. Church history—12th century. 5. Crusades—Philosophy. 6. Jerusalem—History—Latin Kingdom, 1099–1244. I. Title.
956.9'442'03

US Library of Congress Cataloging in Publication Data
Gateway to the Heavenly City: Crusader Jerusalem and the Catholic West (1099–1187) / Sylvia Schein.
p. cm. – (Church, Faith and Culture in the Medieval West)
Includes bibliographical references and index.
1. Jerusalem—History—Latin Kingdom, 1099–1244. 2. Jerusalem—History—Religious aspects—Public opinion. 3. Public opinion—Europe. 4. Christians—Europe—Attitudes. I. Title. II. Series.
D183.S35 2003
956.94'4203–dc21 2003052117

ISBN 13: 978-0-7546-0649-9 (hbk)

Contents

A Room of Her Own: Sylvia Schein's Contribution to Medieval History vii
Acknowledgements x
Foreword xi
Abbreviations xiv

Introduction 1

1 Jerusalem: Goal of the First Crusade 9

2 The Conquest – A Divine Act 21

3 'Inheritance of the Lord': Justifications of Christian Rule in Jerusalem 35

4 Rome, Babylon and Jerusalem: Papal Attitudes to Jerusalem 49

5 From the 'City of the Holy Sepulchre' to the 'City of the Humanity of Christ' 63

6 The City of the Old and the New Testaments 91

7 Jerusalem in the Believer's Plan of Salvation 109

8 Jerusalem – Centre of the World and Scene of the Last Days 141

9 'The Terrible News': The Reaction of Christendom to the Fall of Jerusalem (1187) 159

Conclusion 189

Bibliography 195
Index 223

Sylvia Schein

A Room of Her Own: Sylvia Schein's Contribution to Medieval History

Rooms have voices. On entering Sylvia's home after her funeral, I found it as she left it, beautifully adorned with carefully chosen art and memorabilia. In the middle of her living room, near a large window overlooking Mount Carmel, stood her desk, an academic sanctum surrounded by books and covered with the papers she was working on when her untimely death deprived us of a historian whose work constituted the centre of her life. Lying on the desk were her books: *Fideles crucis* (1991), which had just appeared in Italian translation, a map of medieval Jerusalem that she was proof-reading for this volume, and the draft of a third, still unnamed, book on medieval women. Three books – three milestones on Sylvia's academic path. On the corner of her desk, as if overseeing her work, stood a large photo of Joshua Prawer, Sylvia's teacher and friend. Whether or not they concur with his theses, all Israeli crusader historians are indebted to Prawer's monumental study of the Latin Kingdom of Jerusalem. Sylvia's academic career started under his auspices and he cast a long shadow over her early path, a path that ended in this room of her own where Sylvia's authentic and authoritative voice resounded clearly.

After completing her Master's thesis with Prawer, Sylvia wrote her doctoral dissertation at Cambridge under the direction of Otto Smail. Her first articles, published between 1979 and 1982, which naturally stemmed from this thesis, highlighted European plans – from 1274 to the fourteenth century – to recover the Holy Land. These articles were concerned more with the longing for Jerusalem than with the city's actual history. Sylvia's interests lay primarily in the spiritual concept of Jerusalem as well as in the actual, though partly utopian, plans to re-conquer Jerusalem and to found a new crusader kingdom. Sylvia transformed these very detailed and down-to-earth blueprints into a lesson in political theory. The good advice for rebuilding Jerusalem constituted in fact a kind of workshop for the ideal state, not only correcting the mistakes which had contributed to the disintegration of the crusader kingdom, but also suggesting a healthier foundation for its successor which, in contrast to the historical kingdom, would be ruled by faith and unity – thus enabling it to outlive its enemies. Sylvia emphasized that, although these plans were never in fact implemented, they were devised by those who firmly believed in the kingdom's future resurrection. One of Sylvia's early original contributions to the field was her emphasis on the Mongols' role in those future plans of conquest. Her article, '*Gesta Dei per Mongolos* 1300: The genesis of a non-event' (*English Historical Review*, 94 [1979], 805–19), highlighted Christian hopes for a Mongol invasion of the Holy Land as a means of restoring Frankish rule. A decade later her vision of the wars in the Middle East as an encounter between civilizations had become a commonplace and Mongol history moved to centre stage. She saw the Mongolian ascent not just as altering the tempo of events, but rather as fostering a change in mentality, one that led to a profound shift in thirteenth-century historiography.

Although undoubtedly initiated under Prawer's influence, Sylvia's article 'The Temple between Mount Moriah and the Holy Sepulchre: The Changing Traditions of the Temple in the Middle Ages', *Traditio*, 50 (1984), 175–95 represents another original contribution to Holy Land studies. Sylvia, here, continued in Prawer's footsteps, but set out on a new path of her own. She argued that the concept of Earthly Jerusalem underwent radical change in the crusader age, and that the crusaders saw themselves as the heirs to biblical Israel. Biblical allusions to the Temple Mount, discredited or relocated to the Holy Sepulchre in the Byzantine and Arab periods, flourished in the Latin Kingdom. Earthly Jerusalem, with its hostels and power struggles, blended with the heavenly city, longed for by pilgrims. Sacred traditions once more began to be associated with the Holy Mount, creating a new bone of contention between Christians, Muslims and Jews. Of all her work, this article is perhaps the most often cited. Her new book, *Gateway to the Heavenly City*, takes this fresh path much further, showing the centrality of both Jerusalems, the earthly and the heavenly, to every aspect of the medieval Christian outlook, whether holy places or eschatological concepts.

Nor was this to be her final word on the subject. Several days after her death I received a ninety-page Hebrew manuscript, 'Jerusalem as the Stage for Christ's Suffering in the Late Medieval Period', scheduled to appear in the Yad Ben-Zvi series on the history of Jerusalem. This chapter demonstrates that the fall of the Latin Kingdom had repercussions not only for political realities but also for spiritual conceptions of Jerusalem. In one sense, like the Hassidic Rabbi, Nachman of Braslav, one can say that in her research Sylvia was always on her way to Jerusalem.

She did not, however, write exclusively about dreams and spiritual concepts. Sylvia was at the same time acutely interested in those people on the byways she pursued – women in particular. She wrote about transvestites and female pilgrims, some holy, others more profane, exemplified by such figures as Brigitta of Sweden and Margery Kempe. Their *itineraria* and their voices were to become a central theme in her later articles. She also researched the legal status of women in the Latin Kingdom and her articles, 'Women in the Latin Kingdom of Jerusalem' in *Gendering the Crusades*, ed. Susan Edgington and Sarah Lambert (Cardiff, 2001), 140–53, and in the Amnon Linder *Festschrift* employ sources from thirteenth-century assizes to demonstrate that women's status in the East was in fact better than in their countries of origin ('*Service de Marriage* and Law Enforcement in the

Latin Kingdom of Jerusalem', *De Sion exibit lex et verbum domini de Hierusalem: Essays on Medieval Law, Liturgy, and Literature in Honour of Amnon Linder*, ed. Y. Hen, Turnhout: Brepols, 2001, 71–80). In 1999 she published an article about the crusader queens entitled 'Rulers and Ruled: Women in the Crusader Period', *Knights of the Holy Land*, ed. Silvia Rozenberg, Jerusalem: Israel Museum, 1999, 60–67 a topic that has since attracted much attention. Sylvia did not believe one could understand medieval women simply by defining their legal status; in order to paint a broader picture, she tried to listen to women's voices by probing literature written by, or for, women.

Sylvia demonstrated a keen interest in encouraging young researchers and always tutored her students generously and conscientiously, including those attached to other universities. When she organized a conference in Haifa about the Latin Kingdom, Sylvia made a point of inviting unknown young researchers to speak alongside eminent scholars. She loved conferences, participating in each and every one of the meetings of the Society for the Study of the Crusades and the Latin East (SSCLE) and attending other international colloquia. Nor was her audience restricted to crusader scholars; she was an active member of the Ecclesiastical Historical Society and of the Society of Medieval Historiography, among others.

We can only hope that the third book on her table, the draft of a history of medieval women, may eventually be published. Notwithstanding her strong interest in women, she was not a militant feminist. In writing about medieval women, Sylvia did not use them as a weapon in modern social debate. She already had a room of her own; fiercely independent, she fought her way alone. She should be remembered not only for her outstanding scholarship, but also for her unique personality, which found expression in her varied historical interests.

Yvonne Friedman

Acknowledgements

The Publishers and Moshe Schein wish to express their thanks to Professor Michael Goodich and to Dr Brenda Bolton for their invaluable help in seeing this book through to publication.

Foreword

This book was written in Munich, Haifa and Jerusalem. My first debt of gratitude is to members and staff of the *Monumenta Germaniae Historica* as well as to the Alexander von Humboldt Stiftung which enabled me to work for a year at the *Monumenta* and other libraries in Europe. I should like also to thank all those who gave me advice and encouragement, in particular the staff of Haifa University Library, as well as Professors A. Grabois, A. Paschovsky and S. Menache for their criticisms and suggestions. The late Dr R.C. Smail was until his death a constant source of support and encouragement.

This book is dedicated to someone to whom I owe more than I can say. His lectures, studies and tours of the Old City of Jerusalem first revealed to me the true Jerusalem and thus inspired this book.

Sylvia Schein

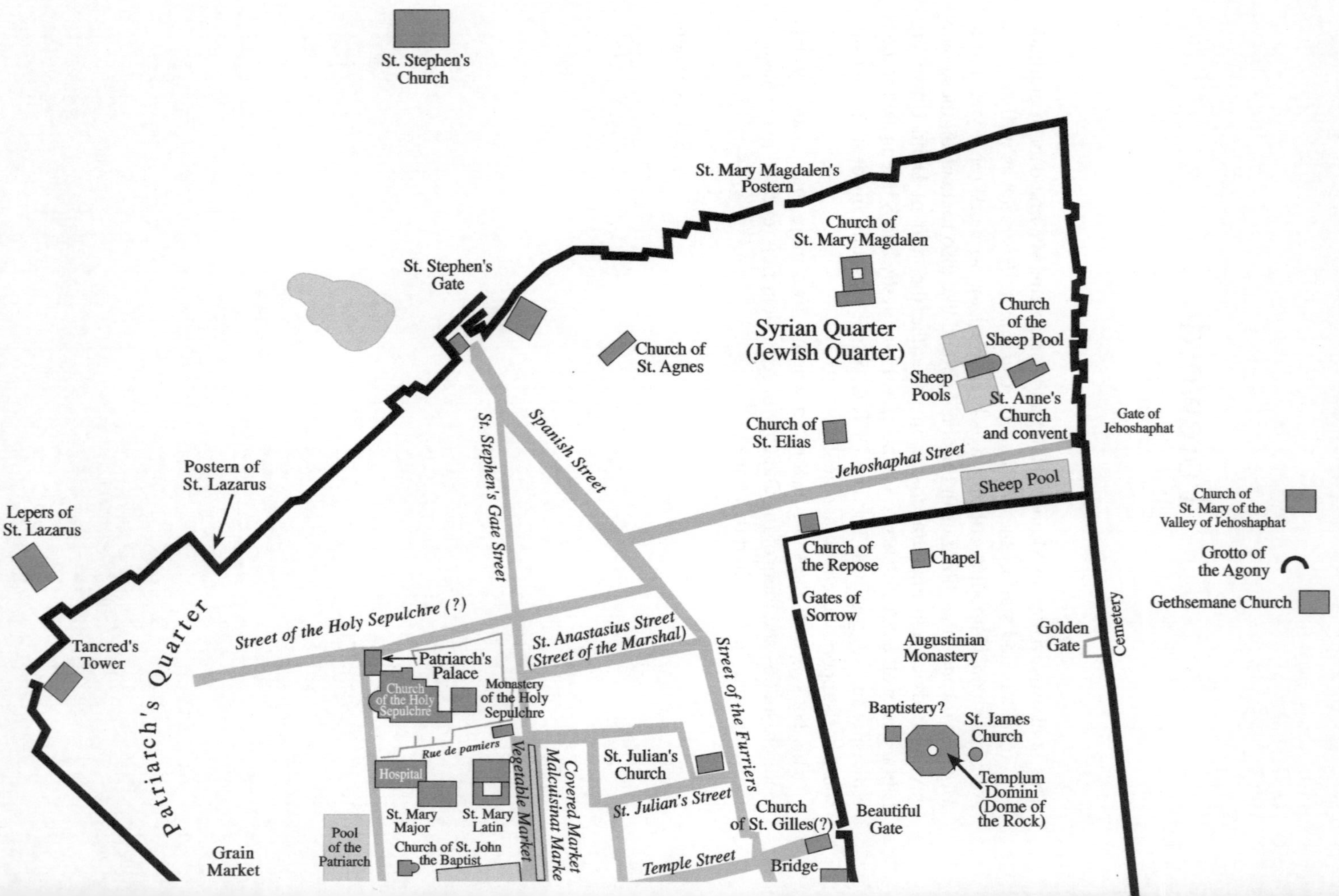
St. Stephen's Church
St. Mary Magdalen's Postern
Church of St. Mary Magdalen
St. Stephen's Gate
Syrian Quarter (Jewish Quarter)
Church of the Sheep Pool
Church of St. Agnes
Sheep Pools
St. Anne's Church and convent
Church of St. Elias
Gate of Jehoshaphat
Spanish Street
St. Stephen's Gate Street
Jehoshaphat Street
Sheep Pool
Postern of St. Lazarus
Lepers of St. Lazarus
Church of St. Mary of the Valley of Jehoshaphat
Church of the Repose
Chapel
Grotto of the Agony
Gethsemane Church
Gates of Sorrow
Street of the Holy Sepulchre (?)
St. Anastasius Street (Street of the Marshal)
Street of the Furriers
Tancred's Tower
Patriarch's Quarter
Patriarch's Palace
Church of the Holy Sepulchre
Monastery of the Holy Sepulchre
Augustinian Monastery
Golden Gate
Cemetery
Baptistery?
St. James Church
Templum Domini (Dome of the Rock)
Rue de pamiers
Vegetable Market
Covered Market
Malcuisinat Marke
St. Julian's Church
St. Julian's Street
Hospital
St. Mary Major
St. Mary Latin
Church of St. Gilles(?)
Beautiful Gate
Pool of the Patriarch
Grain Market
Church of St. John the Baptist
Temple Street
Bridge

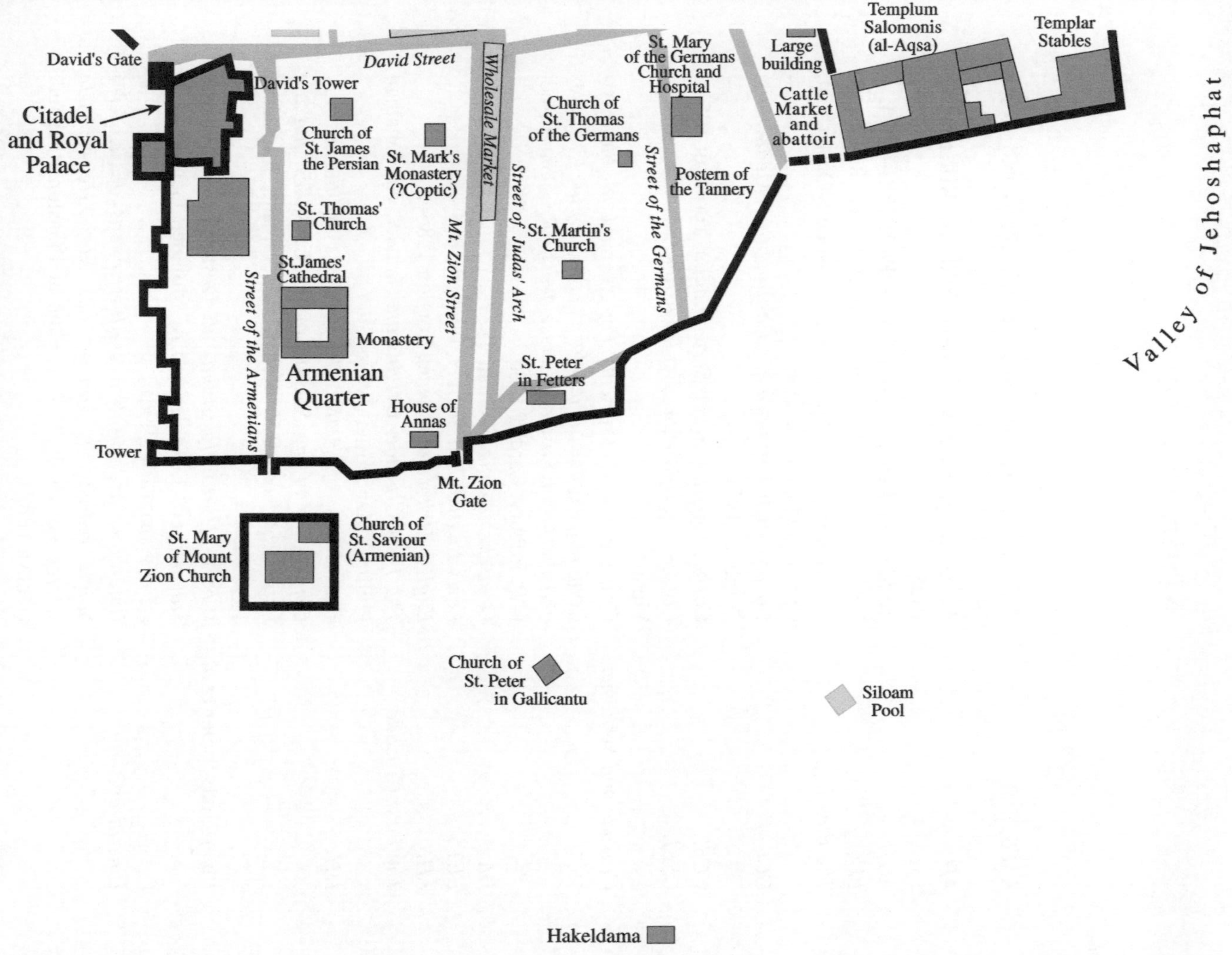
David's Gate
Citadel and Royal Palace
David's Tower
David Street
Church of St. James the Persian
St. Mark's Monastery (?Coptic)
Wholesale Market
St. Thomas' Church
St.James' Cathedral
Monastery
Armenian Quarter
Street of the Armenians
Mt. Zion Street
Street of Judas' Arch
House of Annas
Tower
Mt. Zion Gate
St. Peter in Fetters
St. Martin's Church
Church of St. Thomas of the Germans
Street of the Germans
St. Mary of the Germans Church and Hospital
Postern of the Tannery
Large building
Cattle Market and abattoir
Templum Salomonis (al-Aqsa)
Templar Stables
Valley of Jehoshaphat
St. Mary of Mount Zion Church
Church of St. Saviour (Armenian)
Church of St. Peter in Gallicantu
Siloam Pool
Hakeldama

Abbreviations

AASS	*Acta sanctorum*, 3rd edn, 62 vols, Paris and Rome 1863–1970.
AH	*Analecta hymnica medii aevi*, XXI, XXXII, Leipzig 1895–1899.
AOL	*Archives de l'Orient Latin*, 2 vols, Paris 1881–1884.
BEC	*Bibliothèque de l'Ecole des Chartes*, Paris 1839–.
The Blessings	*The Blessings of Pilgrimage*, ed. Robert Ousterhout, Urbana and Chicago 1990.
CCCM	*Corpus Christianorum. Continuatio Medievalis*, Turnhout 1951–.
CCSL	*Corpus Christianorum. Series Latina*, Turnhout 1953.
CSEL	*Corpus Scriptorum Ecclesiasticorum Latinorum*, Vienna 1866.
Crusade and Settlement	*Crusade and Settlement. Papers read at the First Conference of the Society for the Study of the Crusades and the Latin East and Presented to R.C. Smail*, ed. P.W. Edbury, Cardiff 1985.
DA	*Deutsches Archiv*, Weimar 1937–.
EETS	*Early English Text Society*, London 1864–.
EHR	*English Historical Review*, London 1886–.
The First Crusade	*The First Crusade. Origins and Impact*, ed. Jonathan Phillips, Manchester 1997.
HH	*The Horns of Hattin. Proceedings of the Second Conference of the Society of the Crusades and the Latin East*, ed. B.Z. Kedar, London and Jerusalem 1992.
Itinera, ed. Tobler *et al.*	*Itinera Hierosolymitana et Descriptiones Terrae Sanctae*, ed. T. Tobler and A. Molinier, Geneva 1879 (repr. Osnabrück 1996).
Itinéraires	*Itinéraires à Jérusalem et descriptions de la Terre Sainte rédigés en français aux XIe, XIIe et XIIIe siècles*, ed. Henry V. Michelant and Gaston Raynaud, Geneva 1882.
JEH	*Journal of Ecclesiastical History*, Cambridge 1950.
MGH.Ldt	*MGH. Libelli de lite*, 3 vols, Hanover 1891–1897.

MGH.Scr.rer.Germ.	*MGH. Scriptores rerum Germanicarum in usum scolarum*, Hanover 1839.
MGH.SS	*Monumenta Germaniae Historica Scriptores*, ed. G.H. Pertz *et al.*, 32 vols, Hanover, Weimar, Stuttgart, Cologne and Munich, 1826–1934.
Oriens Pontificius I	*Papsturkunden für Templer und Johanniter. Archivberichte und Texte*, ed. Rudolf Hiestand, *Vorarbeiten zum Oriens Pontificius* I, Göttingen 1972.
Oriens Pontificius II	*Papsturkunden für Templar und Johanniter. Neue Folge*, ed. Rudolf Hiestand, *Vorarbeiten zum Oriens Pontificius* II, Göttingen 1984.
Oriens Pontificius III	*Papsturkunden für Kirchen im Heiligen Lande*, ed. Rudolf Hiestand, *Vorarbeiten zum Oriens Pontificius* III, Göttingen 1985.
Outremer	*Outremer. Studies in the History of the Crusade Kingdom of Jerusalem Presented to Joshua Prawer*, ed. B.Z. Kedar *et al.*, Jerusalem 1982.
Peregrinationes Tres	*Peregrinationes Tres. Saewulf, John of Würzburg, Theodoricus*, ed. R.B.C. Huygens with a study of the voyages of Saewulf by John H. Pryor, *CCCM*, 139.
PG	*Patrologia Graeca*, ed. J.P. Migne, 167 vols, Paris 1857–1912.
PL	*Patrologia Latina*, ed. J.P. Migne, 217 vols plus 4 vols indexes, Paris 1844–1864.
PPTS	*Palestine Pilgrims' Text Society*, 13 vols and index, London 1896–1907 (repr. New York 1971).
RBPH	*Revue Belge de Philologie et d'Histoire*, Brussels 1992.
RHC.Hocc	*Recueil des Historiens des Croisades. Historiens Occidentaux*, 5 vols, Paris 1844–1895.
RHC. Hocc	*Les Assises de Jérusalem*, 2 vols, Paris 1841–1843.
RHGF	*Recueil des Historiens des Gaules et de la France*, ed. M. Bouquet *et al.*, 24 vols, Paris 1737–1904.
RIS	*Rerum Italicarum Scriptores*, ed. L.A. Muratori, 25 vols, Milano 1723–1751.
RIS.NS	*Rerum Italicarum Scriptores, NS*, ed. G. Carducci *et al.*, Città di Castello and Bologna 1990–.
ROL	*Revue de l'Orient Latin*, 12 vols, 1893–1911.
RRH	*Regesta regni Hierosolimitani*, ed. R. Röhricht, Innsbruck 1893; *Additamentum*, 1904.
RS	(*Rolls Series*) *Rerum Britannicarum Medii Aevi Scriptores*, 99 vols, London 1858–1911.
W.T.	*Willelmi Tyrensis Archiepiscopi Chronicon*, ed. R.B.C. Huygens, *CCCM*, 63–63A.

Hail then, holy city, sanctified by the Most High for his own tabernacle in order that such a generation might be saved in and through you! Hail, city of the Great King, source of so many joyous and unheard of marvels! Hail mistress of nations and queen of provinces, heritage of patriarchs, mother of apostles and prophets, source of the faith and glory of the Christian people! … Hail promised land, source of milk and honey for your ancient inheritants, now become the source of healing grace and vital sustenance for the whole earth!

Bernard of Clairvaux, *De Laude Novae Militiae*, cap. V.

Introduction

During the first millennium of the Christian era, Jerusalem suffered many upheavals. Ruined and its temple burned in AD 70, it continued as a Jewish city until 132 when a second revolt of the Jews was launched against the Romans. Once this revolt was quelled, Caesar Aelius Hadrian refounded Jerusalem as a Roman city with a new name – Colonia Aelia Capitolina. It was the embracing of Christianity by Emperor Constantine I (315) that turned Aelia Capitolina, almost overnight, from a provincial city in the Roman Empire into one of the most important in the new Christian Empire and as a place sanctified by the presence of Christ himself. From that time onwards, public and state resources were mobilized to rebuild Jerusalem as a Christian city. As a result, during the Byzantine period (325–638), Jerusalem, with its beautiful new churches and the complex of Constantine's great Church of the Holy Sepulchre at its very heart, became a major pilgrimage centre for pilgrims from both the Byzantine Empire and the Christian West. When the Moslems conquered the city in 628 (to remain under their rule until the crusader conquest in 1099), this disaster was seen as one of the events of the Last Days.

From the very beginning, Jerusalem held a unique position in Christian spirituality. During the first millennium of its existence, Christianity, caught on the horns of the dilemma of the Heavenly versus the Earthly City, invested the term 'Jerusalem' with a great variety of meanings.[1] It was the Holy City of the historical Israel and central to God's providential care for His people. Its pivotal position in sacred history, symbolized by the history of David, Solomon and the Temple, was fulfilled through the Life and Passion of Christ. It was in Jerusalem that Christ had preached, suffered, died, was buried and rose again from the dead; its holy places preserved the memory of his presence and made it a Holy City, the very existence of which inspired material contact with that sacred history.[2] Since the key events of sacred history had taken place in Jerusalem, the city was

---

1 J. Prawer, 'Jerusalem in the Christian and Jewish Perspectives of the Early Middle Ages', *Settimane di studio del centro Italiano di studi sull'alto medioevo. XXVI. Gli Ebrei nell'Alto Medioevo*, Spoleto 1980, pp. 739–82 and below pp. 141–58.

2 For example, St Jerome claimed that material contact with the holy places of Jerusalem generates a better understanding of the Holy Scriptures: 'As those who saw Athens will better understand Greek histories ... so a clearer understanding of Scripture for whoever will see Judaea with his own eyes.' Hieronymus, *Praefatio in librum paralipomenon*, *PL*, 29, col. 401.

considered to be the head of all churches, the mother of Christians, the cosmological centre of the world – its so-called navel – as well as the focus of the events of the Last Days, the place of the Second Coming and the Last Judgement. Moreover, the term 'Jerusalem' derived special significance from the mystical concept of the Heavenly Jerusalem that so dominated Christian literature. On the basis of Paul's epistles and the Book of Revelation, the Earthly Jerusalem was seen as a figure or likeness of Heavenly Jerusalem. The Heavenly City of Jerusalem, with its gates of sapphire, its walls and squares bright with gold and precious stones, as described in the Book of Revelation (21:2–26) and in Tobit (13:20–23), was considered a meeting place for those who had been scattered. It was the goal of the great pilgrimage of peoples (Isaiah 2:2; Tobit 13:14), the place where God resided among His people, the resting place of the righteous, the city of paradise.

Yet there was a paradox for, at the same time, the Earthly Jerusalem in the mountains of Judaea was also the site of Christ's Passion, the city of the Jews, the slayers of Christ, and therefore a place to be despised. As such, it was considered an antithesis to the Heavenly Jerusalem, which was the true country (*patria*) of all Christians. As a result, the prevalent image of Jerusalem was often a vague, hazy mixture of the sacred and the profane, at one and the same time the site of Christ's earthly Passion, his Holy Sepulchre, the figure of the Heavenly Jerusalem; simultaneously it was the *Civitas Dei* and the *Civitas Perfida.*[3]

This ambiguous attitude to the Earthly Jerusalem evolved basically from its portrayal in Holy Scripture.[4] Declaring itself heir to the Old Testament, Christianity had inherited from Scripture three notions of Jerusalem, three different cities, representing different stages of history; and it had also inherited the tense and highly controversial relationship between these three cities: 'Earthly Jerusalem', 'Future Jerusalem' and 'Heavenly Jerusalem'.

The Earthly Jerusalem of the Old Testament was the glorified, eternal capital of the Children of Israel, the city of Solomon's temple. In the Gospels, this Jerusalem became the city of Christ's life, Passion and Resurrection. According to Christ, it was 'the place where one ought to worship' (John 4:20). Paul, whose intention after the Passion was to go up to Jerusalem to celebrate the Pentecost there, addressed those who sought to prevent him, saying: 'What are you doing, weeping and breaking my heart? For I am ready not only to be imprisoned but even to die at Jerusalem for the name of the Lord Jesus' (Acts 21:13). This basically positive

---

3 See below, n. 7.

4 For Jerusalem in the Scriptures, see K. Schmid, 'Aachen und Jerusalem: Ein Beitrag zur historischen Personenforschung der Karolingerzeit', *Das Einhardkreuz. Vorträge und Studien der Münsteraner Diskussion zum arcus Einhardi*, ed. K. Hauck, Göttingen 1974, pp. 207–48; Bianca Kühnel, *From the Earthly to the Heavenly Jerusalem: Representations of the Holy City in Christian Art of the First Millennium*, Rome–Freiburg–Vienna 1987 (*Römische Quartalschrift für Christliche Altertumskunde und Kirchengeschichte*, 42, *Supplementheft*), pp. 17–59.

attitude of the New Testament to Earthly Jerusalem by no means, however, dominated. Running counter to it we find a completely different attitude, involving the complete rejection of the earthly city and its replacement by Heavenly Jerusalem. Earthly Jerusalem, as the place of Christ's passion, was the city which 'did not know the time of [Christ's] visitation' (Luke 19:44), the city which had 'killed the prophets and stoned those who are sent to her' (Matthew 23:37). As such it became also the subject of Christ's curse or prophecy: 'For the days shall come upon you, when your enemies will cast up a bank about you and surround you, and hem you in on every side, and dash you to the ground, you and your children within you, and they will not leave one stone upon another in you; because you did not know the time of your visitation' (Luke 19:43–4).

Four times in the Gospels Christ repeats the saying: 'I am going to destroy this temple made by human hands and in three days build another not made by human hands' (Mark, 14:58; 15:29; Matthew 27:40; John 2:19). Thus it follows that with Christ's Death, Resurrection and Ascension, Earthly Jerusalem forfeited its validity in the history of redemption; God had removed his blessing from the city (Matthew 23:38–9) and the disciples were told to flee from it (Luke 21:21–2). In the epistles of Paul, Earthly Jerusalem serves not only as a symbol of the Old Testament but also the place in which the New Testament would be implemented and fulfilled. According to the author of the Epistle to the Hebrews: 'Here we have no lasting city but we seek the city which is to come' (Hebrews 13:14); or 'But what you have come to is Mount Zion and the city of the living God, the Heavenly Jerusalem where the millions of angels have gathered for the festival, with the whole church in which everyone is a first born son and a citizen of heaven' (*ibid.*, 12:22–3).

Earthly Zion, the meeting place for the tribes of old Israel (Psalms 122:3–4) was being replaced by Heavenly Zion as a meeting place for the new Israel (Revelation 14:1). Like Abraham the Christian lives in this world, where there is no abiding city, like a foreigner in a strange country, looking forward 'to a city founded, designed and built by God' (Hebrews 11:8–10). The superiority of this heavenly city is typologically emphasized by Paul in his Letter to the Galatians 4:24–6 through its confrontation with the historical and earthly Jewish Jerusalem. The promised, expected, and at the same time already existing city is Heavenly Jerusalem, linked to but also separated from Earthly Jerusalem by the Cross of Christ. This denial of the Earthly Jerusalem and its antithetical relation to the heavenly city is clearly presented in Paul's Letter: 'Now this is an allegory: these women are two covenants. One is from Mount Sinai, bearing children for slavery; she is Hagar. Now Hagar is Mount Sinai in Arabia, she corresponds to the present Jerusalem, for she is in slavery with her children. But the Jerusalem above is free and she is our mother.' This Heavenly Jerusalem, as visualized and described in the Book of Revelation, is superior to its earthly counterpart; it will descend from on high to replace the latter after its disappearance (Revelation 21:1). It represents the Christian realization of Isaiah's prophecy: '... I create new heavens and a new earth, and the former things shall not be remembered or come into mind ... For behold, I create Jerusalem a rejoicing' (Isaiah 65:17–18). It is the place where God dwells in complete and eternal fellowship with those who have enrolled as its citizens and who have their names inscribed in the book of life: 'The throne of God and of the Lamb will be in its place in the city; his servants will worship him, they will see him face to face' (Revelation 22:3–4).

In addition to Earthly and the Heavenly Jerusalem, the Scriptures present yet another, the 'Future Jerusalem', at one and the same time a celestial and terrestrial city, given by God to his people in compensation for the old one. According to Baruch, 'the future Jerusalem' will be reconstructed once in the Messianic age; thereafter it will be perfected for ever, in glory, when God renews His creation (the 'Heavenly Jerusalem') (Baruch 32:2–6). According to Zechariah: 'Thus says the Lord: I will return to Zion and will dwell in the midst of Jerusalem and Jerusalem shall be called the faithful city, and the mountain of the Lord of Hosts, the holy mountain' (Zechariah 8:3). Both Isaiah and Tobit (13:16–17) describe the future Jerusalem in terms similar to the description of Heavenly Jerusalem in Revelation (21:9–21). It will be built with precious stones: 'O afflicted one, storm-tossed and not comforted, behold, I will set your stone in antimony and lay your foundations with sapphire. I will make your pinnacles of agate, your gates of carbuncles and all your wall of precious stones' (Isaiah 54:11–12).

By adopting the Pauline view, which relegated Earthly Jerusalem to oblivion, the writings of the Church Fathers presented the historical city in the mountains of Judea as a place either to be despised or, at least, to be ignored. Many of the great spiritual figures of Early Christendom claimed that the centre of the Church was neither the Temple with its Holy of Holies, nor yet the New Temple built (325–35) by Constantine the Great on the site of the Lord's Passion and His Sepulchre but Christ himself. It was not the Holy City that constituted a holy space or 'area', but the new community, the body of Christ, in other words, the Church. Already Hilary of Poitiers (*c.* 315–67), almost a century before Augustine, distinguished between two churches or *ecclesiae*, both identified with Jerusalem, thus establishing the idea of Jerusalem as *ecclesia.*[5] Consequently, in the official exegesis of the period, 'Jerusalem' was historically mainly 'the city of the Jews'. Allegorically it signified the Holy Church or the Holy Soul. Anagogically it was the 'Vision of Peace' (*visio pacis*), the heavenly city whose prime characteristics were true freedom and perfect peace. And finally, in the tropological sense, it was the true home of the Christians – the Heavenly Jerusalem. This was the true Jerusalem, united with that which was in heaven; the place where the perfect Christian life was lived.

---

5 J.P. Pettorelli, 'Le thème de Sion, expression de la théologie de la rédemption dans l'oeuvre de Saint Hilaire de Poitiers', *Hilaire et son temps. Actes du Colloque de Poitiers, 29 Septembre–30 Octobre 1968*, Paris 1969, pp. 213–33.

However, this marked tendency towards what has been called a 'deterritorialization' of the concept of holiness,[6] the, so to speak 'official', relegation of Earthly Jerusalem to a subservient position vis-à-vis its celestial counterpart, the spiritualization of Jerusalem into a symbol of the Church, never became the generally accepted or uniform view of Christendom. Such ideas never expressed the general and commonplace sentiment of believers.[7] The historical facts, the life and death of Christ, created religious facts such as the Passion, Burial, Resurrection and Ascension, and combined to create holy places in Jerusalem as early as the fourth century and even before. The attraction of the holy places of Jerusalem, especially the sites of the Crucifixion and the Holy Sepulchre, was too powerful to be suppressed by learned exegesis. From the expedition of the so-called 'Bordeaux Pilgrim' (333) to the great mass pilgrimages of the eleventh century, pilgrims of all classes flocked to Jerusalem. Some, like Lethbald of Burgundy, believed that the souls of those who followed Christ in their body in order to reach His holy places would follow Him to Paradise. Others came to Jerusalem to cleanse their sins by penance, fasting and the hardships of the long and tedious journey. Most of the pilgrims believed that their reward would not only be a remission of their sins, but also the attainment of the Heavenly Jerusalem. For them, and in fact for some Europeans who never reached the Holy City, Earthly Jerusalem – no less than Heavenly Jerusalem – was the 'Holy City of God'. On a more popular level, the attempts to spiritualize Jerusalem on the one hand, and the status of the earthly city as a major centre of pilgrimage on the other, produced an image which was often a nebulous mixture of earthly topography and celestial vision. Pilgrims and exegetes of the tenth and eleventh centuries often claimed that they left for Earthly Jerusalem moved by their love of the heavenly city; they were well aware of the *Urbild–Abbild* relation between the two cities.[8]

---

6 For this concept see R.J.Z. Werblowsky, 'Jerusalem: Holy City of Three Religions', *Ex Oriente Lux*, 23 (1973–4), pp. 430–33.

7 Prawer, 'Jerusalem in Christian and Jewish Perspectives', pp. 750–82; R. Konrad, 'Das himmlische und das irdische Jerusalem im mittelalterlichen Denken', *Speculum historiale. Festschrift J. Spörl*, ed. C. Bauer *et al.*, Munich 1965, pp. 523–40; A. Bredero, 'Jerusalem in the West', *Christendom and Christianity in the Middle Ages. The Relations between Religion, Church and Society*, trans. Reinder Bruinsma, Grand Rapids, Michigan 1994, pp. 259–71; S. Mähl, 'Jerusalem in mittelalterlicher Sicht', *Die Welt als Geschichte*, 22 (1962), pp. 11–26.

8 E.D. Hunt, *Holy Land Pilgrimage in the Later Roman Empire, AD 312–460*, Oxford 1982; B. Kötting, *Peregrinatio religiosa. Die Wallfahrten in der Antike und das Pilgerwesen in der alten Kirche*, Münster 1950; J. Sumption, *Pilgrimage. An Image of Medieval Religion*, London 1975; J. Wilkinson, *Jerusalem Pilgrims before the Crusades*, Jerusalem 1977; S. Runciman, 'The Pilgrimages to Palestine before 1095', *A History of the Crusades*, 3 vols, vol. 1, Cambridge 1951–1954, pp. 68–78. For Lethbald of Burgundy see L. Lalanne, 'Des pèlerinages en Terre Sainte avant les croisades', *BEC*, 7 (1845–6), p. 11; Kenneth G. Holum, 'Hadrian and St Helena: Imperial Travel and the Origins of Christian Holy Land Pilgrimage', in *The Blessings*, pp. 66–81; P.W.L. Walker, *Holy City, Holy Places? Christian Attitudes to Jerusalem and the Holy Land in the Fourth Century*, Oxford 1990. For the

The aim of this study is to examine the place of Jerusalem in the Christian spirituality of the twelfth century. The subject has up until now been dealt with in various articles on the more general subject of Christian attitudes to Jerusalem during the Middle Ages,[9] in studies on pilgrimages[10] and in studies on the Latin Kingdom and the crusades and particularly crusade propaganda.[11] As a result, one is aware in general of a multitude of attitudes to Jerusalem through the Middle Ages, but not – except for the early Middle Ages[12] – of the fluctuations in those attitudes and the specific characteristics of various periods. For example, it has been suggested rather vaguely that a transformation took place following the crusader conquest of Jerusalem.[13] Other studies, however, trace the roots of twelfth-century attitudes to the period preceding the First Crusade.[14]

Urbild–Abbild relation between the two cities see K.L. Schmidt, 'Jerusalem als Urbild und Abbild', *Eranos*, 18 (1950), pp. 207–48.

9 See the studies of Konrad, Bredero and Mähl cited above in n. 7. See also P. Raedts, 'Jerusalem: Purpose of History or Gateway to Heaven? Apocalypticism in the First Crusade', *Church, Change and Revolution*, ed. J. van den Berg *et al.*, Leiden 1991, pp. 122–38.

10 See above, n. 8.

11 C. Erdmann, *The Origin of the Idea of Crusade*, Eng. trans. M.W. Baldwin and W. Goffart, Princeton, NJ, 1977, pp. 295–305; P. Alphandéry and A. Dupront, *La Chrétienté et l'idée de croisade*, 2 vols, Paris 1954–1959, vol. 1, pp. 18–221; E.O. Blake, 'The Formation of the Crusade Idea', *Journal of Ecclesiastical History*, 21 (1970), pp. 11–32; J. Prawer, *Histoire du Royaume Latin de Jérusalem*, 2 vols, Paris 1969–1970, vol. 1, pp. 123–36; H.E.J. Cowdrey, 'Pope Urban II's Preaching of the First Crusade', *History*, 55 (1970), pp. 177–89; A. Dupront, 'La spiritualité des croisés et des pèlerins d'après les sources de la première croisade', *Convegni del Centro di Studi sulla Spiritualità Medievale*, 4 (1963), pp. 451–83; B. McGinn, '*Iter Sancti Sepulchri:* The Piety of the First Crusades', *The Walter Prescott Webb Lectures: Essays on Medieval Civilization*, ed. R.E. Sullivan *et al.*, Arlington, Texas 1978, pp. 33–71; E. Siberry, *Criticism of Crusading 1095–1274*, Oxford 1985, pp. 37–8, 96, 82–3; H.E. Mayer, *The Crusades*, Eng. trans. J. Gillingham, 2 ed. London 1987, pp. 78–100; J. Riley-Smith, *The First Crusade and the Idea of Crusading*, London 1986, pp. 21–2; D.H. Green, *The Millstätter Exodus*, Cambridge 1966, pp. 262–70; Colin Morris, 'Picturing the Crusades: The Uses of Visual Propaganda, *c.* 1095–1250', *The Crusades and their Sources: Essays Presented to Bernard Hamilton*, ed. John France and William G. Zajac, Aldershot 1998, pp. 79–101; Penny J. Cole, '"O God, the Heathen have come into your inheritance", (Ps. 78:1). The Theme of Religious Pollution in Crusade Documents, 1095–1188', *Crusaders and Muslims in Twelfth-Century Syria*, ed. Maya Shatzmiller, Leiden 1993, pp. 84–111; *idem, The Preaching of the Crusade to the Holy Land, 1095–1270*, Cambridge, Mass. 1991.

12 For the early Middle Ages see above, nn. 1, 7.

13 Prawer, 'Jerusalem in Christian and Jewish Perspectives', pp. 737–50; P. Alphandéry, 'Les citations bibliques chez les historiens de la première croisade', *Revue de l'Histoire des Religions*, 99 (1929), pp. 139–57; A.H. Bredero, 'Jerusalem in the West', pp. 259–71.

14 See Kühnel, *From the Earthly to the Heavenly Jerusalem*, pp. 114–19, 123–65; Riley-Smith, *The First Crusade*, p. 21; Marcus Bull, *Knightly Piety and the Lay Response to the First Crusade. The Limousin and Gascony c. 970–c. 1130*, Oxford 1993, *passim.*

Consequently, though it is by now generally accepted that the Latin occupation of Jerusalem (1099–1187), as well as the crusader movement, were bound to influence the position of Jerusalem in Christian spirituality, the character of this influence has still to be made precise. Therefore, this study aims, first, to analyse the influence of the crusader conquest and the new role of Jerusalem as the capital of a Christian crusading kingdom on the attitude of Christendom to the Holy City. Second, it sets out to define, through comparison with the previous and subsequent periods, the particular character of Jerusalem's sanctity in the context of the major trends of twelfth-century spirituality. It will therefore be our task, *inter alia*, to trace the degree of continuity or change in different periods; to determine the influence of the major trends of twelfth-century spirituality, such as ideals of *Imitatio Christi* and *Vita Apostolica*, the new emphasis on historical and literal meanings in scriptural exegesis, the new (or revived) consciousness of self and inner life, and the growing importance of the cult of the Virgin Mary. We shall be concerned also with the impact of the emerging idea of crusade, accompanied and occasioned by the change in the status of the warrior, from mere soldier to Christian Knight (*miles Christi*). The new concept of crusade as aimed at the recovery of Jerusalem, the crusading movement as charged, among its other aims, with the defence of the Holy City, raises the question of the influence that dissemination of the crusading idea and crusade propaganda had on the attitude to Jerusalem. The place of Jerusalem in crusade propaganda is yet another topic for discussion.

Twelfth-century history comprises intellectual, spiritual and literal innovations, as well as changes in underlying political, social and institutional structures resulting, for example, in the consolidation of the feudal monarchies and the papal theocracy within the *Societas Christiana*, with its organizational and specialized machinery for governing it, a process that made the papacy the focal institution of Europe. As such, the papacy, the initiator of the First Crusade, became the *spiritus movens* and the main pillar of the crusading movement, protector of the Latin States in the Levant and particularly of their capital, Jerusalem. Any examination of attitudes to the Holy City must take all of these changes and transformations into consideration. Other factors to be considered are the general increase in pilgrimage and, with the establishment of the Latin Kingdom and the opening of regular lines of communication with the West, the emergence of the Holy Land as a focal centre of pilgrimage.

The sources show that in the period 1099–1187 Jerusalem loomed large in all spheres of life, ranging from international relations to personal salvation. Jerusalem appears in almost every form of contemporary literary and artistic expression, including liturgy, sermons, religious and profane poetry, *chansons de geste*, drama and the visual arts. In particular, the vernacular, orally transmitted lyrics were seen, heard and often read by the various strata of European society; at a time when literacy was socially still very limited, the impact of sermons or ecclesiastical art on the individual's image world was tremendous. Our sources are therefore most diverse: historiography including annals and chronicles, scriptural exegesis, theological treatises, apocalyptic literature, correspondence, crusade *exitatoria*, descriptions of the Holy Land and itineraries, charters, sermons, liturgy, drama, poetry, *chansons de geste*, architecture, iconography and cartography.

Dealing with the attitude of Christendom to Jerusalem during the crusaders' rule of the Holy City, the *terminus a quo* of this study is the role of Jerusalem as the goal of the First Crusade (1096–1099); the *terminus ad quem* is Christendom's reaction to the loss of Jerusalem (1187). At the same time, however, allusions are made to both the previous and following periods in order to define, through comparison, the particularity of attitudes to Jerusalem in the period under consideration.

The first chapter deals with Jerusalem as the aim of the First Crusade; the second with the reaction to the conquest of the Holy City and the perceptions of its significance. Chapter 3 examines the attitudes of the papacy to Jerusalem on both the ecclesiastical–hierarchical and the spiritual levels as well as attitudes to Rome versus Jerusalem, assessing their potential as major centres of sanctity. Chapter 4 deals with the various arguments used during the twelfth century to justify Christian's right first to conquer, then to rule, and from 1187, to reconquer the Holy City. Chapters 5 and 6 describe the new character of Jerusalem's sanctity as the City of the Old and New Testaments. Chapter 7 discusses the attitude to the earthly Jerusalem as the gateway to the heavenly one, namely, the meritorious position of the earthly city as opposed to its celestial counterpart – the Heavenly Jerusalem. The subject matter of Chapter 8 is the concept of Jerusalem as the centre of the world as well as the place where the events of the End of the World will take place; it also examines the presentation of Jerusalem in the various apocalyptic visions of the twelfth century. It is the examination of the reaction to the fall of Crusader Jerusalem in 1187 – the *terminus ad quem* of this study – that is the theme of the last chapter, Chapter 9.

Chapter 1

Jerusalem: Goal of the First Crusade

On 15 July 1099, Jerusalem was conquered and restored to Christendom, the culmination of the four-year odyssey of the First Crusade (1096–1099). As early as the twelfth century, the deliverance of Jerusalem from the yoke of the infidels became the most common theme associated with the history of the crusades.[1] Yet some historians are still hesitant as to whether the conquest of Jerusalem was merely a result or the original and ultimate goal of the First Crusade.[2] The difficulty in establishing the original goal of the First Crusade stems mainly from the fact, ironic in itself, that Pope Urban II's speech at the Council of Clermont (27 November 1095), proclaiming the crusade, has survived only in various chronicle versions, most of them written a decade after the conquest of Jerusalem and thus influenced by what became the climax of the First Crusade.[3]

There are several indications that Jerusalem was the original goal of the First Crusade. First and foremost, it should be remembered that Urban II, while still only cardinal and papal legate, was close to Pope Gregory VII. Gregory, in 1074, attempted to organize an expedition of 50,000 men who, under his leadership, would both assist the Eastern Churches to withstand the Seljuk onslaught and liberate the Holy Sepulchre. True, what Gregory had in mind was not the same thing as the crusade preached by Urban II twenty-one years later. Yet it is clear that in the two decades preceding the First Crusade the idea of conquering Jerusalem was in the air.[4] Second, sources other than the various versions of Pope Urban's speech – namely, decrees of the Council of Clermont, the letters Pope Urban wrote during his

---

1 Prawer, 'Jerusalem in Christian and Jewish Perspective', p. 739.

2 Mayer, *The Crusades* (1987), pp. 8–13, still hesitates, but is far more ready to accept Jerusalem as the goal of the crusade than in his previous version of his work: *idem*, *The Crusades* (1972 edn), pp. 8–13; and see on this problem Housley, 'Jerusalem and the Development of the Crusade Idea, 1099–1128', *HH*, pp. 27–32; John France, 'Les origines de la Première Croisade. Un nouvel examen', *Autour de la Première Croisade. Actes du Colloque de la 'Society for the Study of the Crusades and the Latin East'* (Clermont-Ferrand, 22–25 juin 1995), ed. Michel Malard, Paris 1996, pp. 43–56.

3 For a summary of the discussion regarding the place of Jerusalem in the speech at Clermont see Riley-Smith, *The First Crusade*, pp. 13–30, esp. pp. 18, 20–22; Robert Somerville, 'The Council of Clermont and the First Crusade', *Mélanges G. Fransen*, 2 vols, vol. 2, Rome 1976 (= *Studia Gratiana*, 20), pp. 325–7; Blake, 'The Formation', pp. 11–13.

4 Riley-Smith, *The First Crusade*, pp. 8, 12; *Kreuzzugsbriefe aus den Jahren 1088–1100*, ed. H. Hagenmeyer, Innsbruck 1901, no. I, p. 136.

preaching of the crusade, reports of sermons he preached in France, as well as charters issued by departing crusaders – all clearly pinpoint Jerusalem as the main goal of the expedition, with the deliverance of the Eastern Churches from Moslem oppression lagging behind.[5] Thus, one of the decrees of the Council of Clermont issued, no doubt, with papal approval as it is repeated verbatim in Urban's letter to Bologna (19 September 1096), refers explicitly to the Church of the Holy Sepulchre: 'Whoever goes from pure devotion and not desire for earthly profit to liberate the Church of God in Jerusalem, his journey will be considered a substitute for all penance.'[6] Also, writing shortly after the Council of Clermont to the faithful in Flanders (December 1095), the pope refers directly to Jerusalem: 'We believe that you, brethren, learned long ago from many reports the deplorable news that the barbarians in their frenzy have invaded and savaged the churches of God in the eastern regions. Worse still, they have seized the Holy City of Christ, embellished by his Passion and Resurrection.'[7] According to a report regarding the pope's visit to the abbey of Saint-Florent of Saumur (January 1096), 'the pope urged people to go to Jerusalem', both publicly and privately.[8] A contemporary report of a sermon preached by the pope at Angers (February 1096) declares that he exhorted people 'to go to Jerusalem'.[9] The Holy City is also presented as the main goal of the crusade by a chronicle of the abbey of St Maixent in Poitou, where Urban II stayed in March 1096: 'By the Pope's order, many men, both noble and base, rich as well as poor, from all lands ... went on the journey to the Holy Sepulchre.'[10] The 'encyclical of Pope Sergius IV', possibly a forgery emanating from the abbey of St Peter of Moissac and probably fabricated at the time Urban II was staying there in May 1096, also refers to the crusade as aimed at the liberation of the Holy Sepulchre.[11] And

---

5 Cowdrey, 'Pope Urban II's Preaching', pp. 177–88; Riley-Smith, *The First Crusade*, pp. 13–25; *idem*, *The First Crusaders 1095–1131*, Cambridge 1997, pp. 56–60; Bull, *Knightly Piety*, pp. 256–8 *et passim.*

6 Somerville, *The Councils of Urban II*, p. 74; *Kreuzzugsbriefe*, no. III, pp. 137–8. Eng. trans. Riley-Smith, *The Crusades*, London 1981, pp. 38–9.

7 *Kreuzzugsbriefe*, no. II, p. 136; Riley-Smith, *Idea and Reality*, p. 38.

8 George T. Beech, 'Urban II, the Abbey of Saint-Florent of Saumur and the First Crusade', *Autour de la Première Croisade*, pp. 60–61.

9 *Fragmentum historiae Andegavensis. Chroniques des comtes d'Anjou et des seigneurs d'Amboise,* ed. L. Halphen *et al.*, Paris 1913, pp. 237–8; Riley-Smith, *The First Crusade*, p. 25.

10 *Chronicon S. Maxentii Pictavensis. Chroniques des églises d'Anjou*, ed. P. Marchegay *et al.*, Paris 1869, p. 412; Cowdrey, 'Pope Urban II's Preaching', p. 181.

11 A. Gieysztor, 'The Genesis of the Crusades: The Encyclical of Sergius IV (1009–1012)', *Medievalia et Humanistica*, V (1948), pp. 3–23; 6 (1950), pp. 3–34. But see now Hans Martin Schaller, 'Zur Kreuzzugsenzyklika Papst Sergius IV.', *Papsttum, Kirche und Recht im Mittelalter. Festschrift für Horst Fuhrmann zum 65. Geburtstag*, ed. Hubert Mordek, Tübingen 1991, pp. 135–53, who argues in favour of its authenticity and dates it 1010.

writing several months later to the faithful of Bologna (19 September 1096) and to the congregation of Vallombrosa (7 October 1096), the pope himself explicitly cited Jerusalem as the goal of the crusade.[12]

During the First Crusade, the faithful too regarded the liberation of Jerusalem, and in particular of the Holy Sepulchre, as their goal. This was apparently a result of the prevalent image of Jerusalem in the early Middle Ages, which specifically depicted the Holy City as the site of the Holy Sepulchre.[13] For example, the great German pilgrimage of 1064–1065, more than seven thousand strong, though motivated by eschatological expectations, was still at its core a pilgrimage to the Holy Sepulchre.[14] Similarly, the participants of the First Crusade saw it as a pilgrimage to the Holy Sepulchre, with its liberation as an additional goal. One of these declared in 1096: 'Considering that God has spared me, steeped in many and great sins, and has given me time to do penance, and fearing that the weight of my sins will deprive me of a share in the heavenly kingdom, I, Ingelbald, wish to seek that sepulchre from which our redemption, having overcome death, wished to rise.'[15] Another participant, Raimond of St Gilles, count of Toulouse, in his charter of endowment to the cathedral of Puy, defined the purpose of his departure as 'the liberation of the Holy Sepulchre'.[16] According to Gerald of Cahors he had gone 'on pilgrimage to wage war on foreign peoples and defeat barbaric nations, lest the Holy City of Jerusalem be held captive and the Holy Sepulchre of the Lord Jesus be contaminated any longer'.[17] Robert II of Flanders went to Jerusalem 'to liberate the Church of God'.[18] Yet another participant, the Gascon, Amanieu of Loubens, similar to Raimond of St Gilles, wished 'to cleanse the place in which the Lord Jesus Christ deigned to undergo death for the restoration of the human race'.[19] In the course of the crusade (spring 1098) its leaders, writing

---

12 *Kreuzzugsbriefe*, no. III, p. 137; W. Wiederhold, 'Papsturkunden in Florenz', *Nachrichten von der Gesellschaft der Wissenschaften zu Göttingen. Phil.-Hist. Kl.*, Göttingen 1901, p. 39; Riley-Smith, *Idea and Reality*, pp. 38–40.

13 For the central position of the Holy Sepulchre in the Christian attitude to Jerusalem in the early Middle Ages see Hieronymus, *Epistulae*, ep. XLVI, pp. 329–44, esp. p. 334; *Breviarius de Hierosolyma*, pp. 109–12; Antoninus Placentinus, pp. 137–9; Arculfus, pp. 185–203; Bernardus Monachus, *Itinerarium*, ed. T. Tobler, *Descriptiones Terrae Sanctae*, Leipzig 1874, pp. 91–3 and below, ch. 5, pp. 63–5.

14 E. Joranson, 'The Great German Pilgrimage of 1064–1065', *The Crusades and Other Historical Essays Presented to Dana Munro*, ed. L.J. Paetow, New York 1928 (repr. New York 1968), pp. 3–14; McGinn, '*Iter Sancti Sepulchri*', pp. 40–55.

15 *Cartulaire de l'abbaye de Saint-Vincent du Mans*, ed. R. Charles and S.M. d'Elbenne, Le Mans 1886, I, col. 69 ; Riley-Smith, *The First Crusade*, p. 23.

16 Somerville, 'The Council of Clermont', p. 326 n. 9.

17 Cited by Riley-Smith, *The First Crusaders*, p. 62.

18 *Ibid.*

19 *Ibid.*, p. 63.

from Antioch, presented themselves as 'pilgrims to the Holy Sepulchre'[20] and in September they invited the pope to lead the crusade himself and bring to an end the war which was Christ's, an expedition ordered by Christ himself and preached by the pope for the 'liberation of the Holy Sepulchre'.[21] This view is also reflected by the fact that the first lordship established by the crusaders in Asia Minor was given to Peter of Aulps to hold 'in fealty to God and the Holy Sepulchre and the princes and the [Byzantine] emperor'.[22] Thus, during the First Crusade the predominant conception of its goal, both in the crusader camp and in Europe, was the liberation of the Holy Sepulchre.[23]

It seems thus that most of the contemporaries of the First Crusade perceived this goal not in terms of the geographical–topographical entity of a city, but rather in liminal terms. The goal itself became more specific – the Holy Sepulchre.[24] This perception of the outcome of the First Crusade may be explained in terms of the holy places and the character of Jerusalem's sanctity in the early Middle Ages. From its inception Christian pilgrimage was directed mainly to holy sites, which consisted of tombs of saints. The saint in heaven was believed to reside bodily in his tomb as indicated, for example, by the inscription on the tomb of St Martin of Tours: 'Here lies Martin the bishop, of holy memory, whose soul is in the hands of God; but he is fully here present as made plain in miracles of every kind.'[25] The tombs of saints were thus, as Peter Brown has argued, 'privileged places where the contrasted poles of heaven and earth met'.[26] As a result Jerusalem pilgrimage, in common with pilgrimages to other centres of sanctity, was *ad limina*, that is, to a church containing a saint's tomb. Thus the pilgrimage to Tours was described as *limina B. Martini*, that to Poitiers as *limina S. Hilarii* and the pilgrimage to Jerusalem was described, from the middle of the seventh century (namely, following the Moslem conquest of Palestine in 638), as *ad sanctum sepulchrum* or *invisere sepulchrum Domini* or *ad adorandum sepulchrum*

---

20 *Kreuzzugsbriefe*, no. XII, p. 154.

21 *Ibid.*, no. XVI, pp. 164–5. For the First Crusade as a pilgrimage see Dupront, 'La spiritualité', pp. 45–83; Riley-Smith, *The First Crusaders*, pp. 66–70. For the goal of the First Crusade in 1098 as the liberation of the Holy Sepulchre see also the letter of the people of Lucca to the faithful in *Kreuzzugsbriefe*, no. XVII, p. 167.

22 Riley-Smith, *The First Crusade*, p. 108.

23 For the same view see *ibid.*, though documented by chronicle accounts completed after the conquest of Jerusalem. *Ibid.*, p. 108 nn. 83, 84.

24 See n. 68 below. Also see Sylvia Schein, 'Jérusalem: Objectif originel de la première croisade?', *Autour de la Première Croisade*, pp. 119–26.

25 Peter Brown, *The Cult of the Saints: Its Rise and Function in Latin Christianity*, Chicago 1981, p. 4.

26 *Ibid.*, p. 3. This was the outcome of both pagan and Jewish practices which have deeply influenced the origins of Christian observances. See Sabine MacCormack, '*Loca Sancta:* The Organization of Sacred Topography in Late Antiquity', *The Blessings*, pp. 7–40; J. Wilkinson, 'Jewish Holy Places and the Origins of Christian Pilgrimage', pp. 41–53.

Domini.[27] Not only those who went voluntarily on pilgrimage[28] but also those who participated in penitential pilgrimages, such as Duke Robert I of Normandy in 1035, and Count Fulk III Nerra of Anjou in 1002–3, 1027–9 and again between 1039 and 1040, defined their goal in terms of the Holy Sepulchre as the place of their intended penitential rituals.[29] Writing to the Archbishop of Tours, Fulk Nerra described his personal motivation as follows: 'Because [I] caused a great deal of bloodshed in various battles and was terrified by the fear of hell [I] went to the sepulchre of the Saviour in Jerusalem ...'.[30]

One by-product of the conquest of Jerusalem was an impassioned literary activity, unprecedented in the historiography of the early and high Middle Ages. Indeed, perhaps no other event until then had inspired a similar quantity of historical writing.[31] The miraculous nature of the event fertilized the historical imagination of its contemporaries and inspired a desire to explain it and describe it. Abbot Guibert of Nogent, who completed his *Gesta Dei per Francos* in 1111, explaining the purpose of his work, argued that 'it would be a scandal not to write about the glory of our time as we ought or at least to the best of our ability'.[32] The Norman Orderic Vitalis, writing between the years 1135 and 1139, explained his reasons for recording the achievements of the First Crusade as follows: 'Many other Latin and Greek writers have treated this memorable subject, and have presented the great deeds of the heroes for posterity in their vivid writings. I too, the last of all who as professed monks follow the Lord in their lives, aspire to include this Christian enterprise for the Lord Jesus in the little book I have begun on ecclesiastical matters,

---

27 *Vita Altmanni Episcopi Pataviensis*, *MGH.SS*, 12, p. 230; Ademarus Cabannensis, *Chronicon*, ed. J. Chavanon, Paris 1867 (*Collection de textes pour servir à l'étude et à l'enseignement de l'histoire*, 20), pp. 189–90; Radulfus Glaber, *Historiarum sui temporis libri V*, ed. M. Prou, Paris 1885, p. 109; *Annales Altahenses Majores*, *MGH.SS*, 3, p. 815; *Gesta Episcoporum Tullensium*, *MGH.SS*, 8, p. 647; E.R. Labande, '"Ad limina": Le pèlerin médiéval au terme de sa démarche', *Mélanges offerts à René Crozet*, 2 vols, ed. P. Gallais *et al.*, vol. 1, Poitiers 1966, pp. 283–91.

28 Antoninus Placentinus, pp. 137–9; Arculfus, pp. 185–203; Bernardus Monachus, pp. 91–3.

29 C. Vogel, 'Le pèlerinage pénitentiel', *Pellegrinaggi e Culto dei Santi in Europa fino alla prima Crociata. Convegni del Centro di studi sulla spiritualità medievale*, vol. 4, Todi 1963, pp. 128–9; L. Halphen, *Le Comté d'Anjou au XIe siècle*, Paris 1906, pp. 213–18; Bull, *Knightly Piety*, pp. 207–9.

30 See the letter in Radulfus Glaber, pp. 2, 4, 5–6. For Fulk's pilgrimages see also Bernard S. Bachrach, 'The Pilgrimages of Fulk Nerra, Count of the Angevins, 987–1040', *Religion, Culture and Society in the Early Middle Ages. Studies in Honor of Richard S. Sullivan*, ed. T.F.X. Noble and J.J. Contreni, Kalamazoo 1987, pp. 205–17.

31 On this phenomenon see Prawer, 'Jerusalem in Christian and Jewish Perspectives', pp. 740–41; J.O. Ward, 'Some Principles of Rhetorical Historiography in the Twelfth Century', *Classical Rhetoric and Medieval Historiography*, ed. E. Breisach, Kalamazoo 1986, pp. 103–65, esp. p. 122; Riley-Smith, *The First Crusade*, p. 135.

32 Guibertus Novigentis, *Gesta Dei per Francos*, *RHC.Hocc*, 4, pp. 119–20.

because I love the brave champions of Christ and delight in praising their valiant deeds ... I will do my best to set out some of the deeds of the pilgrims to Jerusalem truthfully and in order.'[33]

A comparison between those Europeans who wrote a decade or even more after the conquest of Jerusalem and the participants of the crusade reveals a significant difference in their attitude to Jerusalem as the aim of the expedition. This is particularly striking in the chroniclers' versions of Urban II's speech at Clermont. Of the participants, Fulcher of Chartres did not mention Jerusalem at all in his version of the speech.[34] The anonymous author of *Gesta Francorum* and Peter Tudebode described the goal of the crusade as *via Domini et Sancti Sepulchri*.[35] The European writers, especially Robert the Monk, Guibert of Nogent and Baldric of Dol – all of whom, it seems, had been present at the Council of Clermont[36] – launched into a paean of praise of the Holy City. It seems that the zenith of the First Crusade and its chief achievement, the conquest of Jerusalem, inspired those who undertook to record and explain its *res gestae* in an attempt to paint a picture of Jerusalem to portray what the Holy Sepulchre meant and what it stood for in the spirituality of the period. As all three of them were Benedictines, their views represented mainly Benedictine spirituality.

Thus, according to Robert the Monk, the pope described Jerusalem as 'the navel of the world, a land fruitful above all others, like a second paradise of delights. The Redeemer of the human race made it famous by his birth, embellished it by his life, sanctified it by his passion, redeemed it by his death, left his seal upon it by his burial.' Now, however, the pope claimed: 'This royal city, placed at the centre of world, is ... held captive by her enemies and enslaved to pagan rites by a people which does not acknowledge God. So she asks and prays to be liberated and calls upon you unceasingly to come to her aid.'[37]

For Baldric of Dol, in contrast to Robert the Monk and other chroniclers of the First Crusade, the sanctity of Jerusalem flowed not only from its being the city where the life, Passion and death of Christ were enacted and his sepulchre found, but also from being the burial place of the Virgin Mary and the site of the Temple. Unlike other chroniclers, Baldric, in his version of Urban II's speech, referred to

---

33 Orderic Vitalis, *Historia Ecclesiastica*, ed. and Eng. trans. M. Chibnall, 6 vols, Oxford 1969–81, vol. 5, p. 7.

34 Fulcher Carnotensis, *Historia Hierosolymitana*, ed. H. Hagenmeyer, Heidelberg 1913, pp. 130–38.

35 *Gesta Francorum et aliorum Hierosolimitanorum*, ed. R. Hill, London 1961, p. 102; Petrus Tudebode, *Historia de Hierosolymitano itinere*, ed. J.H. and L.L. Hill, Paris 1977 (*Documents relatifs à l'histoire des croisades*, 12), pp. 31–2.

36 Cowdrey, 'Pope Urban II's Preaching', p. 177; Somerville, 'The Council of Clermont', p. 332; *idem*, 'The Council of Clermont (1095) and the Latin Christian Society', *Archivum Historiae Pontificiae*, 12 (1974), pp. 63, 75–82.

37 Robertus Monachus, *Historia Jherosolimitana, RHC.Hocc*, 3, p. 729; Riley-Smith, *Idea and Reality*, p. 44.

the sanctity not of Jerusalem alone but also to that of its physical environment, the entire Holy Land. The very use of the term 'Holy Land' (*terra sancta*) is an innovation; in fact Baldric is one of the first authors to apply it to all of Palestine. In the period preceding the First Crusade the commonly used term was 'the Holy Places' (*loca sancta*), a general concept without geographical localization; it was the crusades and the foundation of the Latin Kingdom that made Palestine the Holy Land of the Christians.[38] Baldric also describes the appalling present condition of Jerusalem and the entire Holy Land, involving Psalms 79:1: 'O God, the heathens have come into thy inheritance, they have defiled thy holy temple; they have laid Jerusalem in ruins.' This prophecy, he argued, has now been fulfilled and one must mourn the land deservedly called 'holy', 'in which there is no one footstep which was not embellished or hallowed by the body or shadow of the Saviour or by the glorious presence of the holy Mother of God or the most beloved company of apostles or the delectable blood spilt by the martyrs Stephen, first of the martyrs ... [and] John the Baptist ...'[39] It is the land of the Children of Israel, whose flight from Egypt prefigured the crusaders; with Jesus as their leader they appropriated that land for themselves and, expelling the Jebusites and other nations, lived in the 'earthly Jerusalem, the type of the Heavenly Jerusalem'.[40]

The use made by Baldric of Dol of Psalm 79:1 is highly significant. This psalm, it should be pointed out, was commonly used to describe disasters, as, for example, by Jerome in his description of the sack of Rome by Alaric in 410.[41] The psalm would become, by the end of the twelfth century, the leitmotif of the reaction to the fall of Crusader Jerusalem in 1187.[42] Within the accounts of the First Crusade the psalm, also inserted by William of Tyre into his version of Urban's speech at Clermont,[43] is used to depict the Moslem presence in Jerusalem as religious pollution, a subject which loomed large in the descriptions of the aims of the First Crusade. Some of the participants in the crusade, as well as authors of the histories of the First Crusade who wrote in the 1110s, presented the aim of the crusade not

---

38 Baldricus Dolensis, *Historia Jerosolimitana*, *RHC.Hocc*, 4, p. 13; Riley-Smith, *Idea and Reality*, p. 50. For the term 'Holy Places' see Antoninus Placentinus, pp. 129, 179; Arculfus, pp. 183, 203; Beda, *De Locis Sanctis*, p. 254; Bernardus Monachus, p. 85. See also Erdmann, *The Origin*, pp. 300–301 and n. 121. The term 'Holy Land' appears however in Augustine's *De Civitate Dei*, 22, 8 in regard to Jerusalem. See Robert L. Wilken, *The Land Called Holy. Palestine in Christian History and Thought*, New Haven and London 1992, p. 125; it appears also in a petition of Sabas and Theodosius to Emperor Anastasius (of *c.* 511) where the monks refer to themselves as 'the inhabitants of this Holy Land', *ibid.*, pp. 168–72.

39 Baldricus Dolensis, p. 14. Riley-Smith, *Idea and Reality*, pp. 50–51. On the theme of the Moslem rule of Jerusalem as religious pollution see Cole, 'O God', pp. 84–111. See also below, ch. 3 pp. 37–9; ch. 9, pp. 163–5.

40 Baldricus Dolensis, p. 14.

41 Hieronymus, *Epistulae*, 127, 12, p. 154; Wilken, *The Land Called Holy*, pp. 224–6.

42 See below, ch. 9, pp. 163–5.

43 *W.T.*, I, 15, p. 133.

only as the liberation of the Holy Sepulchre but also the purification of the Holy Sepulchre as well as the other sacred places of Jerusalem from Moslem defilement; thus some of the contemporaries of the conquest of Jerusalem would present it as an act of purification.[44] Raimond of Aguilers reported that after the fall of Jerusalem, the Egyptian [Fatimid] Sultan boasted that he would recover the Christian holy places, destroy the manger, Golgotha and the Holy Sepulchre, and, unearthing Christian relics, hurl them into the sea from where they could never be found.[45]

The subject of purification as the aim of the First Crusade received, however, even more compelling expression in the crusade histories of Baldric of Dol, Robert the Monk, Guibert of Nogent and Albert of Aachen. Thus Baldric of Dol describes the state of Jerusalem under the Moslem rule as follows: 'that very city in which, as you know, Christ himself suffered for us, since our sins demanded it, has been overwhelmed by the filth of the pagans and, I say it to our shame, led away from the service of God ... To what use now is put the church of Blessed Mary ... ? What of the Temple of Solomon, not to mention the fact it is the Lord's, in which the barbaric races worship their idols, which they have placed there against the law and against religion? We will not recall the Lord's Sepulchre, because some of you have seen with your own eyes to what abomination it has been handed over.'[46] Robert the Monk even presents the pope as claiming that: 'the Holy Sepulchre of Our Lord and Saviour ... is in the hands of unclean races and ... the Holy Places ... are now treated dishonourably and are polluted irreverently by their unclean practices.'[47] Guibert of Nogent made the pope address the participants of the Council of Clermont as follows: 'You, dearest brothers, must take the greatest pains to try to ensure that the holiness of that City and the glory of his Sepulchre will be cleansed, for the Gentiles by their presence continually sully them in so far as they can.'[48] Albert of Aachen has Christ appear before Peter the Hermit and asks him to exhort the believers 'to purge the holy places of Jerusalem and to restore the worship of the saints'.[49] Pondering upon the sanctity of Jerusalem,

---

[44] Cole, 'O God', pp. 88–94 and below, ch. 3, pp. 37–9; ch. 9, pp. 163–5.

[45] Raimond d'Aguilers, *Le Liber*, eds. J.H. and L.L. Hill, Paris 1969 (*Documents relatifs à l'histoire des croisades*, 9), p. 155.

[46] Baldricus Dolensis, p. 13. Eng. trans. Riley-Smith, *Idea and Reality*, p. 50.

[47] Robertus Monachus, p. 729. Eng. trans. Riley-Smith, *Idea and Reality*, p. 43.

[48] Guibertus Novigentis, *Gesta Dei*, p. 138; Riley-Smith, *Idea and Reality*, p. 45. See also on this subject Gilo, *Historia Gestarum viae nostri temporis Hierosolymitane*, *RHC.Hocc*, 5, p. 699; Gilo, *The Historia Vi Hierosolymitane of Gilo of Paris and a Seconde Anonymous Author*, ed. and Eng. trans. C.W. Grocock and J.E. Siberry, Oxford 1997, pp. 4–5, 8–11.

[49] Albertus Aquensis, *Historia Hierosolymitana*, *RHC.Hocc*, 4, p. 273. For the Moslem presence in Jerusalem and particularly in the Holy Sepulchre as pollution see also such European contemporary chroniclers as: William of Malmesbury, *De Gestis Regum Anglorum Libri Quinque*, ed. W. Stubbs, *RS*, 90/2, pp. 423, 434; Hugo Floriacensis (de S.

Guibert of Nogent, in his account of Urban's sermon at Clermont, also described it as the city of the life, the passion and burial of Christ.[50] He, however, introduces two additional motifs. The sanctity of Jerusalem, he claimed, surpassed its being the city of Christ. It derived from both its past, that is, before Christ, and from its future. In his words: 'If, indeed, one reads in the sacred and prophetic writings that this land was the inheritance and the holy temple of God before the Lord walked and appeared there, how much more holy and worthy of reverence must we consider it because when the God of majesty was incarnate there, was nurtured, grew up and in his physical nature walked and travelled from place to place?'[51] And he added: 'Just suppose that Christ had never died nor was buried, nor had lived at any time in Jerusalem. If none of these things had in fact occurred, you ought still to be moved to help the land and the city by this thought alone: that "the law will come out of Zion and the word of the Lord from Jerusalem"' (Isaiah 2:3).[52] The sanctity of Jerusalem is thus rooted, according to Guibert of Nogent, in both its pre-Christian, biblical past as the city of the Children of Israel, as well as in its past as the city of Christ.

Additionally, according to Guibert, Jerusalem is destined to play a particular part in the future, at the end of the world, in the Last Days. Guibert's version of Urban II's speech at Clermont made the pope point this out clearly: 'God wishes some regions of the East to be restored to the faith against the approaching times of Antichrist. For it is clear that Antichrist will wage war neither against Jews nor against Gentiles, but, according to the etymology of his name, he will attack Christians. And if Antichrist finds no Christian there, that is, in Jerusalem, just as today when it is thought that there is scarcely a single one in that place, there will be no one to resist neither him nor any whom he may rightly attack. According to Daniel and to Jerome,[53] his interpreter, he will pitch his tents on the Mount of Olives ... the end of the world is already near although the Gentiles have not been converted to the Lord ... but first before the coming of Antichrist a renewal of the Christian empire in those regions is necessary according to the prophecies, either by means of you or by means of those whom God chooses, so that the head of all the evil ones ... should discover some support of the faith against which he may fight.'[54] In this account of the events of the Last Days, Guibert clearly relied on

---

Maria), *Modernorum Regum Francorum Actus*, *MGH.SS*, 9, p. 394; Sigebertus Gemblacensis, *Chronographia*, p. 367. See also on this subject J. France, 'The Text of the Account of the Capture of Jerusalem in the Ripoll Manuscript, Bibliothèque Nationale (Latin) 5132', *EHR*, 103 (1988), p. 648.

50 Guibertus Novigentis, *Gesta Dei*, p. 138.

51 *Ibid.*, pp. 137–8.

52 *Ibid.*, p. 138.

53 For Jerome see his *Commentariorum in Isaiam*, *CCSL*, 73A, p. 327; *idem*, *De Antichristo in Danielem [IV]*, *CCSL*, 75B, pp. 914–44.

54 Guibertus Novigentis, *Gesta Dei*, pp. 138–9.

the highly popular and widely diffused treatise of Adso of Montier-en-Der, the so-called *Tractatus de Antichristo* (*c.* 950).[55] Like Adso, he insisted that Christian rule of Jerusalem was a necessary prelude to the beginnings of the Last Days and the advent of Antichrist to Jerusalem, since the liberation of Jerusalem from the Gentiles was associated in Luke's Gospel (21:24) with the Last Days.

Thus, Jerusalem as the goal of the crusade figured in the various versions of Urban II's speech at Clermont mainly as the city of the life, passion and burial of Christ, as well as the city of the Children of Israel, the city of the Virgin Mary, the apostles, the saints and the martyrs. As such, it was the place 'from which we have received the grace of redemption and the source of all Christianity',[56] 'the mother of all churches'.[57] But it is also the centre of the world as well as the centre of the events of the Last Days, the scene of the last, crucial battle between Antichrist and Christ. The crusade for its liberation was directed, according to Guibert of Nogent, towards the rescue of 'the cross, the blood and the Sepulchre'; or, in his version of the pope's words at Clermont: 'If you consider that you ought to take great pains to make a pilgrimage to the graves of the apostles [in Rome] or to the shrines of any other saints, what expense of spirit can you refuse in order to rescue, and to make a pilgrimage to, the cross, the blood, the Sepulchre?'[58] Robert the Monk described the goal of the crusade as the deliverance of that 'royal city' from the pagans,[59] while Baldric of Dol referred to Christ's command to follow him as a leader and 'wage war for your own rights over Jerusalem and throw out the Turks'.[60] Particular importance was attached to the liberation and especially to the purification of the Holy Sepulchre. Robert the Monk had the pope exhorting the crusaders 'to take the road of the Holy Sepulchre'.[61] Guibert of Nogent referred to 'the glory of His Sepulchre' and the need to ensure that it would be cleansed of the presence of the gentiles,[62] and Baldric of Dol quoted the pope as describing the place of the Lord's burial as precious, desirable, 'a place beyond compare'.[63] Similarly, other chroniclers of the First Crusade, and European chroniclers in

---

55 Adso's treatise is a version of a sibylline apocalypse, composed in Mesopotamia between the years 644 and 678 and transmitted to Byzantium. As early as the seventh or the beginning of the eighth century it was transmitted to the West and put into writing by Adso *c.* 950. See Adso Dervensis, *De Ortu et Tempore Antichristi*, *CCCM*, 45, pp. 22–30, esp. p. 28. See also *ibid.*, pp. 43–9, 161–6 for two eleventh-century versions. On the apocalypse itself see P. J. Alexander, 'Byzantium and the Migration of Literary Works and Motifs. The Legend of the Last Roman Emperor', *Medievalia et Humanistica*, 2 (1961), pp. 47–68.

56 Guibertus Novigentis, p. 138.

57 Baldricus Dolensis, p. 11.

58 Guibertus Novigentis, p. 138.

59 Robertus Monachus, p. 729.

60 Baldricus Dolensis, p. 15.

61 Robertus Monachus, p. 728.

62 Guibertus Novigentis, p. 138.

63 Baldricus Dolensis, p. 13.

general, such as Albert of Aachen,[64] the anonymous writer of the *Historia Peregrinorum*[65] and Peter the Deacon,[66] referred to its goal in terms of the Holy Sepulchre. The so-called 'Charleville Poet' in his rhymed *Historia Vie Hierosolimitanae* (written before 1118) reported that the crusade was undertaken against those who were 'violating the sacrosanct rights of the Holy Sepulchre'.[67] It was an 'expedition to the Holy Sepulchre' (*iter/via Sancti Sepulchri*) or an 'expedition to Jerusalem' (*iter/via Hierusalem*).[68] In this context, writing between the years 1120 and 1130, that is, almost three decades after the conquest of Jerusalem, the Englishman William of Malmesbury presented a rather unconventional interpretation of the goal of the First Crusade. He described it in the framework of the ancient and prolonged struggle between Christianity and Islam, in the course of which Islam conquered two out of the three parts of the world: Asia and Africa. The First Crusade, he argued, was intended to block and prevent any further expansion of Islam, the liberation of Jerusalem, 'which ought

---

64 Albertus Aquensis, p. 274.

65 *Historia Peregrinorum, RHC.Hocc*, pp. 173, 175.

66 Petrus Diaconus, pp. 755, 765.

67 The Charleville Poet in Gilo (ed. Grocock and Siberry), pp. 2–3.

68 For the crusade as aimed at the liberation of the Holy Sepulchre: *Chronicon S. Petri Amiciensis, RHGF*, 12, p. 377; Hugo de S. Maria, *Itineris Hierosolimitani Compendium, RHC.Hocc*, 5, p. 365; Lupus Protospatarius, *Annales Barenses, MGH.SS*, 5, p. 62; *Chronicon Malleacense, Chroniques des églises d'Anjou*, ed. P. Marchegay *et al.*, Paris 1869, p. 418; *Annales Reicherspergenses, MGH.SS*, 17, p. 449; Caffaro de Caschifelone, *De Liberatione Civitatum Orientis Liber, RHC.Hocc*, 5, pp. 58–9; *Historia Peregrinorum, RHC.Hocc*, 3, pp. 173–5, 178; Radulfus Cadomensis, *Gesta Tancredi in Expeditione Hierosolymitana, RHC.Hocc*, 3, p. 664; Hugo Floriacensis, p. 394; *Chronique de Zimmern*, ed. H. Hagenmeyer, *AOL*, 2 (1884), p. 22; *Gesta Adhemari episcopi Podiensis, RHC.Hocc*, 5, p. 354; *Chronicon S. Andreae Castri Cameracensii, MGH.SS*, 7, pp. 544–5; Gaufredus Grossus, *Vita B. Bernardi de Tironio, RHGF*, 9, p. 167; *Chronica Monasterii Casinensis, MGH.SS*, 34, pp. 475–8. For the crusade as aimed at the conquest of Jerusalem: *Annales Augustani, MGH.SS*, 3, pp. 134–5; *Annales Rosenveldenses seu Annales Harsfeldenses, MGH.SS*, 16, p. 101; *Annalista Saxo, MGH.SS*, 6, p. 728; *Gaufridi Dictamen de Primordiis Ecclesiae Castaliensis, RHC.Hocc*, 5, p. 348; Gilo, *Historia de via Hierosolymitana, RHC.Hocc*, 5, p. 798; Ivo Carnotensis, *Epistolae, PL*, 162, no. 93, cols 113–14; Sigebertus Gemblacensis, *Chronographia, MGH.SS*, 6, pp. 367–8; Adelgot von Magdeburg, ed. W. Wattenbach, 'Handschriftliches', *Neues Archiv*, 7 (1882), pp. 624–6, esp. pp. 625–6; *Annalium Hildesheimensium Continuatio A. 1041–1108, MGH.SS*, 3, p. 106; *Annales Sancti Albani Moguntini, MGH.SS* 2, p. 246; *Annales Sancti Dionysii Remenses, MGH.SS*, 13, p. 83; *Annales Sancti Disibodi, MGH.SS*, 17 p. 16; *Annales Sancti Pauli Virdunensis, MGH.SS*, 16, p. 501; *Annales Seligenstadenses, MGH.SS*, 17, p. 31; Henry of Huntingdon, *De captione Antiochiae, RHC.Hocc*, 5, p. 374; *Walteri Vita Karoli Comitis Flandriae, MGH.SS*, 12, p. 540; Anselmus Cantuariensis, *Opera Omnia*, ed. F.S. Schmitt, Edinburgh 1946–61, 4, epis. CXCV, p. 85. For other goals connected with Jerusalem see: *Notitia Lemovicensis II, RHC.Hocc*, 5, p. 352 (expulsion of the accursed people from Christ's inheritance); Ekkehardus Uraugiensis, *Hierosolymita, RHC.Hocc*, 5, p. 14 (conquest of

to be open to Christians only' and of the Holy Sepulchre, 'that singular assurance of our faith', was part of this grandiose plan that he calls 'the expedition to Jerusalem'.[69]

To sum up: the goal of the First Crusade was twofold – the liberation and the purification of Jerusalem, and particularly that of the Holy Sepulchre. However, in the two first decades of the twelfth century this image of Jerusalem as the 'City of the Holy Sepulchre' had undergone a meaningful transformation. Under the pen of the European chroniclers of the First Crusade, Jerusalem as a city of Christ overshadowed the Holy Sepulchre. It was the city of the life, passion and burial of Christ as well as the city of the Children of Israel, a city whose sanctity was rooted in both: in its pre-biblical past as the city of the Children of Israel, as well as in its past as the city of Christ.

Jerusalem and the Promised Land); *idem*, *Chronicon Universale*, *MGH.SS*, 6, p. 212 (conquest of Jerusalem as the Promised Land, the mother of the faith and the redemption); *Chronicon S. Huberti Andaginemsis*, *MGH.SS*, 8, p. 615 (the crusade as aimed at the liberation of the Promised Land). For the crusade as aimed to rebuild the destroyed holy places see the report of the Chaize-le-Vicomte chronicler in Beech, 'Urban II', pp. 60–61.

69 William of Malmesbury, *De Gestis*, l. 4, pp. 393–8, 399.

Chapter 2

The Conquest – A Divine Act

Both the participants in the First Crusade and Europeans at large presented the conquest and liberation of Jerusalem as a miracle and a divine act. The crusaders on the march saw themselves fighting the 'war of Christ' (*bellum Christi*) as the 'army of God' (*exercitus domini*) or 'army of Christ' (*exercitus Christi*); their crusade was the 'expedition of God and Christ' (*Dei et Domini Jhesu Christi expeditio*). Christ was described as their king, their lord and their commander.[1] Archbishop Manasses of Rheims writing in November or December 1099 to Bishop Lambert of Arras described the liberation of Jerusalem as 'the result, we believe, of divine majesty rather than human might'.[2] A poem composed for the occasion makes use of Psalms 95:1. 'O come, let us sing to the Lord; let us make a joyful noise to the rock of our salvation' as well as Psalms 118:24: 'This is the day which the Lord has made, let us rejoice and be glad in it.' The poet praised the glory of the Lord, who had saved the 'City of David' from the pagans and wrested Jerusalem from the hands of the 'descendants of Hagar'.[3] A sermon attributed to Fulcher of Chartres, composed in the first decade of the twelfth century in commemoration of the liberation of Jerusalem and intended for use during the celebrations of the day of conquest (15 July), described the event in the words of Isaiah 66 (verses 10–11, 18–20) as well as Psalms 118:23 (cf. also Matthew 21: 42): 'This is the Lord's doing; it is marvellous in our eyes.'[4] Another participant, Raimond of Aguilers, presented the conquest as the act of God moved by prayers in His Church, thus fulfilling the promise he had made to the Fathers.[5]

Pope Paschal II also perceived the conquest of Jerusalem as a divine act. Writing on 28 April 1100 to the victorious army in the East, he declared that 'the Lord has certainly renewed his miracles of old' and described the conquest of Jerusalem as the

---

1 Riley-Smith, 'The First Crusade and St. Peter', pp. 58–63.

2 *Kreuzzugsbriefe*, no. XX, pp. 175–6; Housley, 'Jerusalem', p. 28.

3 'Nomen a Solemnibus trahit Solemniacum', *Carmina Burana*, ed. A. Hilka, *et al.*, Heidelberg 1930, no. 52, pp. 104–5. This poem was composed in the middle of the twelfth century for the celebration of the Feast of the Liberation of Jerusalem, in the abbey of Solignac in the neighbourhood of St Martial. See G. Spreckelmeyer, *Das Kreuzzugslied des lateinischen Mittelalters*, Munich 1974 (*Münstersche Mittelalter-Schriften*), 21, pp. 184–94 and here below, n. 53.

4 C. Kohler, 'Un sermon commémoratif de la prise de Jérusalem par les croisés, attribué à Foucher de Chartres', *ROL*, 8 (1900–1901), pp. 160–64.

5 Raimond d'Aguilers, p. 152.

fulfilment of Leviticus 26:11: 'And I will make my abode among you.'[6] Robert the Monk, expressing similar sentiments, had the crusaders declare in 1098: 'Jerusalem will be ours not through the gift of men but through the justice of heavenly judgement. For this judgement proceeds from the severity of God – Jerusalem will be ours.'[7] Baldric of Dol was also of the opinion that the conquest of Jerusalem should be ascribed not to crusaders but rather to God, who worked through them.[8] For Ekkehard of Aura the conquest was a result of: 'Deo conviatore et cooperatore'.[9]

European chroniclers writing in the three decades following the event reflect the influence of the historians. The Benedictine Hugh of St Fleury argued that 'it is clear that only God could have rendered such a victory through the Christian people'. He claimed that 'one should ascribe the victory not to the crusaders but to God who operated through them. It is clear that God is the only one who could achieve such a triumph through the Christian people.'[10] Another chronicler, the so-called *Annalista Saxo*, wrote that the expedition was 'not only a human one but a divine one'.[11] The chronicler of Monte Cassino argued that: 'In our times in the East Christ operated through pilgrim–knights (*peregrini milites*) ... who recovered the Holy City of Jerusalem and the Holy and Venerable Christ's Sepulchre!'[12]

As the liberation of Jerusalem was considered a major event in the history of Christendom, its contemporaries sometimes compared it with other similar events. Baldric of Dol compared it to Joseph of Arimathea taking Christ down from the Cross.[13] This was Baldric's version of a sermon preached beneath the wall of Jerusalem in the summer of 1099: 'Give heed to Christ, who today is banished from that city and is crucified; and with Joseph of Arimathea take him down from the Cross; and lay up on the sepulchre of your hearts an incomparable treasure, that desirable treasure; and forcefully take Christ away from these impious crucifiers. For every time those bad judges, confederates of Herod and Pilate, make sport or enslave your brothers, they crucify Christ. Every time they torment them and kill them they lance Christ's side with Longinus.'[14] Robert the Monk likened the conquest to two other miraculous acts of God: the Creation and the Redemption of mankind on the Cross. Hence his rhetorical question: 'But apart from the mystery of the healing cross, what more marvellous deed has there been since the creation of the world than what was done in modern times in this journey of our own men to Jerusalem?'[15]

6 *Kreuzzugsbriefe*, no. XXII, p. 178.

7 Robertus Monachus, p. 792.

8 Baldricus Dolensis, p. 101.

9 Ekkehardus, *Hierosolymita*, p. 23.

10 Hugo Floriacensis, p. 394.

11 *Annalista Saxo*, p. 728.

12 *Chronica Casinensis auctore Petro diacono*, *MGH.SS*, 7, p. 755.

13 Baldricus Dolensis, p. 101; Riley-Smith, *Idea and Reality*, pp. 49–53.

14 Baldricus Dolensis, p. 101.

15 Robertus Monachus, p. 723; Riley-Smith, 'An Approach to Crusading Ethics', *Reading Medieval Studies*, 6 (1980), p. 12. For the perception of the First Crusade as a miraculous

Raimond of Aguilers associated the liberation of Jerusalem not with Christ's Crucifixion, but with his Resurrection. For him the day of Jerusalem's liberation signified the enfeeblement of the pagans, the strengthening of Christianity and the renewal of the Christian faith; this day was a gift, an answer to the Church's prayers. Therefore, he testified: 'On this day we sang the Office of Easter, because on this day he who rose through his might from the dead, revived [his church] through his grace.'[16] Thus, like Christ through the Resurrection, the hosts of the First Crusade gave new life to the Church through the liberation of Jerusalem. Such comparisons of the conquest of Jerusalem with major events in Holy History reveal the extent to which the fate of Jerusalem was linked with Christ. Some, however, like the author of the *Chanson d'Antioche*, compared it with the conquest of Jerusalem by Titus and Vespasian, both instances symbolizing vengeance for Christ's Passion.[17] Almost a century later, the fall of the Holy City to Saladin would be interpreted as a veritable second Crucifixion of Christ.[18]

As the purification of the holy places of Jerusalem from Moslem pollution was one of the aims of the First Crusade, its conquest was presented not only as an act of liberation but, by some, as an act of purification as well. Thus, Fulcher of Chartres described Tancred purifying the Temple of the Lord from the Saracens' practices of 'rule of idolatry' there. Some of the historians of the First Crusade, such as Robert the Monk and Baldric of Dol, also presented the crusaders' occupation of Jerusalem as a divinely sanctioned act of purification.[19] Gilo (writing before 1120) referred to the time when 'the city had been purified'.[20] Writing in *c.* 1135 Orderic Vitalis argued in his description of the massacre of the Moslems that: 'They [the crusaders] pursued them and cut them down with such great hatred, because they had polluted the Temple of the Lord and the Church of the Holy Sepulchre, and because they shamefully defiled the Temple of Solomon and other churches for their own illicit uses.'[21] A decade later, in December 1145

---

and permanent demonstration of divine power, see Prawer, 'Jerusalem', pp. 742–50; Riley-Smith, *The First Crusade*, pp. 91–119, 139–52.

16 Raimond d'Aguilers, p. 151.

17 *La Chanson d'Antioche*, ed. S. Duparc-Quioc, 2 vols, Paris 1917–1918 (*Documents relatifs à l'histoire des croisades*, 11–12) vol. 1, pp. 28, 29. The *Chanson*, composed by a participant of the First Crusade, has survived only in a version reworked *c.* 1180 by the poet Graindor of Douai. The comparison mentioned is probably of Graindor himself and reflects the influence of the twelfth-century *Venjance Nostre Seigneur.* See Duparc-Quioc in *La Chanson d'Antioche*, vol. 2, pp. 99–100, 143–4; *Venjance Nostre Seigneur*, ed. Loyal A.T. Gryting, *The Oldest Version of the Twelfth-Century 'Venjance Nostre Seigneur'*, Ann Arbor, Michigan 1949 (repr. 1997), *passim.* For the conquest of Jerusalem as avenging Christ's Passion upon the Jews see below, ch. 3, pp. 42–4.

18 See below, ch. 9, pp. 162–3.

19 Fulcher Carnotensis, pp. 359–60; Robertus Monachus, p. 870; Baldricus Dolensis, p. 869; Cole, 'O God', pp. 94–7.

20 Gilo, *Historia ... of Gilo of Paris*, p. 249.

21 Orderic Vitalis, IX, 15, p. 172.

and again in March 1146, in *Quantum predecessores*, Pope Eugenius III evoked the theme of the conquest as the freeing of both the city and the Holy Sepulchre from Moslem defilement.[22]

Pondering the significance of the conquest of Jerusalem, chroniclers described the event mainly in biblical terms and moreover interpreted it as the literal, historical fulfilment of prophecies or omens of both Old and New Testaments. The general idea was that 'God has magnified his mercy by fulfilling in us what he promised in ancient times'.[23] There was, however, no agreement as to which prophecies were applicable. Guibert of Nogent wrote an exposition of Zechariah 12:1–10, in which the prophet predicts an apocalyptic siege of Jerusalem. Commenting on Zechariah 12:3: 'and all the nations of the earth will come together against her', Guibert argued that 'this is not to be understood as allegory but with the eyes of the soul it is to be looked on as recent history reported'.[24] Raimond of Aguilers described the event through a paraphrase of the book of Revelation: 'In the Temple of Solomon, that is, the Dome of the Rock, and the portico, crusaders rode in blood to the knees and bridles of their horses', thus echoing Revelation 14:20: '... and the blood flowed from the wine press, as high as a horse's bridle ...'.[25] Robert the Monk made particular use of Isaiah's great prophecy about Jerusalem (60:9–11, 15–16), identifying the sons of Jerusalem coming from afar with the crusaders. Thus, he declared, the liberation of Jerusalem was already foretold by the prophet, who said: 'For the coastlands shall wait for me, the ships of Tarshish first, to bring your sons from afar ...' (Isaiah 60:9). In his words: 'When it pleased God he led the Frankish people from the ends of the earth and by means of them wished to liberate Jerusalem from the unclean people. Isaiah had long ago prophesied this when he said: "I may bring your sons from afar ... The sons of the pilgrims will build up your walls and their kings minister to you." These many other things we find in the books of prophecy, which are relevant to this liberation done in our days.'[26] He argued that 'now we see in that event the promise which God made through the mouth of Isaiah the prophet'; what Isaiah said of the church spiritually (*spiritualiter*) was now fulfilled in reality (*realiter*), namely, that the gates of Jerusalem, formerly closed, 'shall be open continually, day and night they shall not be shut' (Isaiah 60:11) for pilgrims.[27] Ekkehard of

---

22 P. Rassow, 'Der Text der Kreuzzugsbulle Eugens III.', *Neues Archiv*, 45 (1924), pp. 302–5; Riley-Smith, *Idea and Reality*, p. 57.

23 Guibertus Novigentis, *Gesta Dei*, 4, pp. 237–40.

24 *Ibid.*, p. 238.

25 Raimond d'Aguilers, p. 150. For a paraphrase of Revelation in the description of the conquest see also *W.T.* 8, 20, p. 412. William, archbishop of Tyre, *A History of Deeds beyond the Sea*, trans. Emily Atwater Babcock and A.C. Krey, 2 vols, New York 1943, vol. 1, p. 371.

26 Robertus Monachus, p. 882.

27 *Ibid.*, pp. 739, 868, 881–2 and see Riley-Smith, *The First Crusade*, pp. 142–3; *idem*, 'An Approach to Crusading Ethics', p. 12.

Aura was more hesitant. In his history of the First Crusade he argued that various prophecies, such as Psalms 102:12–28, Isaiah 66:10–14, Ecclesiastes 31:15 and so on, though they refer *per anagogem* to 'our mother Jerusalem which is above', can yet be of value in their historical or verbal sense to simpler people.[28]

The surprising suggestion of the chroniclers that what had previously been regarded as allegorical or anagogical should from this point be treated as literal is hard to explain. It has recently been suggested that it resulted from a 'shock caused by the realization that passages in Scripture which had always been thought to have been susceptible only to allegorical interpretation, were suddenly being literally fulfilled.'[29] In fact, this was a complete revolution in biblical interpretation and cannot be explained in terms of 'shock treatment'. And it was even more of a revolution in regard to the very terms 'Jerusalem' and 'Zion'. Before the First Crusade, the two terms were commonly used in exegetical contexts to indicate 'the Church' or 'Christianity'.[30] This most traditional of Christian interpretations was now reverting to its historical and most literal meaning. Thus, for example, when Guibert of Nogent quoted the verse 'For out of Zion shall go forth the law and the word of the Lord from Jerusalem' (Isaiah 2:3), the word 'Zion' no longer referred to the Church as in the past, but to the city in the mountains of Judaea, because it was from that city that the Law and the Word of the Lord came forth.[31]

This transformation or, perhaps more correctly, this revolution may be explained first and foremost in the context of the general trend in scriptural exegesis at this time, which inclined towards a rather literal interpretation.[32] Thus, on the eve of the First Crusade (*c.* 1096), an anonymous monk of Saint-Père de Chartres, drawing up a charter for Nivelo, the Castellan, wrote that the latter was 'departing on pilgrimage to Jerusalem which until now has been enslaved with her sons'.[33] Thus, from the pen of an anonymous scribe, the allegorical Jerusalem of the Epistle to the Galatians (4:25) became the historical Jerusalem, enslaved by the Moslems. Similarly, Ivo of Chartres, *inter alia* a master at the university of Paris and one of the greatest authorities on scriptural exegesis, referred in a letter written

---

28 Ekkehardus, *Hierosolymita*, p. 38; Prawer, 'Jerusalem in Christian and Jewish Perspectives', p. 749.

29 Riley-Smith, *The First Crusade*, p. 143.

30 See H. de Lubac, *Exégèse médiéval. Les quatre sens de l'Ecriture*, Paris 1959, vol. 1, pp. 645–7; Prawer, 'Jerusalem in Christian and Jewish Perspectives', pp. 770–82.

31 Guibertus Novigentis, *Gesta Dei*, p. 138. For the interpretation of 'Zion' in the early Middle Ages see Hilarius Pictaviensis, *Tractatus Super Psalmos*, *CSEL*, 22, pp. 854–5; Pettorelli, 'Le thème de Sion', pp. 213–33.

32 See B. Smalley, *The Study of the Bible in the Middle Ages*, Oxford 1952, pp. 83–263; D. Luscombe, 'Peter Comestor', *The Bible in the Medieval World. Essays in Memory of Beryl Smalley*, ed. K. Walsh *et al.*, Oxford 1985 (*Studies in Church History. Subsidia*, 4), pp. 111–13.

33 *Cartulaire de l'abbaye des Saint-Père de Chartres*, ed. B.E.C. Guérard, *Collection des Cartulaires de France*, 2 vols, Paris 1840, vol. 2, p. 428.

in 1101–02 to Patriarch Daimbert of Jerusalem following the conquest of Jerusalem, using the same words of the Epistle to the Galatians, praising God that the Earthly Jerusalem was no longer a slave (*ancilla*) but a free woman (*libera*).[34]

Second, it should be remembered that to describe major events by means of Old Testament examples was not uncommon in European historiography of the early Middle Ages. For example, chroniclers describing the Hungarian raids on Southern Germany between the years 900 and 926 referred to the invaders by the good Old Testament name of 'Agareni'. Otto the Great was compared with David, Solomon, Hezekiah and Josiah. Arnoul, count of Flanders, identified himself with Judas Maccabaeus.[35] Bonizo of Sutri, writing in *c.* 1085–86, also compared a certain *miles dei* to Judas Maccabaeus.[36] Manegold of Lautenbach (*c.* 1085), in order to justify the papal fight against Emperor Henry IV, refers to the Maccabees and other Old Testament examples.[37] Ranger of Lucca, writing *c.* 1096, describes the contest between Gregory VII and Henry IV as similar to that between David and Goliath.[38] The *res gestae* of the First Crusade, culminating in the siege and conquest of Jerusalem, could be and indeed were described by its historians in biblical terms, rather than in relation to the New Testament. As M.D. Chenu remarked with regard to twelfth-century history, the Old Testament furnished a series of *exempla* and prefigurations. They were normative and typological actions fitted to provide models for human conduct. They also constituted a textual continuum between Scripture and the present – a present that was dynamic and important.[39] The events of the First Crusade were so miraculous and so powerfully relevant to scriptural prophecies that they had to be interpreted as its literal fulfilment. That the First Crusade is described in terms of the New Testament is therefore not surprising as wars, fights and battles were often so described in the

---

[34] Ivo Carnotensis, *Epistolae*, no. 93, cols 113–14.

[35] J. Dunbabin, 'The Maccabees as Exemplars in the Tenth and Eleventh Centuries', *The Bible in the Medieval World. Essays in Memory of Beryl Smalley*, ed. K. Walsh *et al.*, Oxford 1985 (*Studies in Church History*, *Subsidia*, 4) pp. 30–41. As early as 614, Zechariah, patriarch of Jerusalem, following his exile to Persia, wrote to his congregation, using Psalms 137:5: 'If I forget you, O, Jerusalem, let my right hand wither.' He also argued: 'You should know, my brothers, that we sit by the waters of Babylon and weep remembering Zion, the Holy Calvary and the venerated sepulchre', thus paraphrasing Psalms 137:1. This shows full identification of the Jerusalem Church with a biblical situation. See A. Linder, 'Jerusalem Between Judaism and Christianity in the Byzantine Period', *Cathedra*, 11 (1979), pp. 139–40 (in Hebrew).

[36] Bonizo de Sutri, *Liber ad Amicum*, *MGH.Ldt*, I, p. 599.

[37] Manegoldus, *Ad Gebehardum Liber*, *MGH.Ldt*, I, p. 236.

[38] Rangerius, *Vita Anselmi Lucensis*, *MGH.SS*, 30/2, p. 1161. And see also on the subject Erdmann, *The Origin*, pp. 118, 236, 242, 249–50.

[39] M.D. Chenu, *Nature, Man and Society in the Twelfth Century*, ed. and Eng. trans. J. Taylor and L.K. Little, Chicago 1968, pp. 146–61; Alphandéry, 'Les citations bibliques', pp. 139–57; Prawer, 'Jerusalem in Christian and Jewish Perspectives', pp. 742–50; Riley-Smith, *The First Crusade*, pp. 91–119, 139–59.

previous centuries. What is surprising is that we also find Jerusalem and its holy places being described, during the twelfth century, in biblical terms.[40]

One of the by-products of this new type of exegesis is a link, or even an entirely original relationship, between the two Jerusalems, the Earthly and the Heavenly. Earthly Jerusalem now merges with Heavenly Jerusalem. According to Baldric of Dol, the leader of the crusade, in his speech before the final assault on Jerusalem, declared: 'The city which you see is the cause of all our labours. This Jerusalem is the likeness (*instar*) of Heavenly Jerusalem; this city has the form of that to which we aspire ... Clearly, if you will consider well and diligently, this Jerusalem which you see, to which you come, at which you are present, prefigures and reflects the heavenly city.' What would happen if the Moslems remained in Jerusalem? 'Obviously,' he says, 'one should fear that the heavenly city should not be closed to us, should not be taken from us, if we in our indolence should be disowned of our house by malicious strangers.'[41] Albert of Aachen called Jerusalem the gateway to the Heavenly Fatherland (*coelestis patria*).[42] Ekkehard described the Holy City as 'paradisus terrestris'.[43] According to Gilo, Count Eustace of Boulogne fought for 'a twofold kingdom, since he was looking forward to both Jerusalems: he fought in the one so that he might have life in the other'.[44] The Benedictine Rupert, abbot of Deutz, presented Jerusalem as the *simile* of the Celestial and claimed that it was in the Earthly City and not in the Heavenly one that God built his Temple and worshipped there.[45]

The presentation of the Earthly Jerusalem as the likeness of the Heavenly City was indeed a revolutionary change in the attitude to the Earthly City as well as in the relationship between the two. Already some eleventh-century pilgrims presented Jerusalem as the gateway to the Celestial Jerusalem.[46] It is however only following the crusader conquest of Jerusalem that Earthly Jerusalem emerged as the figure of the Heavenly Jerusalem. As Peter Raedts has pointed out: 'From now on the Heavenly City was not only foreshadowed by the Church and its liturgy, but also, and perhaps foremost, by the city in Palestine.'[47]

The miraculous, even divine, nature of the crusader conquest of the Earthly Jerusalem was also a central theme of the celebrations of the event in Jerusalem in July 1099, as well as of the Feast of Liberation of Jerusalem. According to

---

40 See below, ch. 6, pp. 91–101.

41 Baldricus Dolensis, pp. 100–101; Prawer, 'Jerusalem in Christian and Jewish Perspectives', pp. 749–50 and below, ch. 7, pp. 118–40.

42 Ekkehardus, *Hierosolymita*, p. 40.

43 Albertus Aquensis, p. 482.

44 Gilo, *Historia ... of Gilo of Paris*, p. 245.

45 Rupertus Tuitiensis (of Deutz), *De Victoria Verbi Dei*, ed. R. Haacke, *MGH.Quellen zur Geistesgeschichte des Mittelalters*, 5, Weimar 1970, p. 190. Rupert died in 1129 but stopped writing in 1124–5.

46 For this subject before the First Crusade see below, ch. 7, pp. 111–15.

47 Raedts, 'Jerusalem: Purpose of History', p. 37 and below, ch. 7, pp. 118–24.

Raimond of Aguilers the conquerors of Jerusalem celebrated their conquest for some six or seven days. The celebrations were terminated on 22 July, the Octave of 15 July, when the first ruler of Jerusalem was chosen by the leaders of the crusade. The celebrations included a procession of clergy and laity alike from the Holy Sepulchre to the Temple of the Lord (Dome of the Rock), singing songs of exultation 'and making offerings and most humble supplications'.[48] Raimond of Aguilers described the sentiments of the crusaders on 15 July as follows: 'How they did applaud, shouting and singing a new song to the Lord (Psalms 32 (33):3; 97 (98):1–2) and their minds offered praises to the victorious and triumphant Lord, which cannot be expressed by words. A New Day (Isaiah 65:17–19; 2 Peter 3:13; Revelation 21:5), a new rejoicing, a new and eternal felicity of the consummation of labour and devotion, urged from everybody new words, new songs. This is the Day which the Lord God has made, let us rejoice and be glad in it (Psalms 117 (118):23–24). On this day [15 July] the Apostles broke out from Jerusalem and dispersed over the whole Universe. On the same day the sons of the Apostles vindicated themselves in the eyes of God and of their ancestors through the city and the fatherland (*patria*). This day, the Ides of July, will be celebrated to the praise and glory of the name of God who, moved by prayers in his church, gave the city and the fatherland which he promised by oath to the Fathers; this oath he fulfilled faithfully and with blessings to their children.'[49] Fulk of Chartres described the day as: 'Oh this day so ardently desired! Oh most memorable times! Oh this deed before all other deeds! Desired indeed, because in the inner longing of the heart it had always been hoped by all believers in the Catholic faith that the place where the Creator of all creatures, God, made man, where He, in His manifold pity for mankind, by his birth, death and resurrection, conferred the gift of redemption, would be restored to its pristine dignity by those believing and trusting in Him ... It was a truly memorable time, and justly so, because in this place everything that the Lord God our Jesus Christ did or taught on earth, as a man living amongst men, has been recalled and renewed in the memory of true believers. And the same work which the Lord has chosen to accomplish through

---

48 Raimond d'Aguilers, pp. 151–2.

49 *Ibid.* For the day of the conquest of Jerusalem as that of the Dispersion of the Apostles see the letter of Patriarch Daimbert to the pope (Laodicea, September 1099) in *Kreuzzugsbriefe*, no. XVIII, p. 171. See also Ekkehardus, *Hierosolymita*, p. 23; *Annalista Saxo*, p. 732. It is worth mentioning that Eusebius quoted Psalms 97 (98):1–2 in the first chapter of the tenth book of his *Historia Ecclesiastica*, commenting: 'As those lines command me let me now obediently sing aloud the new song, because after those terrifying darksome sights and stories I was now privileged to see and celebrate such things as in truth many righteous men and martyrs of God before us desired to see on earth and did not see.' Eusebius described here the victory of the Church under Constantine the Great as the realisation of prophecies like those of Psalms 37:33–36 and Psalms 46:8–9. *Historia Ecclesiastica,* 10, 1; *Eusebius Pamphili, Ecclesiastical History*, Eng. trans. Roy J. Defferrari, Washington, DC, 1955, pp. 239–40.

His people, His dearly beloved children and family, chosen, I believe for this task, shall respond and continue memorable in the tongues of all nations until the end of time.'[50]

The celebrations of the conquest by its conquerors in July 1099 would appear to have influenced the liturgy of Jerusalem. From 1099, the 'Feast of the Liberation of Jerusalem' commemorated this miraculous event not only in the crusading kingdom but also in the West. The original festivities were characterized by a public liturgy of joy and thanksgiving. On the vigil of the feast, Vespers was held on the evening of 14 July, and the celebrations continued through Matins and Lauds to the Mass proper for 15 July. On 15 July a solemn procession made its way from the Holy Sepulchre to the Temple of the Lord 'et inde divertentes convertimur contra illum locum ubi civitas capta fuit. Facta statione, fit sermo ad populum'. The sermon was based on Fulcher of Chartres' description of the city and narrative of the siege. The sequence, namely the poem 'Rejoice Jerusalem' (*Jherusalem Laetare*) included the following verses: 'Jerusalem the earthly / The origin of the Celestial / Rejoice in the New Feast / Jerusalem be praised' (*Jherusalem terrestris / principium celestis / laetare novis festis / Jherusalem exulta*). The poem stressed the divine dimension of the victory, manifested in its miraculous nature. Similarly as other eyewitnesses of the conquest of Jerusalem, the poet described the crusaders as the followers of their supreme *dux* and 'unique emperor'; God himself had granted this victory to the crusaders, described as *nova militia*. This view of God as the Victorious Lord of Hosts, Liberator and Saviour of Jerusalem was highlighted also through several psalms and antiphonal texts included in the entire service. The verses used most often are Zechariah 9:9: 'Rejoice greatly, O daughter of Zion! / Shout aloud, O daughter of Jerusalem! / Lo, your king comes to you; / triumphant and victorious is he, humble and riding on an ass ...' (quoted four times); Isaiah 60:1: 'Arise, shine; for your light has come, / and the glory of the Lord has risen upon you' (also four times); Isaiah 2:3: 'It shall come to pass in the latter days / that the mountain of the house of the Lord / shall be established at the highest of the mountains, / and shall be raised above the hills, and all the nations shall flow to it; / and many peoples shall come, and say: / Come let us go up to the mountain of the Lord; / to the house of the God of Jacob; / that he may teach us his way / and that we may walk in his paths. For out of Zion shall go forth the law, and the word of the Lord from Jerusalem' (twice); Isaiah 66:10: 'Rejoice with Jerusalem, and be glad for her, / all you who love her; rejoice with her in joy; / all you who mourn over her' (twice). The office of Vespers was dominated by the emotions of separation and longing of people exiled from Jerusalem: 'By the waters of Babylon, there we sat down and wept, when we remembered Zion ... How shall we sing the Lord's song in a foreign land? If I forget you, O, Jerusalem, let my right hand wither' (Psalms 137:1, 4–5). The liberation is perceived as the return

50 Fulcher Carnotensis, p. 304.

of the exiles: 'When the Lord restored the fortunes of Zion we were like those who dream' (Psalms 126:1). Based upon texts such as the description in Revelation of the Heavenly Jerusalem (Revelation 21:1–3; 21, 25) and the hymn *Urbs beata Hierusalem*, used since the eighth century in the liturgy for dedication of churches, the liturgy of 'In Festivitate Sancte Hierusalem' echoes the emotions of the conquerors like Raimond of Aguilers and present the earthly Old Testament's Jerusalem as prefiguring the Christian Jerusalem. Moreover, it presented the earthly Christian Jerusalem as the *figura* of the heavenly one. By using in the liberation office the *topos* of dedication, the liturgist assimilated the earthly Jerusalem into a new church. In so doing, he was expressing the emotions of both Raimond of Aguilers and Fulcher of Chartres, who described the liberation of Jerusalem in terms of a canticle, a new song of praise to the Lord.[51]

While the original service was characterized by a public liturgy of joy and thanksgiving, two later additions, which were introduced sometime between 1099 and 1149, transformed the whole character of the feast. The two additions were commemorative services, one for the crusaders killed in the battle for Jerusalem, and the other for Godfrey of Bouillon, who died on 17 July. Further changes took place following the dedication of the Crusaders' Church of the Holy Sepulchre on 15 July 1149; the celebrations of dedication were from then on combined with the liberation service. The post-1149 festivities included a solemn procession led by the patriarch from the Holy Sepulchre to the Church of the Temple of the Lord, the Dome of the Rock. Here, at the southern entrance, prayers were recited including the following: 'Grant that we, who rejoice in the liberation of your city of Jerusalem, will be worthy of inheriting the Heavenly Jerusalem.' Then the procession continued across the Temple Esplanade to the area outside the eastern wall of the city – and the entire Temple Mount – to the graveyard of the crusaders who had given their lives for the conquest of the city.[52] The liturgical celebrations also included sermons composed for the occasion. One of the best-known sermons was, as mentioned above, based on the writings of Fulcher of Chartres; another, composed in Catalonia or Provence and copied at the Catalonian monastery of Ripoll in the late twelfth century, includes a description of the conquest of Jerusalem. This was

[51] For the edition of the text of *In Festivitate Sancte Hierusalem* from London, British Library, Additional Ms. 8977, fols 134r–135r and its interpretation see A. Linder, 'The Liturgy of the Liberation of Jerusalem', *Medieval Studies*, 52 (1990), pp. 110–31. Linder dated the text to the first half of the twelfth century, between 1106–27 and 1149 (*ibid.*, pp. 122–3). For the poem *Hierusalem Laetare* see also below, n. 53. For the *Urbs beata Hierusalem* see *AH*, LI, 110.

[52] Linder, 'Liturgy of the Liberation of Jerusalem', pp. 110–12; C. Kohler, 'Un ritual et un bréviaire du Saint Sépulchre de Jérusalem (XIIe–XIIIe siècles)', *ROL*, 8 (1900–1901), pp. 427–8, 436–7; Johannes Wirziburgensis, *Descriptio Terrae Sanctae, Peregrinationes Tres*, pp. 139–40; J. Prawer, *The Latin Kingdom of Jerusalem: European Colonialism in the Middle Ages*, London 1972, p. 185.

based on Raimond of Aguilers and included four exhortations – two praising the liberators, one against the Jews and another praising Jerusalem, beginning with the words 'Laetare Hierusalem' as well as the poem composed for the occasion, the 'Rejoice Hierusalem' (*Jherusalem Laetare*).[53]

In the West the feast was still celebrated in 1146 or 1147, when Gerhoh of Reichersberg completed his commentary on the book of Psalms. Commenting upon Psalm 33, he described the conquest of Jerusalem (relying on Robert the Monk) and remarked that this great victory was still being celebrated.[54] The liturgy for the festival included the singing of the hymn 'Laetare Jherusalem' (that is Isaiah 66:10: 'Rejoice with Jerusalem, and be glad for her all you who love her, rejoice with her joy') as introit and the reading of Matthew 21 as *Evangelium.* As both were part of the liturgy of the feast in the Jerusalemite Church (as described by John of Würzburg, who stayed in the city in *c.* 1160), one may assume that this liturgy was uniform throughout the Christian world.[55]

The First Crusade and the conquest of Jerusalem were likewise commemorated in the stained glass medallions of the so-called 'Crusading Window' in the abbey church of Saint-Denis, installed either between 1146 and 1148 during the abbacy of Suger, or under his successor Odo of Deuil (*c.* 1158). Two of the medallions depict the siege and capture of Jerusalem with the inscription: 'Jerusalema a Francis expugnata'; another of the medallions has nine seated figures representing martyred crusaders. The figures are seated in groups of three; four of them hold palms and the central three receive crowns from above and, like their six already crowned companions, gesture to signify their agreement to be crowned in martyrdom. Like the liturgical celebration of the feast of the Liberation of

---

53 'Account of the capture of Jerusalem', pp. 640–57. For the poem *Hierusalem Laetare* (composed shortly after the conquest of Jerusalem and inspired by the liturgy of 'Laetare Jerusalem' Sunday) see also *Poésies populaires latines*, ed. M.E. du Méril, Paris 1843, pp. 255–60. Another poem was composed for the same feast in the abbey of Solignac in the middle of the twelfth century, 'Dressed in Feast was Solignac called' (*Nomen a Solemnibus trahit Solemniacum*). It described the conquest as a victory over Dagon, Amalek and the descendants of Hagar. Jerusalem is presented as the most noble city, the first city of the king (Christ), where he was crucified; it is also the holy city of the tabernacle, the ark of the Covenant and the Temple. Inspired by Isaiah 24:23, the author exulted in its glory which surpasses that of all the other cities of the world 'with such a clear light that it outdoes that of the moon and the sun'. See *Carmina Burana*, no. 52.

54 Gerhoh Reichersbergensis, *Commentarius in Psalmos*, *MGH.Ldt*, 3, p. 431. For the celebration of the feast see also *Historia Peregrinorum*, *RHO.Hocc*, 3, p. 224. The *Historia* was composed *c.* 1140. It is impossible to gauge the popularity of the feast in the West; possibly it was limited to monasteries closely connected with the Holy Land, especially with the Church of the Holy Sepulchre, monasteries such as Ripoll or Saint-Denis. See A.R. Brown and M.W. Cothren, 'The Twelfth-Century Crusading Window of the Abbey of Saint-Denis: *Praeteritorum enim Recordatio Futurorum est Exhibitio*', *Journal of the Warburg and Courtauld Institutes*, 49 (1986), pp. 20–21 and n. 90.

55 See above, n. 52. See also Johannes Wirziburgensis, *Peregrinationes Tres*, p. 139.

Jerusalem, the 'Crusading Window' of Saint-Denis was intended to inspire viewers to ponder the glorious crusading victories of the past and their relevance to the present and future life of the Church. It was to remind them that the struggle to defend Christ's faith and the land where He once lived would be rewarded on earth with victory and in heaven by the crown of martyrdom, which was an eternal possession; those who pursued Earthly Jerusalem would achieve her heavenly counterpart.[56]

One of the immediate results of the crusader conquest of Jerusalem, as revealed by sources dating from *c.* 1099–1135, was a change in the attitude to the Earthly Jerusalem. First of all, a complete revolution in biblical interpretation occurred with regard to the term 'Jerusalem', now restored to its historical and most literal meaning. Similarly, biblical prophecies regarding Jerusalem were, as part of the accounts of the conquest of Jerusalem, interpreted as the literal, historical fulfilment of prophecies or omens of the Scriptures. It thus follows that the term 'Jerusalem', as well as prophecies referring to Jerusalem, interpreted in the past as allegorical or anagogical, were henceforth treated in the authoritative exegesis, as literal. The revolution in the attitude to the Earthly Jerusalem was particularly striking in the light of the new relationship between the Heavenly and the Earthly Jerusalem. Relegated to oblivion in the centuries previous to its conquest by the learned scriptural exegesis, the Earthly Jerusalem was to be presented as the likeness of Heavenly Jerusalem and as the gate to Jerusalem above.

The event itself, the conquest and the restitution of Jerusalem and its holy places, and above all the Holy Sepulchre, to the Christian faith, was represented as God's will, as the act of God realized through his believers. It was also seen as a miracle, as an event to be compared to such major events as the Creation, the Crucifixion and the Resurrection. Within the accounts of its conquest Jerusalem appeared as *regnorum mundi caput*,[57] *urbs regalis*,[58] *terra repromissionis; mater nostrae redemptionis ac fidei; paradisus terrestris*[59] *quasi orbis umbelicus.* Jerusalem was, above all, the place consecrated by Christ's Incarnation, his Miracles, Passion, Resurrection and Ascension.[60] It was also described as 'the inheritance and the holy Temple of God before the Lord walked and appeared there ... the place where the blood of the Son of God, holier than heaven or earth, poured out and where his body, dead to the fearful elements, rested in the grave ...'.[61]

---

[56] See Brown and Cothren, 'The Twelfth-Century Crusading Window', pp. 1–40. For martyrdom and Jerusalem see below, ch. 7, pp. 120–1.

[57] Rupertus Tuitiensis, p. 338.

[58] Hugo Floriacensis, p. 394.

[59] Ekkehardus, *Hierosolymita*, p. 40.

[60] See, for example, Hugo Floriacensis, p. 394; Ekkehardus, *Hierosolymita*, p. 40.

[61] Guibertus Novigentis, *Gesta Dei*, pp. 137–8. Eng. trans. Riley-Smith, *The First Crusade*, p. 146.

The entire mood of the conquerors, as reflected in both the descriptions of celebrations of their victory and in the liturgy of the Feast of the Conquest of Jerusalem, was one of a new beginning, a new age; it is a new day, a new rejoicing, a new song. The great victory of the crusaders and its celebrations in Jerusalem were described through Psalms 97 (98):1–2: 'O sing to the Lord a new song, for he has done marvellous things! ...'; Psalms 117 (118):23: 'This is the Lord's doing; it is marvellous in our eyes'; Isaiah 65:18: 'But be glad and rejoice for ever in that which I create; for behold, I create Jerusalem a rejoicing, and her people a joy.' The main theme of the celebrations as well as the liturgy of the Feast of the Liberation of Jerusalem was however Isaiah 66:10: 'Rejoice with Jerusalem, and be glad for her all you who love her ...'.

Chapter 3

'Inheritance of the Lord': Justifications of Christian Rule in Jerusalem

The conquest of Jerusalem by the armies of the First Crusade was the end result of Christianity's claim for physical domination of the city; it was hailed as a miracle and a manifestation of Divine Will. This claim was hardly new, and there are indications of its existence as far back as the tenth and eleventh centuries, though disguised until 1074 (when Gregory VII considered the reconquest of Jerusalem) as an eschatological inevitability.[1] Adso of Montier-en-Der wrote in his widely read and highly influential *De Ortu et Tempore Antichristi* (*c.* 950) that the Antichrist would wage war in Jerusalem not against Jews or Gentiles, but against Christians. The presence of Christians in Jerusalem, perhaps even their domination, appeared in Adso's treatise as a precondition for the beginning of the events of the End of the World, for the arrival in Jerusalem of Antichrist, the precursor of the end.[2] Adso and eleventh-century versions of his treatise described the last Roman emperor who, after gaining victory over all his rivals, would rule the entire Roman Empire. On the eve of the arrival of Antichrist he was to come to Jerusalem and deposit his crown and sceptre on the Mount of Olives.[3] As early as the tenth century, this legendary figure was associated in the West with the struggle of Christianity against Islam. This identification became still more pronounced in the eleventh century, and in some versions of Adso's prophecy, propagated in the West under the title *Prophecy of the Last Emperor*, the emperor defeated the Moslems on the eve of his arrival in Jerusalem. Accordingly, Benzo of Alba, in 1086, advised the Emperor Henry IV to conquer Rome, then Constantinople and Jerusalem; thereafter duly honouring Christ's Sepulchre and

---

1 For a discussion of the problem of the claim for domination of Jerusalem as predating the First Crusade, see Prawer, 'Jerusalem in Christian and Jewish Perspectives', pp. 741–50, 800–802, *versus* Bredero, in *ibid.*, pp. 797–800. Bredero did not present any textual evidence in favour of his view that the political claim to Jerusalem predated the First Crusade. Riley-Smith, *The First Crusade*, pp. 48–9, 55, 108, 146, assumed its existence on the eve of the crusade, but produced textual evidence dating from 1098 onwards, emanating mostly from chroniclers who wrote in the first decade following the conquest of Jerusalem.

2 Adso Dervensis, p. 28. For this treatise see O. Prinz, 'Eine frühe abendländische Aktualisierung der lateinischen Übersetzung des Pseudo-Methodios', *DA*, 41 (1985), pp. 1–23 and below, ch. 8, pp. 147–54.

3 Adso Dervensis, pp. 25–6, 46–7, 53, 71–2.

other holy places, he would be crowned 'ad laudem et gloriam viventis in secula seculorum'.[4] This was actually the plan of the emperor's foe, Pope Gregory VII, who in 1074 considered leading an expedition of *fideles Sancti Petri* to liberate the Eastern churches as well as the Holy Sepulchre. In a letter dated 1074 to Emperor Henry IV, Pope Gregory wrote that 'already fifty thousand men are preparing, if they can have me as their leader and prelate, to take arms against the enemies of God and push forward to the Sepulchre of the Lord under His supreme leadership'.[5]

However, only after the conquest of Jerusalem were the justifications of Christians' rights to the Holy City clearly defined for the first time. During the First Crusade, its leaders explained their right to the conquered territories in terms of the 'reconquest' of what rightfully belonged to the Christians. This clearly showed that the Augustinian idea of 'Just War' (*bellum iustum*), which the papacy had since the ninth century associated with the war against the heathen, was by then already linked with the crusade. The conquests were seen as just because they represented nothing other than the recovery of territories rightfully belonging to Christendom which, before being seized by the Moslems, had been part of the Christian Roman Empire.[6] In accordance with this theory, writing in 1098 from Antioch to the ruler of Damascus, the leaders of the crusade declared that they planned to conquer only those lands that had once belonged to the Greeks.[7] The crusaders justified the conquest of Antioch, however, not only by the theory of recovery but in the light of its status as the first see to have been created by St Peter. They were simply reclaiming what 'belongs to God and Christians, for the blessed Peter converted it long ago to the faith of Christ by his preaching'; they were recovering the inheritance of St Peter (*haereditas beati Petri*), an inheritance granted to Peter by Christ himself.[8]

The conquest of Jerusalem and the right of Christendom to dominate it were, however, justified differently than for other territories. Robert the Monk, for example, invoked the concept of recovery in regard to the Holy Land in general:

---

4 Benzo Albensis, *Ad Heinricum IV Imperatorem Libri IV, MGH.SS,* 11, pp. 603, 605–6; C. Erdmann, 'Endkaiserglaube und Kreuzzugsgedanke im 11. Jahrhundert', *Zeitschrift für Kirchengeschichte*, 51 (1932), pp. 394–414.

5 Gregorius VII, *Register*, ed. E. Caspar, *MGH.Epistolae Selectae in usum scholarum*, 2/1, p. 166; H.E.J. Cowdrey, 'Pope Gregory VII's "Crusading" Plans of 1074', *Outremer*, pp. 27–40; Riley-Smith, *The First Crusade*, p. 78; McGinn, '*Iter Sancti Sepulchri*', p. 43.

6 For the Augustinian concept of *bellum iustum* and its history in the early Middle Ages, see F.H. Russell, *The Just War in the Middle Ages*, Cambridge 1975, pp. 16–54. See also Riley-Smith, *What were the Crusades?*, London 1977, pp. 18–33.

7 *Kreuzzugsbriefe*, no. XV, p. 160; Riley-Smith, *The First Crusade*, p. 108.

8 Writing in July 1098 from Antioch to the West, Anselm of Ribemont recalled that a delegation dispatched to Kerbuqa, the governor of Mosul, referred to the right of the Christians to the 'inheritance of St Peter'. *Kreuzzugsbriefe*, no. XV, p. 160. See also *Gesta Francorum*, p. 65; Petrus Tudebode, pp. 62–3, 108; Riley-Smith, 'The First Crusade and St Peter', pp. 41–63 esp. pp. 50–53.

'The land ... is not theirs, although they have possessed it for a long time, for from the earliest times it was ours and your people [that is, the Saracens] attacked it and took it ... and so it ought not to be yours just because you have held it for a long time for by heavenly judgement it is now decreed that what was unjustly taken from the Fathers should be mercifully returned to the Sons.'[9] However, like other chroniclers of the First Crusade, Robert made a clear distinction between Jerusalem and other territories. The conquest of Jerusalem was legitimate not by virtue of the idea of recovery but because it was Christ's personal possession and His inheritance. Thus, Robert the Monk argued, Pope Urban II had declared at Clermont that the Holy Places of Jerusalem were 'given by God as a possession to the children of Israel', and had exhorted the believers 'to take the road to the Holy Sepulchre, rescue that land from a dreadful race and rule over it yourselves ...'.[10]

Other historians of the First Crusade justify the conquest, as already mentioned above,[11] claiming that the Moslems' presence in Jerusalem represents a pollution from which the City has to be cleansed. Ekkehard of Aura cites a speech in which the participants of the crusade of 1101 were told: 'You are the people of the inheritance of Christ (*Christi hereditas*) ... He made you His to sacrifice voluntarily your blood and that of your brothers in order to purify Jerusalem, sanctified by His death.'[12] This need to cleanse Jerusalem is mentioned also by Albert of Aachen. God, says Albert, appeared in a dream to Peter the Hermit and commanded him to exhort the believers 'to cleanse the Holy Places of Jerusalem'.[13] Both William of Malmesbury and the anonymous author of a mid-twelfth-century Catalan or Provençal description of the conquest of Jerusalem, composed for the celebration of the 'Feast of the Liberation of Jerusalem', presented the Moslem rule of Jerusalem as usurpation and sacrilege. William of Malmesbury, who saw the First Crusade as part and parcel of the six centuries'-long struggle between Christianity and Islam, declared that the Saracens 'usurp even the sepulchre of our Lord ... and sell to our pilgrims admissions to that city, which ought ... to be open to Christians only'.[14] William of Tyre, the greatest historian of Outremer, writing in the 1170s, also raised the same point, arguing that the violation of such a sacred place as Jerusalem justified the massacre of its non-Christian population, not only as a Just War but also

9 Robertus Monachus, p. 792 and see also *ibid.*, p. 750. For the same idea of 'recovery' see also Guibertus Novigentis, p. 204; Baldricus Dolensis, p. 74; Riley-Smith, *First Crusade*, pp. 145–6 and see also Caffaro, *De Liberatione*, p. 63.

10 Robertus Monachus, p. 728.

11 See above, ch. 1, pp. 15–17.

12 Ekkehardus, *Chronicon*, p. 222. See also Baldricus Dolensis, p. 14; Blake, 'The Formation', p. 22. For the crusade of 1101 see Riley-Smith, *First Crusade*, pp. 120–34.

13 Albertus Aquensis, p. 272. For the role of Peter the Hermit in the First Crusade and its historiography see E. Blake and C. Morris, 'A Hermit Goes to War: Peter and the Origins of the First Crusade', *Studies in Church History*, 22 (1985), pp. 79–107.

14 William of Malmesbury, pp. 394–5. 'The Text of the Account of the Capture of Jerusalem', p. 647.

as the execution of a divine commandment.[15] Clearly, Jerusalem, the scene of Christ's Passion and Resurrection and in particular the site of His Sepulchre, the liberation of which was the goal of the First Crusade, required different treatment in regard to the justification of its conquest compared with other territories. The Holy City was Christ's personal possession and His inheritance, to be cleansed from the pollution of 'unclean races' and ruled by His believers. And indeed, as Penny Cole has pointed out, since the First Crusade the conviction that the Moslems were polluters of Christ's sacred things and the violators of His people loomed large in twelfth-century writings. According to Cole, this conviction 'served a real purpose: it was versatile to the point where any Moslem action against a Christian position, such as Edessa, could be interpreted as unprovoked aggression against Christ; it was essential to the idea of the crusade as a cleansing operation and a divinely mandated war of reprisal, and it justified the crusaders' violence, however excessive or wanton, against Moslems'.[16]

This concept, however, served yet another purpose, namely the justification of the crusaders' conquest of Jerusalem and their rule in the Holy City. Thus Fulcher of Chartres interpreted the crusaders' capture of Jerusalem as a stage in a comprehensive rite of purification. Condemning the Moslem presence in the Temple of the Lord (*Templum Domini*), the place of Christ's presentation as a pollution, Fulcher argued that the presence of Moslems there, similarly to anywhere else in the Holy City, was an outrage. Jerusalem stood in a unique relationship with God, representing His domain sanctified for Christians by the saving life and redeeming death of his Incarnate Son. In Fulcher's opinion, in occupying the Holy City the Moslems had violated divine right.[17] And two other clerics and participants of the First Crusade, Raimond of Aguilers and Peter Tudebode, advanced similar ideas. Raimond had William Hugh and the clergy call upon the crusaders to pray to God to judge those who occupied Jerusalem shamefully, defiled the site of his Passion and Burial, and who were trying to impair the operation of the grace of redemption by withholding access to Christ's sacred places.[18]

In the writings of the three French Benedictine authors of the histories of the First Crusade, namely Robert the Monk, Baldric of Dol and Guibert of Nogent, the theme of pollution became directly linked with Psalms 78 (79):1: 'O God, the heathen have come into inheritance: they have defiled thy holy temple; they have laid Jerusalem in ruins.' Consequently Robert presented the crusaders' occupation of Jerusalem as a divinely sanctioned act of purification. For Baldric of Dol, it was the gentile pollution and maltreatment of the holy places that had first led Pope

---

15 *W.T.*, I, 15, pp. 131–5. Eng. trans., I, pp. 88–93.

16 Cole, 'O God', p. 111.

17 Fulcher Carnotensis, pp. 286–7: Cole, 'O God', pp. 89–91. On 'Temple of the Lord' see below, ch. 6, pp. 91–107.

18 Raimond d'Aguilers, p. 145; Cole, 'O God', pp. 93–4 and see also Petrus Tudebode, p. 99.

Urban II to proclaim a crusade; the pope was instructed by God, who decreed that the polluters of his sanctuaries should be eliminated. Guibert of Nogent similarly presented the goal of the First Crusade, that is, the conquest of Jerusalem as ensuring 'that the holiness of that city and the glory of his Sepulchre will be cleansed, for the gentiles by their presence continually sully them in so far as they can'.[19] Also, yet another of the chroniclers of the First Crusade, Albert of Aachen, saw the aim of the crusade in terms of purgation; according to him Christ spoke in a dream to Peter the Hermit encouraging Peter to preach the crusade: 'for the purification of the holy places of Jerusalem'.[20]

From the 1140s this theme was used in crusade propaganda as both justification of the Frankish rule in Jerusalem and as the *causa iuxta* of the crusades dispatched to the crusader kingdom. In his encyclical, *Quantum predecessores* of 1 March 1146, Pope Eugenius III evoked the example of the First Crusade and interpreted its achievement as freeing Jerusalem and the Lord's Sepulchre from 'the filth of the pagans'; he warned the believers that the 'bravery of the fathers' could be proved 'to be diminished in the sons'.[21] Both Bernard of Clairvaux and Peter the Venerable of Cluny used the subject in their crusade propaganda. Bernard in his letter to the duke and people of Bohemia (1147) informed them that 'evil people have begun to occupy this land of the new promise and unless someone resists them they will be feasting their eyes upon the sanctuary of our religion and will try to stain that bed, on which for our sake slept our life in death; they will profane the Holy Places, the places, I say, purpled with the blood of the immaculate lamb'.[22] Peter of Cluny, in his sermon of 1146 on the theme of the Holy Sepulchre, praised the first crusaders for 'having purged with the swords of piety the place and the house of heavenly purity from the fifth of the impious'.[23]

Another vindication of the crusaders' capture and domination of the Holy City appealed to the concept of the 'Promised Land'. Chroniclers of the First Crusade alternatively used the terms 'inheritance of Christ', 'possession of God' and 'Promised Land' to fortify the Christian claim to Jerusalem.[24] The

---

19 Robertus Monachus, p. 870; Baldricus Dolensis, pp. 11–14; Guibertus Novigentis, pp. 137–8. See also Orderic Vitalis, 5, pp. 96, 100, 172, who wrote under the influence of the histories of Fulcher and Baldric, Cole, 'O God', pp. 94–101.

20 Albertus Aquensis, p. 273.

21 Eugenius III, *Quantum praedecessores*, pp. 302–5; Riley-Smith, *Idea and Reality*, pp. 57–9; Cole, 'O God', p. 101.

22 Bernardus Claraevallensis, *Opera*, vol. 8, n. 458, p. 435; Riley-Smith, *Idea and Reality*, p. 97.

23 Petrus Venerabilis, *Sermo in Laudem Sepulchri Domini*, p. 247. For this sermon see below, ch. 5, p. 71.

24 Ekkehardus, *Hierosolymita*, pp. 14, 32–3: 'Christi haereditas'; Guibertus Novigentis, p. 137: 'haereditas Dei'; Robertus Monachus, p. 728: 'possessio Dei'; Monachus Anonymus Scaphusensis, *De Reliquiis Sanctissime Crucis et Domini Sepulchri*, *RHC.Hocc* 5, p. 335: 'haereditas Dei'; *Chronicon S. Huberti Andaginensis*, *MGH.SS*, 8, p. 615: 'terra repromissionis'.

description of Jerusalem and the holy places, as well as all of Palestine, as the 'Promised Land', dates from the fourth century. Egeria (*c.* 381–384) referred to 'Palaestina' as the 'Promised Land'.[25] Jerome, writing in 398, confined the term to Jerusalem, interpreting God's promise to Abraham, 'Go from your country and your father's house to the land that I will show you' (Genesis 12:1), as a promise to Christ. Later, however, in 414, he referred to the whole of Palestine as the 'Promised Land' because of the Passion and Resurrection of Christ.[26] The concept of Jerusalem as the 'Promised Land' thus predated the First Crusade. Employment of that term, together with 'inheritance of Christ', used to justify Christian rule in the Holy City, was a by-product of the crusader conquest. It was during the First Crusade and the first subsequent decade that these two theological concepts became, so to speak, the two main formulae defining Christendom's right to the political domination of Jerusalem. For example, Raimond of Aguilers, describing the conquest of Jerusalem, claimed that on 'this day ... God ... returned to His People His City as well as His Homeland which He had pledged to the Fathers'.[27] Thus, to the chroniclers of the First Crusade, the words 'Promised Land' were far more than a mere *façon de parler* or a liturgical formula. The Holy Land, with the city of Jerusalem at its centre, was the land promised by God to His Chosen People Israel and thus the legacy of Christendom. The crusaders were the Chosen People of the New Israel, heirs to the historical Israel. When they conquered Jerusalem, they merely took what was properly Christ's and their own.

Another justification of the conquest of the Holy City – and one in which the conquerors themselves believed – was the miracle of the 'Holy Fire'. Celebrated on Holy Saturday, the feast of the Holy Fire had apparently been one of the festivals peculiar to Jerusalem since the ninth century; it was certainly the most renowned event before the crusader conquest.[28] It was a miraculous sign of the unique

---

25 Egeria, *Itinerarium*, *CCSL*, 175, p. 53.

26 Hieronymus, *Epistulae*, no. XLVI, p. 330; no. CXXIX, pp. 162–75, esp. p. 169. For the term in the early Middle Ages see also Eucherius, *De Situ Hierosolimae*, *CCSL*, 175, pp. 239–40 quoting Hieronymus, *Epistulae*, ep. XLVI; *Commemoratorium de Casis Dei*, p. 303; Bernardus Monachus, p. 315. The term 'Promised Land', it should be noted, does not appear literally in the promise to Abraham (Genesis 12:7; 13:14–15), in the Old Testament, or in the Epistle to the Hebrews 11:9: 'By faith he [Abraham] sojourned in the land of promise, as in a foreign land ...'.

27 Raimond d'Aguilers, p. 151; Y. Katzir, 'The Conquests of Jerusalem, 1099 and 1187. Historical Memory and Religious Topology', *The Meeting of Two Worlds. Cultural Exchange between East and West during the Period of the Crusades*, ed. V.P. Goss *et al.*, Kalamazoo 1986, p. 107.

28 For the festival see A.S. Tritton, 'The Easter Fire at Jerusalem', *Journal of the Royal Asiatic Society* (1963), p. 249; M. Canard, 'La Destruction de l'église de la resurrection par le Calife Hakim et l'histoire de la descente du feu sacré', *Byzantion*, 35 (1965), pp. 16–43; McGinn, '*Iter Sancti Sepulchri*', pp. 33–71, esp. 33–8. It is first recorded by the pilgrim Bernard the Monk (*c.* 870). See Bernardus Monachus, p. 92.

character of Jerusalem as well as a sign of God's approval of its inhabitants.[29] Like many early medieval miracles, it was a localized manifestation of divine power. Unlike these miracles the 'Holy Fire' possessed a universal significance, as a result of its association with Jerusalem, the centre of the world,[30] at the key moment of the liturgical year, when the central event of the history of salvation was reenacted.[31] Its annual repetition was deemed a visible expression of divine providence. So powerful was its appeal that Baldric of Dol had Pope Urban II mention it at Clermont.[32]

Consequently, no greater sacral disaster could be imagined than the failure of the Holy Fire. When it failed to appear on Holy Saturday of the year 1101 at its appointed time and place, this was interpreted by the crusaders as a divine sign and moreover an indication of doubt about the providential status of the new Kingdom of Jerusalem. It is clear from the events of the Holy Fire ceremony of 1101 that the appearance of the Holy Fire was perceived as God's approval of crusader rule in Jerusalem. Bartold of Nangis' version of Fulcher of Chartres' *Historia Hierosolymitana*, written before 1109,[33] described those events: 'When the entire church [of the Holy Sepulchre] was filled by an innumerable crowd waiting for the divine act of grace', the fire failed to appear, despite prolonged prayers and supplications. When Easter Day dawned and, to the disappointment and distress of all those crowded into the church and awaiting 'the mercy of God', the light had still not appeared, 'a sorrow lamentation and fear greater than any since the earliest days of the Christian religion seized all, so much so that some almost fell into despair ...'. Considering the failure of the miracle as a sign of God's punishment for the sins of His people, Patriarch Daimbert ordered a procession to be held. The Latin clergy, together with the king and his retinue and most of the people, went barefoot, in a penitential procession, to the Temple of the Lord. This was where God had promised Solomon to listen to his prayers (1 Kings 8:34–51; 2 Chronicles 6:25–40), to beg forgiveness for the sins that had hindered the appearance of the Holy Fire. As the procession was returning, the patriarch was informed that the long-expected light had finally appeared in one of the lamps of the Holy Sepulchre. Daimbert entered the tomb with the keys that he had been carrying during the procession, probably to avoid any suspicion of fraud. Seeing the miraculous fire he immediately announced

---

29 Bartolfus de Nangis, *Gesta Francorum Jherusalem Expugnantium*, *RHC.Hocc*, 3, pp. 385–7; McGinn, '*Iter Sancti Sepulchri*', p. 36.

30 For Jerusalem as the centre of the world see below, ch. 8, pp. 141–3.

31 McGinn, '*Iter Sancti Sepulchri*', p. 36; B. Ward, *Miracles and the Medieval Mind*, London 1982, pp. 120–22.

32 Baldricus Dolensis, p. 13.

33 Fulcher Carnotensis, pp. 831–7. For this interpolation (Codex L) see Hagenmeyer's edition of Fulcher's *Historia*, pp. 831–4. This was printed in the *RHC.Hocc* edition of Fulcher as the version of Bartolf of Nangis, *RHC.Hocc*, 3, pp. 321–418n. The story of the events of 1101 was known to Guibert of Nogent, who wrote his *Gesta Dei per Francos* in the years 1104–1108. See Guibertus Novigentis, pp. 255–6.

his discovery to all. According to an eyewitness: 'when all of us who had gone forth tearfully crying out "Kyrie Eleison" saw this, we rejoiced with great joy. We then exulted in the miracle even more than we had been cast down in sorrow.'[34]

Both the descent of the Holy Fire and the failure of the miracle were interpreted as a sign of God's approval of crusader dominion in Jerusalem for the rest of the twelfth century. The Genoese Caffaro, also apparently an eyewitness to the failure of the Holy Fire in 1101 (though he did not compose his *De Liberatione Civitatum Orientis* until *c.* 1155), argued that 'God makes miracles for unbelievers, not for the faithful, and while the Holy City was under the rule of unbelievers it was good and rightful that God converted unbelievers to the faith through miracles. Now, therefore, when [Jerusalem] is ruled by believers, miracles are no longer needed.'[35] Peter the Venerable of Cluny, too, regarded the Holy Fire as being relevant to the Christian–Moslem polemic. He pointed out, writing on the eve of the Second Crusade, that the miracle had occurred in Jerusalem precisely during the five-hundred-year rule of the 'Saracens, enemies of God' a sure proof of God's preference for Christianity over Islam.[36] An Icelandic pilgrim who visited Jerusalem in the middle of the century held a diametrically opposite view. He maintained that the Holy Fire had appeared just at the time the Christians were ruling the Holy City: 'The roof of the temple [of the Lord's sepulchre] is open to the church where the fire descends on Easter evening on the candles standing there, when Christian people rule the city.'[37]

Attempts to justify the legitimacy of Christian rule in Jerusalem were directed during the twelfth century not only at the Moslems but at the Jews as well. The fact that Peter the Venerable included the miracle of the Holy Fire in *Adversus Judeorum* is symptomatic of this way of thinking.[38] True, the goal of the crusaders was to liberate Jerusalem from the Moslems, but the Jewish past of Jerusalem loomed too large to be either forgotten or ignored. The capture of Jerusalem was seen as additional proof of the victory of *Ecclesia* over *Synagoga*; it was yet another punishment meted out to the Jews who, together with the Moslems, were now forbidden to settle in Jerusalem. The crusader conquest of Jerusalem was compared to the destruction of the city by Titus and Vespasian, and both events were considered to be just punishment for the Jews. It thus reflected an early medieval Christian notion, handed down from Eusebius, that the defeat of the Jews was the penalty for

---

34 *RHC.Hocc*, 3, pp. 385–7; Sylvia Schein, 'Between Mount Moriah and the Holy Sepulchre', *Traditio*, 60 (1985), pp. 183–4; McGinn, '*Iter Sancti Sepulchri*', pp. 33–5.

35 Caffaro, p. 61.

36 Petrus Venerabilis, *Sermo in Laudem Sepulchri Domini*, p. 252; *idem*, *Adversus Judeorum*, *CCCM*, 58, pp. 122–3.

37 B.Z. Kedar and C. Westergard-Nielsen, 'Icelanders in the Crusader Kingdom of Jerusalem: A Twelfth-Century Account', *Mediaeval Scandinavia*, 11 (1978–79), pp. 204–7. For the Holy Fire as a sign of God's eternal love for Jerusalem see also the poem 'Nomen a Solemnibus trahit Solemniacum', *Carmina Burana*, no. 52.

38 Above, n. 36.

their failure to accept Christ as the Messiah; that with the defeat and the destruction of their Temple, God had indicated his acceptance of a new Chosen People. Thus, for example, the pseudo-encyclical of Sergius IV (1009–12), a crusade *exitatorium* fabricated in Moissac on the eve of the First Crusade, proclaimed that the Christians would prevail in their battle for Jerusalem, as had Titus and Vespasian in their day.[39]

Another source, an anonymous description of the conquest of Jerusalem composed for the 'Feast of the Liberation of Jerusalem', gave an account of Moslem rule in Jerusalem, of Christ's inheritance, and of the usurpation of His Holy Places, including an exhortation addressed to the Jews. Beginning with 'Oh you miserable Jews' (*Vos autem miseri Judei*), the anonymous author referred twice to Christ's prophecy about the destruction of Jerusalem (Matthew 24:2), reminding the Jews that Jerusalem was destroyed by the Romans in order to avenge Christ's crucifixion. This was the reason that they had been expelled from their sanctuary and were still vainly awaiting their Messiah, that the Promised Land had recently been restored not to them but to the Christians, and that they should have no hope of returning to the city where they had killed Christ. This right to return was granted by God to His faithful and not to those who wished to obliterate Christ forever from the city which had been destroyed because of them. The 'True Israel' and 'Sons of Abraham' were not the Jews, now scattered all over the world as a punishment for rejecting Christ, but the Christians, who therefore deserved to hold their possessions, namely, Jerusalem and the entire Promised Land. The Jews, claimed the anonymous writer, should now recognize the evil of Antichrist and the goodness (*bonitas*) of Christ, inasmuch as the Holy Sepulchre of God was being glorified by those who recognized it and who had expelled the Saracens and the Turks from the sanctuary.[40]

The same idea appeared in an interpolation to the *Chanson d'Antioche*, the greatest of the vernacular epics of the First Crusade, composed by the poet Graindor of Douai (*c*. 1180). He argued that when Christ was hanging on the cross between the two thieves, one of them, the good thief, had commented: 'It would be most just, moreover, if you should be avenged on these treacherous Jews by whom you are so tormented'; Christ turned to him and said: '... the people are not yet born / Who will come to avenge me with their steel lances / ... And my land [will be] conquered and my country freed. / A thousand years from today they [the Jews] will be baptized and raised / And will cause the Holy Sepulchre to be regained and adored.'[41] Otto of Freising claimed – wrongly as it so happened – in

---

39 Gieysztor, 'The Genesis of the Crusades', pp. 33–4; Seidel, 'Images of the Crusades', pp. 377–91 but see now Schaller, 'Zur Kreuzzugsenzyklika Papst Sergius IV.', pp. 135–53, who argues that the encyclical is authentic and dates from 1010. For Eusebius see his *Historia Ecclesiastica*, III, 5.

40 'The Text of Account of the Capture of Jerusalem', pp. 647, 648, 652–3.

41 *Chanson d'Antioche*, I, pp. 25–9. See also *ibid*., pp. 68, 79, 223, 363, 383; *ibid*., II, pp. 99–100, 143–4; Riley-Smith, *The First Crusade*, pp. 55–6 and n. 120.

his work, *The Two Cities*, that Jerusalem had been conquered by the crusaders on the eighth day (the day of Resurrection following the Sabbath), to indicate that Jewish law and the seventh day had been superseded.[42]

From among the justifications that evolved after the conquest of Jerusalem, the most popular, during the twelfth century, was that describing Jerusalem as Christ's inheritance (*hereditas*) or patrimony (*patrimonium*), namely, according to feudal law, allodial property as opposed to a fief. As such, Jerusalem was regarded as Christ's personal property, to be protected for him by his family, the Christians. For example, William of Tyre declared that almost every syllable of the books of both Old and New Testaments taught that the Redeemer 'designed to call that part of the Earth, that is, the Holy Land, or rather that little spot, His heritage (*hereditas*)', although 'the earth is the Lord's and the fullness thereof, the world and those who dwell therein' (Psalms 24:1; 1 Corinthians 10:26). Although from the beginning Christ consecrated to Himself the entire land, He still adopted the Holy City as particularly His own, as testified by the prophet, who said: 'the Lord loves the gates of Zion more than all the dwelling places of Jacob' (Psalms 87:2). He chose it, in very truth, for He who Himself chose her testifies, saying: 'And out of Jerusalem, the city that I have chosen, shall come a saviour for you' (1 Kings 2:36; Zechariah 9:9).[43]

Bernard of Clairvaux, a contemporary of William of Tyre, while preaching the Second Crusade on behalf of Pope Eugenius III, declared in a letter of 1147 to the duke of Bohemia and his subjects that: 'The earth has been shaken and has trembled [Psalms 68:8], because the Lord has caused His land to lose territory. His land, I repeat, where He was seen and in which He lived among men for more than thirty years, His land, which He honoured by his birth, embellished by His miracles, consecrated with His blood and enriched by His burial ... Evil men have begun to occupy this land of the new promise, and unless someone resists them they will feast their eyes upon the sanctuary of our religion ...'.[44] Writing in 1146 to the Eastern Franks and the Bavarians, Bernard asked rhetorically: 'But what is on our minds, brothers? Has the hand of the Lord been shortened or has it been made powerless to save, just because He has called poor little worms to guard His inheritance and restore it to Him ...?'[45] Again, in a letter written sometime after 7 May 1150 to Peter of Cluny following the failure of the Second Crusade, Bernard declared that the king of the Heavens lost a part of His land, indeed, of His inheritance.[46] Following the defeat of Hattin and the fall of the Latin Kingdom, the justification for the right of Christendom to dominate Jerusalem became the

---

42 Otto Freisingensis, *Chronica sive Historia de Duabus Civitatibus*, *MGH.SS*, 20, p. 314. Jerusalem was actually conquered on a Friday, 15 July 1099.

43 *W.T.*, I, 15, p. 131. Eng. trans., I, pp. 88–9.

44 Bernardus Claraevallensis, *Opera*, VIII, no. 458, p. 435.

45 *Ibid.*, no. 363, p. 313.

46 *Ibid.*, no. 521, p. 483.

causae iuxtae for its recovery. Pope Gregory VIII announced the disaster in his famous encyclical *Audita tremendi* (October–November 1187), opening with the words of Psalms 78(79):1: 'Oh God, the heathen are come into thy inheritance'.[47] A description of the Holy Land, composed in the 1190s and emanating from the kingdom of Jerusalem, describes its barons as being always prepared to fight 'for the inheritance of the Lord'.[48]

On a more popular level, poets and troubadors, addressing themselves to a noble audience, tended to describe the relationship between Christ, Jerusalem and Christendom in feudal terminology. As early as the eve of the Second Crusade, the author of the poem 'Knights, now is this a time of Grace' (*Chevalier, mult estes guariz*), wrote: 'Knights, now is this a time of Grace / For God has brought to you a claim / Against the Turks and the Moors / Who treated Him dishonourably / They wronged Him and seized His fief.'[49] Another anonymous poet (*c.* 1189) compared the love of God to the vassals' love for their lord, who does not deserve to be deserted.[50] Conon of Bethune, in a poem composed on the eve of his departure on the Third Crusade ('Alas, O Love'), described God as 'being besieged in His saintly inheritance'.[51] The so-called 'Maistre Renaut', in his 'In order to comfort the people' (*Pour lou peuple rescomforteir*), presented the recovery of Jerusalem, 'the land of Our Lord' and the 'Promised Land', as a service which those who had received the sacrament of baptism owed to the Son of God.[52]

Such terminology, borrowed from feudal law, was also invoked by Pope Innocent III (while preaching the crusade) to define the relationship between Christ and the faithful as that between the feudal lord, the seigneur, and his vassals. The recovery of Jerusalem and the entire Holy Land was the duty of the vassals, who were obliged to help their lord recover his fief, which had been unjustly wrested from him by his enemies. Christ is thus described as a feudal lord of all Christians, who owe him their lives and possessions, and have a duty to fight for him as vassals serving their lord. Writing in 1208 to the faithful in Lombardy,

47 Gregorius VIII, *Audita tremendi*, p. 5, and see below, ch. 9, pp. 164–6.

48 *Tractatus de locis et statu sancte terre ierosolimitane*, ed. G.M. Thomas, *Sitzungsberichte der philosophisch-historischen Klasse der K. Bayrischen Akademie der Wissenschaften*, 3 (1865), p. 157; Riley-Smith, 'Peace Never Established: The Case of the Kingdom of Jerusalem', *Transactions of the Royal Historical Society*, 28 (1978), pp. 95–6 and see below, ch. 9, p. 167.

49 *Chansons de croisade*, ed. J. Bédier *et al.*, Paris 1909 (repr. Geneva 1974), no. I, p. 8; C. Morris, 'Propaganda for War', p. 95. For the feudal terminology as a problem for theologians see Riley-Smith, 'Crusading as an Act of Love', *History*, 65 (1980), pp. 180–82.

50 'Vos qui ameis de vraie amour', *Chansons de croisade*, no. II, pp. 20–22. The poet claims that those who desert the Lord in time of need deserve to be 'forjuge', again a term taken from feudal law, meaning banishment from their fiefs.

51 'Ahi, Amour', *ibid.*, no. III, p. 34; Morris, 'Propaganda for War', p. 97.

52 *Chansons de croisade*, no. VII, pp. 78–81.

Innocent argued as follows: 'What will a king do to his trustless and faithless vassals who are not only responsible for the loss of his kingdom but refuse to help him in his efforts to recover it? Is there a greater crime than this that Christ was banished from the land He purchased with the price of His blood, not because of His fault but theirs, and they did not rise to aid Him to recover it. For it was the Lord Himself who let His land fall into the hands of unbelievers, solely in order to give Christians an opportunity to serve Him in their efforts to give it back to Him.'[53] Proclaiming the Fifth Crusade (*Quia maior*, 19–29 April 1213), the pope warned his listeners: 'For if any temporal king is thrown out of his kingdom by his enemies, when he regains his lost kingdom surely he will condemn his vassals as faithless men, and for these bad men will devise unimagined torments, with which he will bring them to a bad end, unless they risk for him not only their possessions but also their persons! In just such a way will the King of Kings, the Lord Jesus Christ, who bestowed on you body and soul and all the other good things you have, condemn you for the vice of ingratitude and crime of infidelity if you fail to come to his aid when he has been, as it were, thrown out of his kingdom ...'.[54]

Far less popular during the twelfth century was the justification of Christian rule in Jerusalem through the Augustinian concept of 'Just War', that is, the non-legitimacy of Moslem rule in territories that had previously belonged to the Roman Empire. Even canonists, or those familiar with their theories, used a different argumentation in regard to Jerusalem. A case in point is Peter Comestor, dean of Troyes from 1147 and Chancellor of Paris from 1168. In the 1150s or 1160s, using Gratian's *Decretum*, Peter wrote to Amalric of Nesle (1157–80), the patriarch of Jerusalem: '... your land embodies the visible reminiscences of the life and death of Our Lord Christ and the most Holy Places. God chose it as His house (Psalms 26:8) and His temple, which He cleansed of its impurities (Matthew 21:12). Therefore, as the pagans are now attempting to sow their impurity in the Holy Places and to obliterate from there the memory of Christ and the name of Christianity, [places] which belong not to them but to Christ (Philippians 2:21), ... those who participate in this struggle [against the pagans, that is, the Saracens] will be considered the martyrs of our time ...'.[55] Similarly the decretist Huguccio (d. 1210), writing on the eve of the Third Crusade, though he had dealt at length with 'Just War' in his *Summa*, argued regarding the Holy Land that since divine law takes precedence over the law of nations (*ius gentium*), the pious have the right to

---

53 Innocentius III, *Opera Omnia*, *PL*, 215, cols 1500–1501. See also *Die Register Innocenz' III*, *2. Band*, *2. Pontifikatsjahr 1199/1200: Texte*, ed. O. Hageneder *et al.*, Rome–Vienna 1979, no. 241, pp. 460–62.

54 *Quia maior*, ed. G. Tangl, *Studien zum Register Innocenz' III*, Weimar 1929, pp. 88–97; Riley-Smith, *Idea and Reality*, pp. 118–24.

55 J. Leclercq, 'Gratien, Pierre de Troyes et la Seconde Croisade', *Studia Gratiana*, 2 (1954), p. 592. For Peter Comestor see D. Luscombe, 'Peter Comestor', *The Bible in the Medieval World*, pp. 109–29. For fighting for Jerusalem as a gateway to salvation see below, ch. 7, pp. 116–25.

expel the impious from their territories and to rule in their stead. Just as the Jews had expelled God's enemies from the Promised Land, so now the Christians legitimately occupied the possessions of those who had offended God. On the whole, Huguccio considers the Holy Land as belonging rightfully to Christians; its occupation by Saracens was an offence to Christianity, not because the Saracens were unbelievers or because the Holy Land was previously under Christian rule, but because of its sanctity.[56]

The justifications of the Christian right to Jerusalem reflect the new status the city acquired during the twelfth century. Following its conquest by the crusaders, the right to dominate Jerusalem was formulated in terms totally different from those applied to other conquered territories, including the rest of the Holy Land. Thus, even canonists who normally used the Augustinian concept of 'Just War' to justify Christian conquests, now argued that it was Jerusalem's sanctity that justified that it should be ruled by Christians. Considered Christ's inheritance or his fief, the Christian rule in Jerusalem was in the twelfth century considered almost as axiomatic, as self-evident truth. Any other rule was regarded as an insult to God, an offence to Christianity, a pollution of its most Holy Places. Paradoxically, following the fall of Jerusalem, there were those who argued that though indeed Jerusalem and the Holy Land were His inheritance, God, however, used to grant it to people chosen by Him to hold it. Thus, according to the English chronicler, William of Newburgh, it was not a perpetual grant, for according to Scriptures: 'The land shall not be sold in perpetuity, for the land is mine, for you are strangers and sojourners with me' (Leviticus 25:23). Similarly, it was stated of Abraham in the Acts of the Apostles 7:5: 'Yet He gave him no inheritance in it, not even a foot's length, but promised to give it to him in possession and to his posterity after him.' The land has always disgorged its inhabitants, who had polluted it with their sins. Therefore, Abraham was not allowed to possess it. It was his successors who received it, only after the Amorites had been banished because of their sins (Deuteronomy 31:4–5). The Children of Israel had been banished from their land by the Babylonians because of their sins. In the same way, the conquest of Jerusalem by Vespasian and Titus had been God's judgement, as stated in Matthew 10:15: 'It shall be more tolerable on the day of judgement for the land of Sodom and Gomorrah than for that city.' Subsequently the unbelievers had trampled on the sanctuary of God (*Sanctuarium Dei*) until the time of Constantine the Great. As a result of Constantine's piety, the land had then been granted as an inheritance and possession to the true seed of Abraham, the Christians. However, they too lost it to the Children of Hagar for their sins. The Holy Land had been given to the crusaders not because of its fertility, for in this respect India was superior to Syria. It had been given to them as the descendants of Abraham, as the land in which the Saviour had offered Himself as a sacrifice for them, so that they could praise Him there. However, whereas the crusaders who had conquered the

---

56 Russell, *The Just War*, pp. 114–22.

Holy Land had been pious men, in the course of time the land became populated with people dedicated to luxury, as well as homicides, actors and mimes.[57]

The concept of the inheritance was so prominent in European minds that the motto of the Third Crusade was to avenge the insult inflicted upon Christ by the occupation of His inheritance by the Moslems. Thus the goal of the Third Crusade as defined by Alain of Lille, in a sermon preached upon the Feast of the Exaltation of the Holy Cross (14 September 1189), was: 'the recovery of the land of our inheritance (*hereditas*), the inheritance of Christ, the dowry of the Virgin'.[58] This idea of inheritance even came to be associated with the goals of crusading in Europe. At the Fourth Lateran Council (1215), the bishop of Livonia, so it was reported, addressed the pope as follows: 'Holy Father, as you have not ceased to cherish, with your holiness' care, the Holy Land of Jerusalem, the land of the Son, so also you ought not to abandon Livonia, the land of the Mother ... For the son loves His Mother and, as He would not care to lose His own land, so too He would not wish to have His Mother's land endangered.'[59]

---

57 William of Newburgh, *Historia rerum Anglicarum*, *RS*, 82/1, pp. 249–65. For the origin of this *topos* see R. Morse, *Truth and Convention in the Middle Ages. Rhetoric, Representation and Reality*, Cambridge 1991, pp. 109–10.

58 Alain de Lille, 'Sermo de cruce domini', ed. M.-T. d'Alverny, *Alain de Lille: Textes inédites*, Paris 1965 (*Etudes de philosophie médiévale*, 52), p. 281.

59 *Henrici Chronicon Lyvoniae*, p. 293; Riley-Smith, 'An Approach', pp. 7–8.

Chapter 4

Rome, Babylon and Jerusalem: Papal Attitudes to Jerusalem

The crusader conquest of Jerusalem created a multi-faceted situation which the papacy would be required to resolve. Two problems presented themselves: on the one hand, that of the institutional status of the city within the hierarchy of the Catholic Church, while on the other, on the theological–spiritual level, the question was raised of the position of Jerusalem as a major centre of sanctity – now once more under Christian rule and the capital of the Latin Kingdom of Jerusalem – versus Rome, the traditional capital of Christendom.

The institutional status of Jerusalem within the hierarchy of the Catholic Church was a most acute question. Since the Council of Chalcedon (451), Jerusalem had enjoyed the status of a patriarchate. On the eve of the Moslem conquest (638) it ranked equally with the other four patriarchates – Rome, Antioch, Alexandria and Constantinople.[1] Up to the First Crusade, Rome enjoyed a special position as the only patriarchate within the Catholic Church and Western Christianity. With the recovery of both Antioch (1098) and Jerusalem, and consequently their return to the institutional framework of the Church, Rome was faced with the challenging task of defining their status in relation to its own. In this context, Rome was guided by one major objective: the establishment of its own primacy and supremacy. The result was the reorganization of the churches of Jerusalem and Antioch according to the principle of *provincia regnum*, as well as the relegation of the rights of the patriarch of Jerusalem to those of a primate, in accordance with the Pseudo-Isidorian principle which equated patriarchs and primates.[2] Antioch, which together with Rome and Alexandria had been regarded as one of the three senior patriarchates and, like Rome, claimed foundation by St Peter himself, retained its patriarchal title; its dignity, however, was diminished when the archbishopric of Tyre was transferred to the jurisdiction of Jerusalem.

1 It should be remembered that Jerusalem's elevation to the status of patriarchate at the Council of Chalcedon (451) was purely informal; however, it appears in the *Novellae* of Justinian as equal to the other four patriarchates. See *Corpus Iuris Civilis, III, Novellae*, ed. R. Schoell and G. Kroll, Berlin 1895, no. cxxiii, c. III; Mansi, VII, cols 177–84; Vancourt, cols 2263–4.

2 Y. Katzir, 'The Patriarch of Jerusalem, Primate of the Latin Kingdom', *Crusade and Settlement*, pp. 169–75.

Moreover, whereas the patriarch of Jerusalem was considered a primate, his counterpart in Antioch was not.[3]

The significance of these papal decisions regarding the patriarchate of Jerusalem can only be properly understood against the background of two developments which changed and transformed the structure of the Church during the period when Jerusalem was under Moslem rule (638–1099). First Rome, mainly on the basis of the Petrine theory as well as the partly forged decretals of Pseudo-Isidore (*c.* 850), theoretically consolidated its supremacy over the other patriarchates and in practice strengthened is primacy within the Church. Second, under the influence of Pseudo-Isidore, a new rank was added to the episcopal order (*ordo episcoporum*), that of 'primate', that is, the bishop of a see created, according to tradition, by the apostles or their immediate successors who thus enjoyed the dignity of *primae sedes*. Tending to equate the offices of primate and patriarch, Pseudo-Isidore, using the equation *provincia–regnum*, argued that primates held a central position in the episcopal order because their sees were capital cities and because metropolitans were directly subordinated to them; primates and lay rulers should administer the same territory from the same city. Despite this equation, however, Pseudo-Isidore distinguished sometimes between the offices of primate and patriarch. The title of patriarch was, in his opinion, reserved for the three patriarchates founded by St Peter – Rome, Antioch and Alexandria; Jerusalem was granted the same rank because St James the Just, the brother of Christ, had been consecrated by the apostles Peter, James and John as its first bishop.[4]

Pseudo-Isidore's theory of Church organization, in an attempt to model it on a secular–political organization, described the primate as a sort of ecclesiastical ruler of a political unit, in keeping with the principle of *provincia–regnum*. Isidore of Seville (636) advanced a different theory of church organization, still influential in the eleventh century – the eve of the crusader conquest of Jerusalem. In his *Etymologiae* he elaborated a theory of the ecclesiastical hierarchy which divided the episcopal 'order' into four grades: patriarchs, archbishops, metropolitans and bishops. Contrary to Pseudo-Isidore, Isidore of Seville, though recognizing the pope's pre-eminence in the Church, did not draw any distinction between the different patriarchs, whose mark of recognition was in his view their apostolic or, specifically, Petrine foundation.[5]

---

3 *Ibid.*, pp. 171–2; J. G. Rowe, 'Paschal II and the relations between the spiritual and temporal powers in the Kingdom of Jerusalem', *Speculum*, 32, pp. 490–97; *idem*, 'The Papacy and the Ecclesiastical Province of Tyre (1100–1187)', *Bulletin of the John Rylands Library*, 43 (1960/61), pp. 160–89; B. Hamilton, *The Latin Church in the Crusader States. The Secular Church*, London 1980, pp. 18–51.

4 *Decretales Pseudo-Isidorianae*, ed. P. Hinschius, Leipzig 1863, pp. 39, 79, 120–21, 185; R.L. Benson, 'Provincia–Regnum', *Prédication et Propagande au Moyen Age. Islam, Byzance, Occident*, Paris 1983, *Penn–Paris–Dumbarton Oaks Colloquia*, 3, pp. 57–60; Katzir, 'The Patriarch', pp. 169–70.

5 Isidorus Hispalensis, *Etymologiarium sive Originum Libri XX*, ed. W.M. Lindsay, Oxford 1911 (repr. 1971), VII, CII, 4–6; Benson, 'Provincia–Regnum', pp. 45–6.

The Reform papacy of the eleventh century adopted the theory of Pseudo-Isidore as an operative plan for the reorganization of the Church. This theory suited the papacy on the grounds that, first and foremost, it acknowledged Roman primacy. Next, the Church's organization as suggested by Pseudo-Isidore was considered to portray that of the primitive church. Third, Pseudo-Isidore's theory served as a useful tool in the hands of the papacy by which it could achieve reorganization of the Church according to the principle of '*provincia* (that is, primate)–*regnum*'.[6] Thus, it was used to justify the granting of the status of primate to the bishops of Lyons (1079), Toledo (1088) and Rheims (1089). Between the ninth and the end of the twelfth century similar status was granted to the bishops of Arles, Carthage, Salonica, Narbonne, Bourges, Magdeburg, Vienne, Canterbury, York, Salerno and other centres; Aquila (Grado) was likewise also elevated from bishopric to patriarchate.[7]

Describing the position of the patriarchate of Jerusalem, Pope Paschal II defined it according to Pseudo-Isidorean theory as that of a primateship, though he did not use that exact word. Writing on 9 June 1110 to Patriarch Gibelin of Jerusalem (1108–1112), he argued, citing the *Dictatus Papae*, as follows: 'The kingdoms of the world change in accordance with the changes of time and conditions; hence it seems necessary that the boundaries of ecclesiastical parochial districts in many regions should also be changed and amalgamated. The boundaries of the churches of Asia were defined in early times, but these divisions have been disturbed by the influx of various races of people with a different faith. But in our times ... the cities of Antioch and Jerusalem, with their environs and the adjacent regions, have been restored to the rule of Christian princes. We must therefore ... in accordance with the movement of the times, rearrange that which needs to be rearranged.' Accordingly, Paschal II granted to the church of Jerusalem all those cities and provinces which had been or would be conquered by the kings of Jerusalem. Through the exercise of patriarchal and metropolitan rights, the patriarch of Jerusalem was granted the rule and disposition of all cities and provinces which Divine Grace has restored to the rule of the aforesaid king or shall design to restore to it in the future. This the pope explained as follows: 'For it is fitting that the Church of the Lord's Sepulchre shall obtain due honour according to the desires of the soldiers of the Faith and that, freed from the yoke of the Turks and the Saracens, it shall be exalted more generously by the hand of the Christians.'[8]

---

6 Katzir, 'The Patriarch', p. 170.

7 G. Le Bras, 'Institutions ecclésiastiques de la chrétienté médiévale', *Histoire de l'Eglise*, ed. A. Fliche *et al.*, vol. 12, Paris 1959, pp. 535–7; H. Fuhrmann, 'Studien zur Geschichte mittelalterlicher Patriarchate', *Zeitschrift der Savigny-Stiftung für Rechtsgeschichte kanonistische Abteilung*, 39 (1953), pp. 127–8; 40 (1954), pp. 112–76; 41 (1955), pp. 1–84. Cf. Gregory VII, *Register*, pp. 201–8.

8 *Oriens Pontificius III*, no. 10, pp. 108–9; Fulcher Carnotensis, p. 742; *W.T.*, XI, 28, p. 538; *RRH*, no. 61. For the validity of the *Dictatus Papae* see C. Morris, *The Papal Monarchy. The Western Church from 1050 to 1250*, Oxford 1989, pp. 111–12.

It was Innocent II who first used the term primate in regard to the patriarch of Jerusalem. Thus, in a letter of 8 March 1138 to Patriarch William of Jerusalem, Innocent stated that the archbishop of Tyre ought to obey the patriarch as his primate.[9] The result of this papal decision was that the boundaries of the patriarchate of Jerusalem were proclaimed as conterminous with those of the kingdom of Jerusalem in conformity with the principle of *provincia–regnum*. Moreover, for the first time in its history, the patriarchate of Jerusalem was subordinated to the bishopric of Rome. At the same time, however, because of its sanctity, and above all as a result of the veneration in which Jerusalem's holy places, and especially the Holy Sepulchre, were held, the patriarchate of Jerusalem was treated as the most important in the East.

The official papal position regarding the status of the patriarchate of Jerusalem versus Rome was summarized in 1199 by Innocent III, one of the greatest popes of the Middle Ages, who was familiar, perhaps more than any other medieval pope, with both canon and feudal law. Contrary to the generally accepted theory included in the influential *Decretum* of Gratian,[10] which ranked the patriarchates in accordance with the dignity of their founders, Innocent III did not distinguish between the patriarchates founded by the apostles Peter, John and James. Two of the patriarchates, he claimed, namely Antioch and Rome, owed their existence directly to St Peter, whereas Alexandria was founded by him indirectly, through his disciple Mark. Constantinople was John's foundation while Jerusalem had been sanctified by James, through his martyrdom. Innocent granted Jerusalem the status of 'mother of the faith' (*mater fidei*) since it was from there that the apostles had gone out to preach the sacraments of the true faith all over the world. Rome had, in his view, the status of 'mother of the faithful' (*mater fidelium*), not because of its priority in time (*ratione temporibus*) but because of its dignity (*ratione dignitatis*). The Gospels, he argues (referring to Matthew 10:2; Mark, 3:16–19; Luke 6:14–16; Acts 1:13), show that though Andrew was converted before Peter, the latter enjoyed higher status than all other apostles, because of his dignity, not his priority. Rome, as 'mother of the faithful', was responsible for all churches. James, the founder of Jerusalem's patriarchate, wished to rule Jerusalem alone and left to Peter the governance (*gubernandum*) not only of the Universal Church but of the whole world (*saeculum*).[11]

The major features of the papal treatment of Jerusalem as a patriarchate, namely the identification of its patriarchate with James and, moreover, the very

---

9 *Oriens Pontificius III*, no. 38, p. 149. See also *ibid.*, no. 91, p. 245; no. 118, p. 293; *Cartulaire du Chapitre Saint-Sépulchre de Jérusalem*, ed. D.G. Bresc-Bautier, Paris 1984 (*Documents relatifs à l'histoire des croisades*, 15), no. 105, p. 224.

10 Gratianus, *Decretum, Corpus Juris Canonici*, ed. A. Friedberg, 2 vols, Leipzig 1879–81, I, dist. XXII, c.II, VII.

11 Innocentius III, *Acta*, ed. T. Haluščynskyj (*Pontificia Commissio ad Redigendum Codicem Iuris Canonici Orientalis. Fontes* III/2), Vatican City, 1949, no. 9, pp. 192–3; Innocentius III, *Register*, II, no. 200 (209), pp. 383–9.

principle of Roman supremacy over Jerusalem and, consequently, the negation of its past status as autonomous patriarchate, were not generally accepted by canonists and sometimes even openly contested. Some, like Gratian, agreed with the formal papal position that 'primates and patriarchs have different names but the same office',[12] but others, like Stephen of Tournai, argued that patriarchates differed from primates because only the former had been founded by the apostles.[13] John Beleth, who composed a treatise on the subject of ecclesiastical offices (*c.* 1165), described the *ordo episcoporum* as consisting of pope, patriarch, cardinal, primate and finally archbishop-metropolitan. Thus, Beleth not only distinguished between patriarch and primate but ranked cardinal after patriarch, though before primate.[14] Huguccio (d. 1210), though he equated patriarchs and primates in regard to their functions, argued that the five patriarchs exceeded the primates in their dignity, as they held the most ancient (*seniores*) sees.[15] This was also the opinion of Sicard of Cremona, who wrote his *Mitrale seu De Officiis Ecclesiasticis Summa* before 1215. Referring to the sees enjoying the status of patriarchate, he listed three: one in Asia, namely Antioch; a second in Africa, at Alexandria; and a third in Europe, that of Rome. Sicard claimed that only recently had the dignity of Alexandria been transferred and henceforth bestowed upon the see of Aquileia while that of Antioch devolved to Jerusalem; this was because both Alexandria and Antioch, where Peter and Mark had preached Christianity, had 'returned like a dog to its vomit' (Proverbs 26:11; 2 Peter 2:22), that is, to the vomit of infidelity. Jerusalem, he declared, was honoured with the dignity of patriarchates, 'for out of Zion shall go forth the law and the word of the Lord from Jerusalem' (Isaiah 2:3).[16] Thus Sicard was admitting that the *ordo episcoporum* was influenced and determined by political factors. Hence, he clearly recognized the papal right to transfer the patriarchal dignity from one see to another and also perceived the status of Jerusalem as directly derived from its past as the source of Christianity.

The principle of Rome's supremacy over Jerusalem was, on the whole, generally accepted, even by the patriarchs of Jerusalem (unlike those of Antioch),[17] but there remained some notable exceptions. Among the opponents was the anti-

---

12 Gratianus, *Decretum*, c. I, dist. XXII, c.VI; dist. LXXX, c. I–VI; dist. XVIX, c. I–IV; Rufinus, *Summa*, ed. J. F. von Schulte, Giessen 1892, dist. XXI, XXII, XCIX; Sicardus Cremonensis, *Mitrale seu De Officiis Ecclesiasticis Summa*, *PL*, 213, col. 70.

13 Etienne de Tournai, *Summa*, dist. XXI, XXII, XCIX.

14 Johannes Beleth, *Summa de Ecclesiasticis Officiis,* CCCM, 41A, 13, 14.

15 Huguccio, dist. XXI, XXII; G. Catalano, *Impero, Regni e Sacerdozio nel Pensiero di Uguccio da Pisa*, Milan 1956, pp. 69–71.

16 Sicardus Cremonensis, *Mitrale*, col. 70.

17 Fulcher Carnotensis, pp. 737–45; Röhricht, *Beiträge*, I, p. 193; R. Hiestand and H.-E. Mayer, 'Die Nachfolge des Patriarchen Monachus von Jerusalem', *Basler Zeitschrift für Geschichte und Altertumskunde*, 74 (1974), pp. 128, 130. For Antioch, see Rowe, 'Pascal II', pp. 470–501; Hamilton, *The Latin Church*, pp. 27–9.

Gregorian Norman Anonymous (probably of Rouen),[18] who, in his anti-papal polemic regarding the Investiture Contest, argued in favour of the supremacy of Jerusalem over Rome. In a treatise composed one year after the crusader conquest of Jerusalem (*c.* 1100), the anonymous author argued that Rome, as a place, had no right to claim any spiritual pre-eminence over any other. The apostle Peter bestowed no more distinction on Rome than on Jerusalem or Antioch; before Rome had received the true faith from Jerusalem, it was an evil city. Of all the earthly churches, Jerusalem deserved the position of supremacy and that of 'mother of all churches' (*mater omnium ecclesiarum*). Though Peter may have called himself the 'prince of the apostles', he had received this post (*magisterium*) from a higher prince, whom he feared and honoured – James, the bishop of Jerusalem. Christ, the head of the Church, had lived in Jerusalem, not in Rome, and it was there that he had composed the sacraments. Jerusalem was also the home of Stephen the proto-martyr, as well as the apostles, and from that city they had proclaimed the laws of God to all other parts of the world. Human beings had founded Rome while Jerusalem owed its existence to Christ and the apostles. Moreover, Jerusalem was the only city which communicated directly with heaven and was therefore named by Paul 'our mother' (Galatians 4:26), whereas John called her the 'bride of the Lamb' (Revelation 21:2). It was also said of her that 'of Zion it shall be said this one and that one were born in her' (Psalms 87:5), and 'For out of Zion shall go forth the law and the word of the Lord from Jerusalem ...' (Isaiah 2:3). Therefore the 'Holy Mother Zion' was the head and owner (*domina*) of all the churches; all the above passages of the Holy Scriptures referred directly to Jerusalem, not to Rome. Now that his kinsmen, the Norman crusaders, had liberated Jerusalem, the Norman Anonymous argued that the supreme position in the Christian Church should be transferred from the bishops of Rome to the patriarchs of Jerusalem.[19]

The polemics of the Norman Anonymous probably did not enjoy much popularity during the twelfth century, as his views regarding the status of

---

[18] For the problem of the identity of the *Anonymous*, also mistakenly known as the 'Anonymous of York', see G.H. Williams, *The Norman Anonymous of 1100 A.D.*, Cambridge, Mass. 1951 (*Harvard Theological Studies* 18), pp. 1–127; N.F. Cantor, *Church, Kingship and Lay Investiture in England 1089–1135*, pp. 91–101; R. Nineham, 'Anonymous of York, the So-Called', *Journal of Ecclesiastical History*, 14 (1963), pp. 31–45.

[19] *Tractatus Eboracenses, II, MGH.Ldt*, 3, pp. 659–60. See also Williams, *Norman Anonymous*, pp. 139, 141–2; W. Ullmann, *The Growth of Papal Government in the Middle Ages*, London 1955, pp. 394–7. The view advocating the religious superiority of Jerusalem over Rome was hardly new. Adhemar of Chabannes had argued as early as the beginning of the eleventh century that 'not from Rome but from Zion went forth "the law and the word of the Lord"...' (Isaiah 2; 3); moreover, Jerusalem was the source of Christianity (*fons Christianitatis*) and the mother of all churches. Ademarus Cabannensis, *De Apostolatu Sancti Martialis*, *PL*, 141, col. 110.

Jerusalem versus Rome did not become a subject of controversy. Similar opinions reappeared only just before the Latin conquest of one of the other historical patriarchates – that of Constantinople in 1204. John Kamateros, patriarch of Constantinople, refused to acknowledge the subjection of his see to Rome, in other words, the principle of Roman primacy in the Church. He contested Rome's claim to supremacy, arguing that the Scriptures reinforced the superiority of Jerusalem to Rome, not only in its sanctity but also through its history in terms of both priority and authority.[20]

Whereas the principle of Roman supremacy over Jerusalem was seldom contested or contradicted, one might say that Rome's attempt to identify the origins of the Jerusalem patriarchate, and to some degree its sanctity, with James (whether James the Just or James the first bishop) met with complete failure. During the crusader period, contemporaries, while identifying Rome with St Peter, associated Jerusalem first and foremost with Christ.[21] Referring to its sanctity in hierarchical terms, they referred first to Christ, then to the Virgin Mary and only then to the apostles and the saints. To all intents and purposes, both apostles and saints played only a secondary and quite marginal part in the concept of the sanctity of Jerusalem in the twelfth century. This way of thinking can be seen, for instance, in the opinions of pilgrims, who often stated their personal reasons for undertaking the Jerusalem pilgrimage. John of Würzburg, who came to Jerusalem *c.* 1160, described his journey as motivated by his love of Christ and his desire to visit the admirable places sanctified by the physical presence of Christ, the Virgin and the apostles.[22] Theoderich, another pilgrim, dealing with the same subject about a decade later, asserted that, as previously testified by Josephus Flavius and Jerome, Jerusalem was the holiest of all cities in the entire world. This holiness flowed not from its own essence but from the presence of God, Christ and the Virgin Mary, as well as the patriarchs, the prophets, the apostles and other saints who had lived, preached and suffered martyrdom there.[23] Such opinions were by no means characteristic of pilgrims alone. In the twelfth century, as in previous centuries, Jerusalem's sanctity was derived mainly from its special link with Christ.[24]

While the papacy acted promptly and resolutely on the institutional level to enforce its supremacy over Jerusalem as a patriarchate, the definition of its position in regard to the sanctification of Jerusalem was a more protracted process, which did not fully develop until mid-century. It was to a large extent influenced and shaped by the Europeans. The papacy, while defining the status of the church

---

20 Innocentius III, *Register*, II, no. 199 (208), pp. 380–81.

21 See below, ch. 5, pp. 66–76. Baldric of Dol, however, presented Jerusalem's dignity as patriarchate as connected with James. Baldricus Dolensis, p. 104.

22 Johannes Wirziburgensis, p. 79.

23 Theodericus in *Peregrinationes Tres*, p. 145.

24 See below, ch. 5, pp. 66–76.

of Jerusalem on the institutional level as evolving mainly from James, could not ignore the reaction to the crusader conquest of the Holy City, nor that city's position as a major centre of sanctity and pilgrimage. Faced with the significant increase in the influx of pilgrims to Jerusalem, themselves often also crusaders, and sometimes passing through Rome,[25] as well as with the task of defending the capital of the crusading kingdom by means of the crusade, the papacy described Jerusalem in what soon became a sort of formula of sanctity: 'the place where His feet had trodden'. In this it followed the Vulgate version of Psalms 132:7 '... let us worship at his footstool' and of Zechariah 14:4: 'On that day his feet shall stand on the Mount of Olives ...'.[26]

This formula first appears in a letter of Paschal II (1113) and again in the letters of Innocent II (1138), Eugenius III (1146), Alexander III (1181), Celestine III (1192) and Innocent III (1198–1199).[27] Thus, Paschal II, writing on 2 January 1113 to Hugh, abbot of the monastery of St Mary in the Valley of Jehoshaphat, speaks of the grace of God which had liberated the place of His Passion and Resurrection, the City of Jerusalem, in which 'God my King from old [is] working salvation in the midst of the earth' (Psalms 74:12); and, as stated by the prophet, 'I will make the place of my feet glorious' (Isaiah 60:13).[28] Alexander III described Jerusalem as the place which 'the Saviour of all glorified and honoured by His corporeal presence and where He chose for the sake of our Redemption to become a man'.[29]

The Holy Sepulchre figures relatively rarely in the papal references to Jerusalem. It does, however, appear in the letters of Pope Paschal II. Writing on 9 June 1110 to Patriarch Gibelin of Jerusalem, Paschal argues that he is granting to the patriarchate of Jerusalem the rights to all cities and provinces conquered by the

---

[25] For the pilgrimage movement in the twelfth century see below, ch. 5, pp. 76–80.

[26] The term *ubi steterunt* appears in Eusebius, *Vita Constantini*, *PG*, 20, 3, 24, in reference to Helena's pilgrimage to Jerusalem. See Wilken, *The Land Called Holy*, pp. 97–8; Paulinus of Nola, *Epistulae*, 49, 14, p. 402 used this term regarding his pilgrimage to Jerusalem: 'No other sentiment draws people to Jerusalem than the desire to see and touch the places where Christ was physically present, and to be able to say from their own experience "We have gone into his tabernacle, and have worshipped in the places where his feet stood".' According to Gregory of Nyssa the Holy Places of Jerusalem had 'received the footprints of Life itself and for this reason they are palpable reminders that God once walked this earth'. Gregorius Nyssenus, *Epistolae*, 3, 2, ed. Maraval, pp. 126–7; Wilken, *The Land Called Holy*, p. 117 and n. 67.

[27] Paschal II, *Oriens Pontificius III*, no. 14, p. 118; Innocentius II, *ibid.*, III, no. 37, p. 147; Eugenius III, *ibid.*, no. 69, p. 191; *idem*, *Epistolae*, *PL*, 163, col. 1394; Alexander III, *Epistolae*, *PL*, 200, no. 1505, col. 1296; Celestinus III, *Oriens Pontificius* III, no. 166, p. 345; Innocentius II, *Register*, I, no. 13, p. 22; no. 343, p. 513; no. 355, p. 531; no. 437, p. 661. For this theme within papal crusade propaganda see Cole, *The Preaching*, pp. 37–97 and *passim*.

[28] *Oriens Pontificius* III, no. 14, p. 118.

[29] *Ibid.*, no. 109, p. 276.

kings of Jerusalem 'for it is fitting that the Church of the Lord's Sepulchre shall obtain due honour according to the desires of the soldiers of the Faith, and that, freed from the yoke of the Turks and the Saracens, it shall be exalted more generously by the hand of the Christians.'[30] Later, significantly, the Holy Sepulchre is mentioned mainly in appeals for aid to the kingdom of Jerusalem. Eugenius III proclaiming the Second Crusade in his encyclical *Quantum praedecessores* of March 1146, and Alexander III echoing him in *Quantum praedecessores nostri* of 14 July 1165, exhorted the believers to come to the aid of Jerusalem as the place where 'Our Saviour wished to suffer for us and left us for His memory the glorious Sepulchre of His Passion'.[31] In *Inter Omnia* of 20 July 1169, Alexander III was moved to write: 'When long ago the land fell under the power of the Saracens because of the people's sins, men of virtue rose up, restored the land to the faith of Christ after wiping out or routing the heathen and, after setting up the standard of faith there, returned the Lord's Sepulchre, famous because much frequented by people, to Christendom.'[32]

Thus it follows that Jerusalem, unlike Rome, was described by the papacy as drawing its sanctity not from a sepulchre or sepulchres, but rather from its history as the place where God had operated as man, the place where His feet had trodden. It was the place where He had worked for the salvation of mankind, and thus the 'place of salvation' and therefore the 'source of Salvation'. Moreover, it was not by mere accident that this was the 'place of Redemption'; it was so by an act of God, who had chosen Jerusalem as the stage for His work of Redemption.[33] Following the fall of Jerusalem, this concept was formalized by Pope Gregory VIII. Summoning Christians to a crusade, he wrote in the encyclical *Audita tremendi* (October–November 1187): 'First the prophets and then the apostles and their followers laboured with all their zeal, so that the worship of God might be established in that land and flow from it to all regions of the world. Moreover – this is the greatest and most unutterable fact – God, through whom all things were made, desiring to be made flesh, wished in his ineffable wisdom and incomprehensible mercy to bring about our salvation there through the infirmity of the flesh, which is to say hunger, thirst, the cross and death, and through his resurrection, according to the saying he "hath wrought salvation in the midst of the

---

30 *Ibid.*, no. 10, p. 109. See also *ibid.*, no. 8, p. 105.

31 Eugenius III, *Quantum Praedecessores*, p. 302; Alexander III, *Epistolae*, no. 260, col. 384. For this bull of Alexander see J.G. Rowe, 'Alexander III and the Jerusalem Crusade. An overview of problems and failures', *Crusaders and Muslims in Twelfth-Century Syria*, ed. Maya Shatzmiller, Leiden 1992, pp. 118–23.

32 Alexander III, *Epistolae*, cols 599–600. See also Rowe, 'Alexander III', pp. 123–5. For references to Jerusalem see also Innocent II in *Oriens Pontificius* III, no. 37, p. 147 and Paschal II above, n. 8.

33 See also Eugenius III, *Oriens Pontificius* II, no. 64, p. 196; Alexander III, *ibid.*, no. 109, p. 276; Innocentius III, *Register*, I, no. 11, p. 19; no. 302, p. 432; no. 336, p. 499; no. 343; p. 513; no. 355, p. 531; no. 406, p. 607; no. 437, p. 661; no. 487, p. 717.

earth" (Psalms 74:12); he deigned to work this through himself, which the tongue cannot speak of nor the heart of man contemplate.'[34]

On the whole, papal references to Jerusalem are notable for the widespread omission of the Holy Sepulchre and its significance as the place of Burial and Resurrection of Christ. This may be understood in terms of the rivalry between Rome and Jerusalem as centres of sanctity and pilgrimage. Rome's major attraction for pilgrims derived from the tombs of the apostles Peter and Paul.[35] It may also be explained, however, within the framework of the general attitude to Jerusalem. As will be demonstrated in the next chapter,[36] the characteristic attitude to Jerusalem in the early Middle Ages, before the First Crusade, namely, the almost exclusive identification of the city with the Holy Sepulchre, had already given way to the concept of Jerusalem as the scene of the life of Christ, in the first decade of the twelfth century. Recent studies have shown that the traffic of ideas flowed in both directions: while the papal *curia* influenced the public it was addressing, it was itself influenced by that same public. The popes came under the spell, so to speak, of popular ideas, much as their own crusading propaganda affected popular opinion. A similar two-way transmission of ideas is discernible in regard to Jerusalem.[37] In the first stage, the papacy was influenced by the general enthusiastic and spontaneous reaction to the liberation of Jerusalem in the First Crusade. In the second stage, papal letters featured Jerusalem almost exclusively as 'the place where His feet had trodden' (Psalms 132:7; Zechariah 14:4) and where God was 'working salvation in the midst of the earth' (Psalms 74:12). This notion influenced the Europeans at large and became the most popular expression of the attitude of those faithful arriving in the Holy City before its fall to Saladin in 1187.

In the twelfth century the subject of Rome versus Jerusalem was dealt with at several very different levels: the ecclesiastical–hierarchical level, on the spiritual one, as a major centre of sanctity, and on the mystical–allegorical level, with the two cities symbolizing Babylon and the Celestial City. Rome was frequently described as Babylon, Paul himself referring to the Roman community of the faithful as 'she who is at Babylon' (1 Peter 5:13), while in the Book of Revelation it appears as 'Babylon the great, mother of harlots and of earth's abominations' (Revelation 17:5). Even after it had emerged in the fourth century as the city of the

---

34 Gregorius VIII, *Audita tremendi*, pp. 7–8. Riley-Smith, *Idea and Reality*, p. 65.

35 On this subject see Sumption, *Pilgrimage*, pp. 217–31.

36 See below, ch. 5, pp. 66–7.

37 On the transmission of crusading ideas and crusading propaganda see C. Morris, 'Propaganda for War: The Dissemination of the Crusading Ideal in the Twelfth Century', *Studies in Church History*, 20 (1983), pp. 88–9; Riley-Smith, 'An Approach to Crusading Ethics', pp. 3–19, esp. p. 15; Cole, *The Preaching*, *passim*. For papal propaganda in general see G. Constable, 'Papal, Imperial and Monastic Propaganda in the Eleventh and Twelfth Centuries', *Prédication et Propagande au Moyen Age. Islam, Byzance, Occident*, Paris 1983 (*Penn–Paris–Dumbarton Oaks Colloquia*, 3), pp. 179–99.

apostles Peter and Paul, their martyrdom and their tombs, Rome was to remain identified with Babylon. Referring to pagan Rome, Augustine portrayed it as the 'other Babylon', whose king was the Devil; Jerusalem was to him the 'Vision of Peace' (*visio pacis*). Thus Augustine contrasted the Heavenly Jerusalem, the city of eternal peace, with Babylon, the city of earthly, temporal peace and of confusion. The imagery of both Revelation and *De Civitate Dei* was too powerful not to influence the medieval mind. Though in the early Middle Ages, Rome was sometimes eulogized as the New Jerusalem, it was more often identified with Babylon.[38] In the twelfth century, however, this dichotomous attitude to the two cities, the identification of Rome with Babylon, thus the personification of evil and confusion, and of Earthly Jerusalem with the Vision of Peace and the prefiguration of the Heavenly Jerusalem, seems to have become less prominent than in previous centuries. Now not only Rome, but also Jerusalem, was sometimes identified with Babylon. The analogy was, however, limited in one respect: though Rome was sometimes Jerusalem, Jerusalem was never Rome. Thus, for example, the Benedictine Sigebert of Gembloux claimed in his letter of 1103 against Pope Paschal II that Rome was Babylon and that the Romans shared with the Jews the guilt of Christ's death; moreover, He had been crucified by the Romans and not by the Jews.[39]

The contrast between the two cities, Earthly Babylon and Heavenly Jerusalem, was the central theme of Otto of Friesing's world history, *Chronica sive Historia de Duabus Civitatibus*, composed in the 1140s. Deeply influenced by Augustine, Otto presented Babylon as the personification of the wickedness of this world, as the worldly city which would fall and be destroyed by means of the Last Judgement. It would then be replaced by 'the Holy City Jerusalem, the City of Christ' as described in the Book of Revelation. Sometimes this Babylon was, for him, Rome, whereas the Earthly Jerusalem was the 'City of Christ', 'the City of God', the 'Holy City'.[40]

For others too, Jerusalem was Babylon. Rahewin, in his history of Frederick I, alluding to the schism between Victor IV and Alexander III and the schism

---

38 Augustinus, *Enarrationes in Psalmos*, *PL*, 137, col. 1761; Konrad, 'Das himmlische … Jerusalem', pp. 523–30. J. Benzinger, *Invectiva in Romam. Romkritik im Mittelalter vom 9. bis zum 12. Jahrhundert*, Lübeck–Hamburg 1968, pp. 1–74. For identification of Jerusalem as *visio pacis*, for the etymological explanation of the Hebrew word 'Jerusalem' as 'city of peace' that goes back to Philo of Alexandria, see Kühnel, *From the Earthly to the Heavenly Jerusalem*, pp. 74–81; Konrad, p. 526 and n. 15.

39 Sigebertus Gemblacensis, *Epistola Leodicensium adversus Paschalem Papam*, *MGH.Ldt*, 2, pp. 451–5; I.S. Robinson, *Authority and Resistance in the Investiture Contest. The Polemical Literature of the Late Eleventh Century*, Manchester 1978, pp. 175–9. Bernard of Cluny in his *De Contemptu Mundi*, *RS*, 59/2, pp. 18, 212, 23 27–33, 31, 47–8, 113, 171–8 has 'Jerusalem or Zion (*Sion*) as the symbol of the Heavenly City whereas Rome is Babylon'; moreover: 'Zion is choked by Babylon (*Sion angitur Babylon*)'.

40 Otto Freisingensis, *Chronica*, pp. 68, 150, 154, 310, 431–6.

between the latter and the emperor (1160), declared: 'One should mourn Rome as Jeremiah mourned over Jerusalem, crying: "The kings of the earth did not believe or any of the inhabitants of the world that foe or enemy could enter the gates of Jerusalem" (Lamentations 4:12). Once our mother the Roman Church was Jerusalem ... Now, however, we say with pain: our "Jerusalem sinned grievously therefore she became filthy" (*ibid.*, 1:8) ...'.[41] For Bernard of Clairvaux, Rome was Jerusalem. He described Rome as a 'princess among the cities and the greatest among nations' (Lamentations 1:1), the place which the Lord had made 'his house of the ruler of all his possessions' (Psalms 105:21; Deuteronomy 32:26), 'where his feet had trodden' (Psalms 132:7) – these were the same terms as those in which he described Jerusalem.[42] Gerhoh of Reichersberg alternated between Rome as Jerusalem, and Babylon and Jerusalem as Babylon. In his commentary to Psalm 33, written in 1146 or 1147, he argued that 'Zion' was Rome which, through St Peter and his martyrdom, as well as its conversion to Christianity, had changed its name from Babylon to Zion.[43] A few years later, in 1166 or 1167, however, disgusted with the Roman Church because of the schism between Victor IV and Alexander III, he claimed that 'the Holy Place, the city of Rome, is the other Babylon'.[44] In his *Libellus de investigatione Antichristi*, written between 1160 and 1162, Gerhoh, who claimed that the three major evils of his time were the disputes between the empire and the papacy, the schism between Victor IV and Alexander III and the failure of the Second Crusade, branded both Rome and Earthly Jerusalem as Babylon. Both were accused of avarice, an accusation which was traditionally levelled only against Rome. Describing the avarice of the inhabitants of Jerusalem, Gerhoh portrayed greed as one of the beasts of the Apocalypse, which had fled from Rome to Jerusalem and then returned.[45]

Thus, whereas both Jerusalem and Rome were presented as Babylon, mainly within a polemical framework, Rome appears less frequently than in previous centuries in the role of the New or Heavenly Jerusalem.[46] In the past, Rome was often eulogized as the *Abbild* of Heavenly Jerusalem, the 'other Jerusalem', *caput mundi*, while Earthly Jerusalem was either rejected completely or spiritualised. This changed with the crusader conquest of Jerusalem.[47] Earthly Jerusalem re-

---

41 *Rahevini Gesta Friderici I Imperatoris*, *MGH.Scr.rer.Germ.*, pp. 320–21.

42 Bernardus Claraevallensis, *Opera*, ed. J. Leclercq *et al.*, Rome 1957–77, no. 238, p. 117; *idem*, *De Laude Novae Militiae*, *Opera*, III, pp. 223–4 and below, ch. 5, pp. 77–8.

43 Gerhoh Reichersbergensis, *Commentarius in Psalmos*, p. 444.

44 *Idem*, *Opusculum ad Cardinales*, *MGH.Ldt*, 3, p. 400.

45 *Idem*, *Libellus de investigatione Antichristi*, *MGH.Ldt*, 3, pp. 377–84; Siberry, *Criticism of Crusading*, pp. 199–200; R.W. Southern, *The Making of the Middle Ages*, pp. 152–4; Benzinger, *Invectiva in Romam*, pp. 91–4.

46 For Earthly Jerusalem as Heavenly Jerusalem in the twelfth century see below, ch. 7, p. 136.

47 Konrad, 'Das himmlische ... Jerusalem', pp. 526–33; Prawer, 'Jerusalem in Christian and Jewish Perspectives', pp. 748–82 and see below, chs 6, 7.

emerged as the centre of the world, as well as both *Abbild* and *Vorbild* of the Heavenly City.[48] At the same time, the contrast between Babylon and Jerusalem was so powerfully rooted in the collective imagination that it became a kind of criterion for the evaluation and description of cities in general. Abbot Philip of Harvengt, for example, compared Paris to biblical Jerusalem, describing it as *Cariat Sepher*, or *Civitas litterarum*, whereas some of his contemporaries, such as Bernard of Clairvaux, presented Paris as Babylon.[49] Bernard of Clairvaux, likewise, compared adulterous women to the daughters of Babylon; nuns and virgins were for him 'daughters of Zion', 'daughters of Jerusalem'.[50] But above all, Babylon and Jerusalem exemplified the opposition between this world and paradise, between the earthly and heavenly cities. Thus, Abelard, in one of the most beautiful poems of the Middle Ages, sang: 'Be ours with minds uplifted / Unto the heights of God, / With our whole heart's desiring, / To take the homeward road / And the long exile over, / Captive in Babylon, / Again unto Jerusalem, / To win at last return / ... Jerusalem is the city / Of everlasting peace, / A peace that is surpassing / And utter blessedness / Where finds the dreamer waking / Truth beyond dreaming far, / Nor is the heart's possessing / Less than the heart's desire.'[51]

It follows thus that most of the faithful members of the Church, seculars, regulars and laity alike, accepted in the twelfth century the supremacy of Rome over Jerusalem within the institutional framework of the Church as a patriarchate, that is on the ecclesiastical–hierarchical level. On the spiritual as well as on the mystical–allegorical level, however, Jerusalem was considered superior to Rome. Both Rome and Earthly Jerusalem were now identified with Babylon; Rome was sometimes presented as Jerusalem. But Jerusalem was never Rome.

---

48 See below, chs 7, 8.

49 Philippus de Harvengt, *Epistolae*, *PL*, 203, col. 31; Le Goff, 'Babylone ou Jérusalem: La ville dans l'imaginaire collectif au Moyen Age', *Critique*, 34 (1978), pp. 554–9.

50 Bernardus Claraevalensis, *Opera*, no. 113, p. 289; *idem*, *Letters*, Eng. trans. B. James, Chicago 1953, no. 116, pp. 174–7.

51 Petrus Abelardus, 'Sabbato ad Vesperas', *The Oxford Book of Medieval Latin Verse*, no. 169, pp. 243–4. English trans. H. Waddell, *Mediaeval Latin Lyrics*, London 1929 (1952 edn), pp. 175–7. For this theme see also Bernard of Cluny, 'Hora novissima', in *The Oxford Book of Medieval Latin Verse*, no. 160, pp. 223–6. The theme of Babylon as *saeculum* appears in a letter written in 1117 by Reimbald of Liège for a pilgrim named Dermont who was on his way from Ireland to Jerusalem and who wanted to urge others to go likewise from Babylon to Jerusalem: 'You see that I have made myself an exile for God ... You see I take flight to God not only in Jerusalem but everywhere'. Reimbald of Liège, 5–7, p. 4; G. Constable, *The Reformation of the Twelfth Century*, Cambridge 1996, p. 310 and see also *ibid.*, p. 269.

Chapter 5

From the 'City of the Holy Sepulchre' to the 'City of the Humanity of Christ'

The crusader conquest of Jerusalem had a tremendous impact on the very essence and character of the overall perception of the city's sanctity. During crusader rule in the Holy City, Jerusalem had undergone a major transformation and with it the very character of its sanctity.

In the early Middle Ages Jerusalem was seen mainly as the city of the Holy Sepulchre.[1] As a matter of fact it seems that the two centuries preceding the First Crusade were characterized by an almost obsessive devotion to the Holy Sepulchre. Not only did that shrine overshadow Jerusalem as a goal of pilgrimage but the marked increase in the volume of pilgrim traffic to the place of the Lord's Burial and Resurrection also inspired the foundation of monasteries and hospices dedicated to the Holy Sepulchre. These could be found on the main pilgrim roads in Italy, such as the monastery of the Holy Sepulchre at Bobbio (founded in 950) and the monastery of Borgo San Sepolcro southwest of Florence (founded in 934).[2] Moreover, that period saw the erection in the Catholic West of many churches whose design imitated the rotunda of the Holy Sepulchre, that is, the Constantine Church of the Resurrection (*Anastasis*). The oldest of these was the chapel of St Michael at Fulda, erected 820–822 by Abbot Eigil, possibly on the advice of Rabanus Maurus, and rebuilt in the late eleventh century.[3] Constructed by Bishop Conrad between 934 and 976 and rebuilt in the thirteenth century was the chapel of

---

1 *Vita Altmanni*, p. 230; Ademarus Cabannensis, *Chronicon*, pp. 189–90; Radulfus Glaber, p. 109; *Lettres de Gerbert (983–997) (=Silvester II)*, ed. J. Havet, Paris 1889, pp. 22–3; *Annales Altahenses Majores*, p. 815; *Gesta Episcoporum Tullensium*, p. 647; Lalanne, pp. 1–23 and above, ch. 1, pp. 10–16.

2 J. Heers, 'Bourgs et faubourgs en Occident: Les pèlerinages et devotions au Saint-Sépulchre', *Jérusalem, Rome, Constantinople. L'image et le mythe de la ville au Moyen Age*, ed. D. Poiron (*Cultures et Civilisations Médiévales*, 5), pp. 205–15.

3 R. Krautheimer, *Studies in Early Christian, Medieval and Renaissance Art*, New York 1969, p. 117; R. Ousterhout, 'The Church of Santo Stefano: A "Jerusalem" in Bologna', *Gesta*, 20 (1981), p. 311; *idem*, 'Loca Sancta and the Architectural Response to Pilgrimage', *The Blessings*, p. 110; Kühnel, *From the Earthly to the Heavenly Jerusalem*, p. 202 and n. 219. As early as 801 the chapel of Charlemagne's palace at Aachen represented the Holy Sepulchre, but it was understood as standing for Heavenly Jerusalem. Konrad, 'Das Himmlische … Jerusalem', pp. 528–30; Kühnel, *From the Earthly to the Heavenly Jerusalem*, pp. 116–19, 131–2.

St Maurice at Constance. Like the chapel at Fulda, it contained a model of the tomb's aedicula.[4] Some hundred years later, in 1033, Meinwerk, the saint and bishop of Paderborn, dispatched Wyno, abbot of Helmarshausen, to Jerusalem to measure the church and the monument of the Holy Sepulchre with a view to building a replica. The church, built at Busdorf outside Paderborn, was consecrated in 1036.[5] Another eleventh-century copy of the Church of the Holy Sepulchre which also originally contained a copy of the tomb of Christ, was the large church of Neuvy St Sepulchre in Central France, founded in the middle of the century (1042) but reconstructed in the next.[6]

Some eleventh-century churches contained an imitation of the tomb of Christ without being themselves a copy of the Church of the Holy Sepulchre. One of them was a pyramid-shaped monument recalling the Holy Sepulchre itself in the precincts of the monastery of Beaulieu, *Belli Locus*, contracted by Fulk Nerra (before 1039). The church of the monastery was granted by Fulk such relics as a stone from the Holy Sepulchre and a piece of the True Cross.[7] Another was that of St Hubert at Ardenne; this was a marble monument representing the Lord's Sepulchre consecrated in 1076 by Bishop Herman of Metz.[8] These copies had for the faithful an icon-like value and were considered a conduit of the tomb's life-giving power, effecting miracles, cures, and aid in salvation. Regarded as relics, the copies of the *rotunda* or of the tomb were believed to have – as a result of the transfer of the measurements of the original site – the same divine powers as those projected in the original building. Moreover, some of the copies were funerary in function, aiding in the salvation of those buried nearby by making spiritually present the life-giving power of the Tomb of Christ. St Michael at Fulda, St Maurice at Constance and the Holy Sepulchre of Cambrai, constructed between 1063 and 1064 and now destroyed, were all funerary chapels. At Paderborn, Bishop Meinwerk was buried near the chapel of the Holy Sepulchre which he had constructed. According to his *Vita*, the completion was rushed so that the chapel might be dedicated before his death, and the purpose of Meinwerk's building of the chapel was 'so that he might reach Heavenly Jerusalem'.[9]

---

4 G. Bresc-Bautier, 'Les imitations du Saint-Sépulchre de Jérusalem (IXe–XVe siècles). Archéologie d'une dévotion', *Revue d'histoire de la spiritualité*, 50 (1974), pp. 324, 331, 334–40.

5 *Vita Meinwerci*, *MGH.SS*, 11, pp. 158–9; Krautheimer, *Studies in ... Art*, p. 117; Riley-Smith, *The First Crusaders*, p. 33.

6 Krautheimer, *Studies in ... Art*, p. 127; Ousterhout, 'Loca Sancta', p. 111.

7 B.S. Bachrach, *Fulk Nerra, the Neo-Roman Consul, 987–1040. A Political Biography of an Angevin Count*, Berkeley 1993, pp. 110, 247–9.

8 *La Chronique de Saint-Hubert dite Cantatorium*, pp. 50, 94–5; Bresc-Bautier, 'Les imitations du Saint-Sépulche', pp. 319–41; Ousterhout, 'Loca Santa', pp. 119–29 and n. 20.

9 *Vita Meinwerci*, pp. 158–9; Bresc-Bautier, 'Les imitations du Saint-Sépulchre', pp. 323–5; Ousterhout, 'Loca Sancta', pp. 108–24.

A popular German song, which probably originated in the eleventh century and was sung during the First Crusade, went as follows: 'In God's name we make our way / And we journey in His grace / Power of God help us today / And the Holy Sepulchre / Where God Himself lay / Kyrie Eleison.'[10] And indeed, as we have shown previously, the *Iter/Via Sancti Sepulchri* was one of the terms most commonly used to designate the First Crusade. The Holy Sepulchre was the crusaders' motto during the four-year-long crusade, as well as the main goal of their expedition.[11] The participants of the crusade saw themselves as 'pilgrims to the Holy Sepulchre' (*peregrinatores S. Sepulcri*). Thus, for example, the leaders of the crusade wrote from Antioch in September 1098 to the pope, inviting him to join them to head their crusade and bring it to its end, that is, the liberation of the Holy Sepulchre.[12] According to one of the participants, the author of the *Gesta Francorum*, the crusaders suffered in the course of the crusade 'for the sake of Christ and the liberation of the road to the Holy Sepulchre'.[13] He reports that during the siege of Jerusalem, in July 1099, it was decided to attack the city with war machines so that the crusaders would be able 'to enter the city and pray at the site of the sepulchre of our Saviour'.[14] Raimond of Aguilers, another participant in the First Crusade and eyewitness to the conquest of Jerusalem, argued that the very sight of the crusaders venerating the Holy Sepulchre justified all the sufferings and hardships of the crusade.[15] Indeed, as shown above, both the participants of the First Crusade and Europeans at large saw the main target of their campaign as the capture of the Holy Sepulchre rather than that of Jerusalem as a whole. Accordingly, the new Latin rulers of Jerusalem defined their status in relation to that holy shrine. After his election as ruler of the kingdom of Jerusalem, Godfrey of Bouillon refused to wear a crown in the city where Christ had worn the Crown of Thorns and adopted as one of his titles 'Advocate of the Holy Sepulchre'.[16] His brother and successor, Baldwin I, who did assume the title of king, also called himself 'Defender of the Holy Sepulchre'.[17]

The centrality of the Holy Sepulchre for the contemporaries of the First Crusade was clearly expressed in the crusading oath. According to one version, the crusader

---

10 U. Müller, *Kreuzzugsdichtung*, Tübingen 1979, no. 8, p. 9; Morris, 'Propaganda for War', pp. 93–4.

11 See above, chs 1, 2.

12 *Kreuzzugsbriefe*, no. XVI, pp. 164–5.

13 *Gesta Francorum*, p. 62.

14 *Ibid.*, pp. 89–92.

15 Raimond d'Aguilers, p. 151.

16 *Kreuzzugsbriefe*, no. XVIII, p. 168; *W.T.*, 9, 9, pp. 431–2. Eng. trans., I, p. 392; *De Situ*, p. 433; Riley-Smith, 'The Title of Godfrey of Bouillon', *Bulletin of the Institute of Historical Research*, 52 (1979), pp. 83–6.

17 *Codice diplomatico della Republica di Genova*, ed. C. Imperiale di Sant'Angelo, 3 vols, Rome 1936–42, vol. 1, no. 15, p. 20; no. 16, p. 22; Riley-Smith, 'The Title of Godfrey of Bouillon', p. 84.

swore 'to journey and visit the Sepulchre of the Lord in Jerusalem ... when a general expedition to the Holy Lord will be proclaimed by the Holy Church of Rome'.[18] This version of the crusading oath was apparently formulated as early as the eve of the First Crusade, whereas the liturgical ceremony for the taking of the crusading cross, the bestowal by the ecclesiastical authorities of the cross upon the crusader, did not develop until well into the twelfth century. A comparison between the two reveals a significant development. In the version of the Bari Pontifical of the twelfth century, the prayer for the departing crusaders describes their goal as 'the place where our Lord Christ ... was born from a Virgin, died, was resurrected and ascended to Heaven ... "to worship in the place where His feet had trodden" (Psalms 132:7)'.[19]

This is symptomatic of the general trend in the development of the concept of the holiness of Jerusalem in the twelfth century. During the crusader period, the nature of the city's sanctity as entirely deriving from, and thus dependent on, the Holy Sepulchre, underwent a significant transformation. A new attitude developed which emphasized the significance of the Holy Sepulchre as the place of resurrection rather than the burial of Christ. Additionally, the very concept of Jerusalem's sanctity was now considerably enlarged. In the wake of a large-scale topographical identification or re-identification of holy traditions, what might be called the 'Holy Geography of the Holy City' was enriched to an extraordinary degree. Jerusalem now appeared in a new and different perspective, namely, as the 'City of the Life of Christ'. Thus, the almost exclusive connection of the city with the Holy Sepulchre as the site of Christ's burial gave way to a new collective concept, according to which Jerusalem was presented as the scene of Christ's lifetime.

This new and expanded concept of Jerusalem as the place where 'His feet had trodden' (Psalms 132:7) parallels a new preoccupation with the humanity of Christ, on the increase since the eleventh century, and forges an undoubted link between the two. The marked feature of the spirituality of the late eleventh and the twelfth century was the general devotion to the humanity of Christ and His life on earth, together with the desire to suffer with and for Christ. Most of the spiritual writings of the eleventh and twelfth centuries stress the desire not only to follow, but also to copy and almost to replicate the life of Christ down to the last detail. The devotion to the humanity of Christ now took on a new importance both as an object of devotion and as an example for imitation. The devotion concentrated especially on the childhood and death of Christ, namely his most human periods, as well as on his poverty and sufferings, although the other aspects of his life on earth were frequently cited. A particular feature of the devotion was the concentration on

---

18 J.A. Brundage, *Medieval Canon Law and the Crusader*, Madison 1969, pp. 116–18 and n. 2.

19 K. Pennington, 'The Rite for Taking the Cross in the Twelfth Century', *Traditio*, 30 (1974), p. 431.

themes of tenderness for the helplessness and the sufferings of Christ as a man. This devotion to the humanity of Christ penetrated every aspect of the life of the Church. It can be seen in the liturgy, in the devotion to the Cross, the Holy Lance and the Five Wounds of Christ, and most clearly in iconography. The figure of Christ on the Cross was transformed from that of a crowned, erect, alive and victorious Saviour of the early Middle Ages to a slumped and mutilated dead body. Artists henceforth depicted the agony on the Cross with growing realism and a moving realization of the extreme limits of human suffering: the dying figure was portrayed as stripped of its garments, the head hanging to one side, with closed eyes, the blood running down the Cross.[20]

The ideas of pilgrimage and crusade were both expressions of the ideal of 'imitation of Christ'. Pilgrims felt that they were re-enacting in their own lives the sufferings of Christ; they were performing an act of personal redemption, just as Christ by His death had made possible the salvation of all men. The pilgrims' life in Jerusalem was conceived as a continuously repeated drama of the life of Christ. For example, a passage attributed to Abelard, in the so-called 'Cambridge Commentary' on the epistles of Paul, compared a visit to Jerusalem to reception of the Eucharist, emphasizing that same compassion and fervour for God that the site and the relics of Christ's Passion inspired in those who travelled to Jerusalem.[21] Theodorich, the pilgrim, who came to Jerusalem in the 1170s, reported that, as in previous centuries, it was common for pilgrims to have themselves flogged at the site of the flagellation pillar, preserved in the chapel of *Lithostrotos* on Mount Zion.[22] Bernard of Clairvaux described the moment of a pilgrim's arrival in the city as follows: 'How sweet it must be for pilgrims after the fatigue of their long journey and their many perils on land and sea to find rest there at last – there where they know their own Lord has rested! I should think that in their joy they no longer feel their weariness, nor regret their expenses, but claim the reward of their labour and the prize of their course (1 Corinthians 9:24) according to their words: "They rejoice exceedingly to have found the tomb" (Job 3:22)'.[23]

As the centre of Christendom in which physical engagement with the *imitatio Christi* was always at its strongest, the crusader city of Jerusalem became a refuge for those who had chosen to realize this ideal in its most extreme form. One such was Teotonicus (1086–1166), monk of the monastery of the Holy Cross near Ombra in Portugal. Teotonicus, who came to Jerusalem twice, spent his time in the

---

20 J. Leclercq, 'De Saint Grégoire à Saint Bernard', *Histoire de la Spiritualité Chrétienne*, ed. L. Boyer *et al.*, vol. 2, Paris 1961, pp. 233–72; Vandenbroucke, 'Nouveaux problèmes: du XIe au XVIe siècles', *ibid.*, pp. 299–307; R.W. Southern, *The Making of the Middle Ages*, London 1953, pp. 231–46; G. Constable, *Three Studies in Medieval Religion and Social Thought*, Cambridge 1996, pp. 169–217.

21 *Commentarius Cantabrigiensis in epistolas Pauli e schola Petri Abaelardi. In epistolam ad Hebraeos*, ed. A. Landgraf, Notre Dame 1945, p. 747.

22 Theodoricus, *Peregrinationes Tres*, p. 172.

23 Bernardus Claraevallensis, *De Laude*, p. 230.

city prostrating himself before the Holy Cross and was later beatified.[24] For some, like the French knight William of Maraval, this meant denial of the aristocratic way of life for an eremitical one in Jerusalem. On the advice of Pope Eugenius III he went on pilgrimage to Jerusalem. Having completed his pilgrimage vow, William asked the patriarch of Jerusalem to have a little hut built for him in the *atrium* of the patriarch's house, in which he lived in solitude and total silence as a pauper for nine years.[25] The Russian abbot of Chernigov, Daniel, who came on pilgrimage to Jerusalem in *c.* 1108, mentioned seeing a stylite on the Mount of Olives near the Tomb of St Pelagia, 'a very spiritual man, terrible in appearance and ancient in days'.[26] The Cistercian St Hildegund or Joseph (she lived disguised as a man in the Cistercian abbey of Schönau until her death in 1188) came to Jerusalem 'to visit the Lord's Sepulchre as well as other Holy Places where His feet had trodden' and stayed for a year in the place of the Templars, namely at the 'Temple of Solomon' on the Temple Mount.[27] Godric, the hermit of Finchale (in his early life a prosperous trader), visited, after his conversion, Rome, Santiago de Compostella and twice – barefoot – Jerusalem. He stayed there for a year, tending the sick in the Hospital of St John.[28]

Others sought salvation by serving the lepers who lived in a house outside the walls of Jerusalem. One of them, Alberic, a noble by birth, kissed each of the lepers every day after Mass and carried the feeble among them on his shoulders. He ate the scraps left by the lepers and when he found himself disgusted by the blood and secretion from their sores mixed in the water with which he washed the lepers' feet, he immersed his face in the bucket and drank from it. He would flog himself so hard that the blood streamed down his body; he prayed so often on his knees that they became thickened with calluses, and he wore a shirt of rough goat-hair. Some of those who came to live in Jerusalem to imitate Christ chose a communal life, remaining for some time in a large cave near the city. Others followed the eremitical model such as the Hungarian priest Cosmas, who shut himself in a cell atop the walls of Jerusalem.[29] According to Reiner of Liège, St Gertrude of Liège lived on the walls for seven years as a recluse, under the

---

[24] A. Jotischky, *The Perfection of Solitude. Hermits and Monks in the Crusader States*, Pennsylvania 1995, p. 165.

[25] *Theobaldi vita S. Guilelmi eremitae*, *AASS*, *Feb. 2*, p. 459; Jotischky, *The Perfection*, p. 165.

[26] Daniel, *Pilgrimage*, *PPTS*, 4, p. 26.

[27] *AASS*, April, vol. 2, pp. 779–81; Bredero, 'Jerusalem in the West', pp. 266–7.

[28] *Reginaldus Dunelmensis*, *Vita S. Godrici Eremitae Finchalae*, *AASS*, May, vol. 5, cols 71, 72; B. Ward, *Miracles*, pp. 76–7.

[29] Jotischky, *The Perfection*, pp. 170–71; B.Z. Kedar, 'Gerard of Nazareth. A Neglected Twelfth Century Writer in the Latin East', *Dumbarton Oaks Papers*, 37 (1983), pp. 65–6, 72–4. For such extreme practices of mortification of the body in the twelfth-century West see G. Constable, 'Attitudes toward Self-Inflicted Suffering in the Middle Ages,' in G. Constable, *Culture and Spirituality in Medieval Europe*, Aldershot 1996, no. 9, pp. 15–34.

patronage of the nuns of the convent of Mary Magdalen, otherwise St Mary 'La Grande'.[30] This explains why Conrad of Montferrat, in a letter to Baldwin of Canterbury written from Tyre in September 1188, lamented that 'the walls of Jerusalem were left deprived of the hermits, its habitants'.[31]

Crusaders, whose ethos combined the ideal of pilgrimage with that of the Christian knight, believed, too, that, by taking the cross, they were literally fulfilling Christ's precepts, 'If any man will come after me, let him deny himself, and take up his cross daily, and follow me' (Luke 9:23; Matthew 16:24), as well as 'one must suffer for the sake of my name' (Acts 9:16). A crusade *exitatorium* of *c.* 1095–1096 proclaimed that 'it is a precept to follow the road of Christ only with the cross'. Pilgrims and crusaders alike saw bearing the cross as imitating Christ's crucifixion by their own torments. According to the pilgrimage tradition – a major source of crusading ideology – the physical *imitatio Christi* was of little value unless accompanied by internal spiritual emulation, which took the form of moral regeneration and was a sign of true repentance. According to the apologists, the true crusader did not merely bear the cross upon his clothes as an outward symbol; it was also imprinted upon his heart, mind or body. So strong was veneration of the cross among the crusaders that some branded or tattooed actual crosses on their bodies; at other times, according to the evidence of crusaders, crosses appeared miraculously upon men's flesh.[32] A marching song composed between the years 1096 and 1099 and used by crusaders, the 'Jerusalem the marvellous' (*Jerusalem mirabilis*), tells the story of Christ's Passion, dwelling upon his Crucifixion. An exhortation to Holy War, this song called upon the believers to sell their fiefs (*honores*) and go to the Holy Sepulchre to acquire the 'Temple of God'.[33] The identification with the torments of the Cross, coupled with a realistic approach to their physical aspects, is well expressed by an anonymous poet in the lines, 'You who love with true love' (*Vos qui ameis de vraie amour*) of *c.* 1189: 'You who love with true love, awake! Sleep no more / ... the day of peace has come / Which give to those who for love of him / Take the cross for their burden / Suffer pain both night and day / Then he will see who truly love him / ... He who was put on the cross for us / Did not love us with a simulated love / He loved us like the Finest friend / And lovingly for us / Carried with

---

30 *Reineri Annales*, *MGH.SS*, 16, p. 659. For this convent of Mary Magdalen see M. de Vogüé, *Les Églises de la Terre Sainte*, Paris 1860 (repr. Jerusalem 1973), p. 265; B. Hamilton, 'Rebuilding Zion: The Holy Places of Jerusalem in the Twelfth Century', *Studies in Church History*, 14 (1977), pp. 112–13.

31 Radulphus de Diceto, *Opera Historica*, *RS*, 68/2, p. 61; Kedar, 'Gerard of Nazareth', p. 72, no. 1.

32 *Kreuzzugsbriefe*, no. XVI, p. 164; Fulcher Carnotensis, pp. 169–70; *Gesta Francorum*, p. 1; Raimond d'Aguilers, p. 102; Petrus Tudebode, p. 31; Robertus Monachus, pp. 727–8, 730, 850; Baldricus Dolensis, p. 16; Ekkehardus, *Hierosolymita*, pp. 15, 19, 34, 39; Radulfus Cadomensis, p. 674; *Historia de translatione*, p. 257; *Historia Peregrinorum*, pp. 172–3. For the *exitatorium* see Gieysztor, 'The Genesis of the Crusades', p. 33.

33 'Iherusalem mirabilis', *AH*, 55b, no. 96.

so much anguish / The Holy Cross very gently / Between his arms, before his breast, / Like a gentle lamb, simple and devout. / Then he was nailed with three nails / Painfully through his hands and through his feet ...'.[34]

Viewing the journey to the Holy City, now more intensely than in previous centuries as the imitation of Christ's Passion, people identified the Holy Sepulchre as the site of the Resurrection rather than of the burial of Christ. This attitude, typical of the crusading period, meant a restoration and a reversion of the shrine to its original significance, dominant during the previous Christian rule in the Holy City, in the Byzantine period. Constantine the Great, in the Roman tradition of theophanies, intended the Anastasis Rotunda to be first and foremost a monument to the theophany of resurrection and only in the second place a sepulchre. Eusebius, Constantine's biographer, recounting the emperor's deeds in Palestine, described the significance of the Anastasis as 'the trophy of the Saviour's victory over death' and referred to Jerusalem as *locus resurrectionis Dominicae.*[35] For Cyril of Jerusalem (347–348), it was the 'city of the resurrection' and for Jerome, the *sepulcrum resurrectionis*. The location of the Church of the Holy Sepulchre, not far from what was then the centre of the city, was intentional: it was to be the new place of salvation, to replace the old centre of Jerusalem – the Jewish Temple. Adopting the Jewish tradition of Jerusalem as the centre of the world and the Temple as the centre of the Holy City, Constantine transferred it to the site of the Resurrection.[36]

Indeed, this was precisely the significance attached to the Holy Sepulchre, both in the West and in the kingdom of Jerusalem, during the period of the crusaders' domination of the Holy City.[37] The Holy Sepulchre's significance as the symbol

---

[34] *Chansons de croisade*, no. II, pp. 20–22; Riley-Smith, *Idea and Reality*, pp. 89–90.

[35] Eusebius, *Vita Constantini*, cols 1086, 1093; Hieronymus, *Epistulae*, 108, p. 316; *Enchiridion Locorum sanctorum*, ed. D. Baldi, Jerusalem 1945, p. 629; Kühnel, *From the Earthly to the Heavenly Jerusalem*, pp. 81–7. Lamenting the capture of Jerusalem by the Persians (614), Sophronius, the patriarch of Jerusalem, described the Anastasis as the place 'where the King of all rose again trampling the power of death'; Wilken, *The Land Called Holy*, p. 230.

[36] Kühnel, *From the Earthly to the Heavenly Jerusalem*, pp. 87–9; and see also below, ch. 6. For the early Middle Ages see Antoninus Placentinus, pp. 138, 163; Arculfus, 45, pp. 187–8; Willibaldus, *Vita seu potius hodoeporicon*, ed. Tobler, *Descriptiones*, pp. 29–30; *Innominatus I*, *PPTS*, 6, pp. 1–2. For the early crusader period (*c*. 1102–3) see Saewulf, *Peregrinatio de Hierosolymam et Terra sancta* in *Peregrinationes Tres*, pp. 64–5, 67–8. Eng. trans. A. Rogers, *PPTS*, 4, pp. 59, 64–7.

[37] For a different view see Y. Friedman, 'The Status of the Church of the Holy Sepulchre in the Crusader Period', *Sefer Z. Vilnai*, Jerusalem 1987, vol. 2, pp. 56–8. For the Holy Sepulchre as the place of the burial of Christ see Baldricus Dolensis, p. 13: '... in that place ... God was laid to rest, ... there he was buried. How precious is that place of the Lord's burial, how desirable, a place beyond compare!' Guibert of Nogent, who refers to Jerusalem as the place where 'Christ lived and suffered', does not mention the Resurrection but the place 'where the blood of the Son of God ... poured out and where his body, dead to the fearful elements, rested in the grave'; Guibertus Novigentis, p. 138.

of the supreme event in the history of the Redemption, the Resurrection of Christ, thus placing it at the centre of the world, is the central theme in Peter the Venerable's sermon *In Laudem Sepulchri Domini* (1147). He described Christ's Sepulchre as the most admirable and adorable of all holy places in the entire world. It is the Holy Temple of the Lord, known as the 'holy of holies'; in this place Christ lay as dead for the sake of the life of human beings. It is clear, he argued, that this place is the centre of the world, the place called by the Saviour the heart of the world. Both the sepulchre of Christ and the city of Jerusalem were situated at the heart of the world – at its centre, just as the human heart was situated at the centre of the human body. This is because the Fathers of the Church, the first Christians, had placed the sepulchre at the centre of the world. The main significance of the Holy Sepulchre as the site of Christ's Resurrection is demonstrated, in Peter the Venerable's view, by the miracle of the Holy Fire. He argued that: 'Evidence for the greatness of the miracle lies in the fact that it does not appear in Nazareth, where Mary was told that she would give birth to a son by the Holy Spirit; nor in Bethlehem where He [Christ] was born; nor in Capernaum, where He suckled as a baby; nor anywhere in Judaea or in Galilee, where He grew up as infant, boy, adolescent and young man, where He studied, lived and preached; but solely in the site of the Lord's Sepulchre ... The Holy Fire appears there not on the Day of the Annunciation, nor on Christmas Day, nor on the day of Christ's transfiguration, nor on the day of Christ's presentation in the Temple; nor on the day of his passion; the Celestial Fire does not appear during the three days and nights when He lay buried in the Sepulchre; but on Holy Saturday, the day of the Resurrection.' The Holy Sepulchre is thus the symbol of the Resurrection of Christ rather than his burial. 'You should remember,' he argues, 'that the Lord does not lie dead in the Sepulchre but rules in Heaven as God.'[38] Another abbot, Gerhoh of Reichersberg, writing and commenting at the same time (1146 or 1147) on Psalm 33, refers to the sepulchre 'from which He was resurrected and ascended to Heaven to rule the world sitting at the right of His Father'.[39]

This absolute identification of the Holy Sepulchre with Christ's Resurrection is one of the most characteristic features of the Franks' concept of its sanctity. One significant manifestation of this concept is that in the course of the twelfth century, the patriarchs of Jerusalem altered their title from *Deo Gratia patriarcha Jherosolimitanus* to *patriarcha sancti Resurrectionis Christi ecclesie*.[40] The perception of the act of Resurrection as central to the history of Redemption and

---

38 Petrus Venerabilis, *Sermo*, pp. 248–50, 253.

39 Gerhoh Reichersbergensis, *Commentarius*, pp. 430–31.

40 *Cartulaire ... du Saint Sépulchre*, no. 54, p. 143; no. 119, p. 242; no. 147, p. 288; no. 148, p. 290; no. 150, p. 293; no. 162, p. 314; Hiestand and Mayer, 'Die Nachfolge des Patriarchen Monachus von Jerusalem', p. 128; H.E. Mayer, 'Die Stiftung Herzog Heinrichs des Löwen für das Heilige Grab', *Heinrich der Löwe*, ed. W.D. Mohrmann, Göttingen 1980, pp. 323–4.

thus also to Jerusalem's sanctity is most prominent in the thought of one of the archbishops of the Church of Jerusalem, and the greatest historian of the Latin Kingdom, William of Tyre. William referred to King Amalric of Jerusalem as 'defender and advocate of the Holy Places of the Passion and Resurrection' (*locorum venerabilium dominice passionis et resurrectionis defensor et advocatus*).[41] In his version of Pope Urban's proclamation of the First Crusade at Clermont (written in the 1170s), William called Jerusalem 'the city where our Saviour, through his teachings, the Passion and the Resurrection, worked "salvation in the midst of the earth"' (Psalms 74:12). In his lengthy description of the city of Jerusalem, he referred to 'the place of the sepulchre of the Saviour' in the terms of *Resurrectionis ecclesia*, without any mention at all of the more common term 'Church of the Holy Sepulchre'.[42] Similarly, other inhabitants of Jerusalem were familiar with – and perhaps even used in daily life – the term 'Church of the Resurrection' rather than that common in the West, 'Church of the Holy Sepulchre'. Moreover, the crusaders, when they rebuilt Constantine's buildings on the site of Christ's sepulchre and Crucifixion, an undertaking which was completed in 1149, decorated the monument of the Holy Sepulchre with inscriptions depicting it mainly as the site of the Resurrection. According to John of Würzburg, some of those inscriptions read as follows: 'The place and guardian testify to Christ's resurrection; also the linen clothes, the angel and redemption' and: 'Here Christ was laid within this sepulchre of rock / His burial heaven's gate to mortals doth unlock'.[43] Moreover, according to *Citez de Jherusalem*, a pilgrims' guide to Jerusalem written *c.* 1187, the deacon chanting the Mass of the Resurrection in the Holy Sepulchre used to turn toward Mount Calvary when he came to the word 'Crucifixum'; he would then face the Sepulchre and say: 'Surrexit, non est hic …'.[44] The motto of the crusader defenders of Jerusalem during Saladin's siege (20

41 *W.T.*, XX, 22, p. 942. Eng. trans., II, p. 378.

42 *Ibid.*, VIII, 3, p. 385. Eng. trans., I, p. 343; IX, 4, p. 425. Eng. trans., I, p. 384. See also 18, 3, p. 812. Eng. trans., II, p. 240.

43 Johannes Wirziburgensis, pp. 89–90.

44 *Citez de Jherusalem*, ed. T. Tobler, *Descriptiones*, p. 204. See also Prawer, *The Latin Kingdom*, pp. 180–81. An Easter trope (from the Introit of the Mass of Easter) from *c.* 900 runs as follows: 'Quem quaeritis in sepulchro, / o Christicolae? / Jesum Nazarenum crucifixum, / o caelicolae. / Non est hic, / surrexit, sicut praedixerat. / ite, nuntiate, / quia surrexit ...' *The Oxford Book of Medieval Latin Verse*, no. 99, pp. 136–7. As early as the eleventh century these tropes were sung by several clerics; the distribution of parts varied slightly from place to place; normally the words of the angel (that is, the question 'Quem queritis ...' and the answer 'Non est hic') were recited by a single priest, and those of the Marys, by two cantors. In the eleventh century this developed into an Easter play, known as *Visitatio Sepulchri*. In the play the sepulchre was represented by the altar, the angels were portrayed by two persons and the three Marys by three clerics. K. Young, *The Drama of the Medieval Church*, 2 vols, Oxford 1933, I, pp. 201–72. For the version of the Easter play as performed in the Rotunda of the Church of the Holy Sepulchre in 1160 see *ibid.*, I, pp. 261–3. The Marys were played by three young clerics dressed as women, each carrying a

September–20 October 1187) was: 'The Holy and True Cross! The sepulchre of the Resurrection of Jesus Christ! Save the City of Jerusalem and its inhabitants!'[45]

A different approach to the Holy Sepulchre was that expressed by Bernard of Clairvaux. In his *De laude Novae Militiae* (1129) addressed to the Templars, he included a long theological excursus in the form of a meditation on the Holy Sepulchre. His central theme was not the mystery of the Incarnation or the Resurrection but that of salvation through the death of Christ; the theme of the Resurrection appeared only as a secondary theme. For him it was mainly the bodily resting of the Lord: 'Somehow the Holy Sepulchre seems to be the most attractive of holy places. I do not know exactly why the devout are more drawn to the place where his dead body rested than to where he lived and worked, or why the remembrance of his death should move us more than that of his life. Perhaps it is because the latter is regarded as more demanding, while the former is more satisfying; or that human weakness is more drawn by his peaceful repose than by strenuous activity; by the security of his death than by the rectitude of his life. The life of Christ has furnished me with a pattern for living, but his death has delivered me from death ... These thoughts and similar ones regarding the Holy Sepulchre present themselves to the reflection of Christians according to the attractions of each. Still, I think that those who are actually able to see even with their eyes the bodily resting place of the Lord must experience the strongest emotions, from which they will receive no little profit. Even though this place is empty now of its sacred contents, it remains full of delightful mysteries for us – for us, I say, because it is really our resting place ... How sweet it must be for pilgrims, after the fatigue of their long journey and their many perils on land and sea, to find rest then at last – there where they know their own Lord has rested! I should think that in their joy they no longer feel their weariness, nor regret their expenses, but claim the reward of their labour and the prize of their course (I Corinthians 9:24) according to the words: "They rejoice exceedingly and are glad, when they find the grave" (Job 3:22).'[46]

At the same time Jerusalem emerged as the 'City of the life of Christ'. This was already one of the meanings of Jerusalem in the Byzantine period. Already Eusebius, describing the pilgrimage of Helena, in his *Vita Constantini*, had her 'render due reverence to the ground which the Saviour's feet had trodden,

---

gold or silver vessel containing ointment. Those impersonating the angels wore amices over their heads and carried candles. Some time between 1099 and 1160, however, the performance was discontinued because of the throngs of pilgrims who came to see it. See also *ibid.*, II, Appendix A, pp. 507–13. In its second stage of development the play was expanded to include Peter and John as central figures. *Ibid.*, I, pp. 307–68. In the final stage, in the latter part of the twelfth century, another scene was added, centred on the risen Christ. *Ibid.*, I, pp. 369–410.

45 *De Expugnatione Terrae Sanctae Libellus*, ed. J. Stevenson, *Radulphi de Coggeshall Chronicon Anglicanum, RS*, 66, p. 242.

46 Bernardus Claraevallensis, *De Laude*, pp. 229–30. Eng. trans. Grenia, pp. 154–63.

according to the prophetic word which says "let us worship at the place whereon his feet have stood" (Psalms 132:7)'.[47] Sulpicius Severus (*c*. 360–420) described the footprints at the Mount of Olives as 'enduring proof of the soil that this place has been trodden by God ...'.[48] Gregory of Nyssa (*c*. 330–395) referred to the holy places of Jerusalem as 'the places that had received the footprints of Life itself' and 'for this reason they are palpable reminders that God once walked this earth; his pilgrimage to Jerusalem aimed at seeing the signs of the Lord's sojourn in flesh'.[49] Paulinus of Nola declared in 431 that: 'no other sentiment draws men to Jerusalem than the desire to see and touch the places where Christ was physically present, and to be able to say from our own experience: "We have gone into His tabernacle and adored in the very places where His feet have stood" (Psalms 132:7).'[50]

This transformation in the character of Jerusalem's sanctity could be explained by the new and general devotion to the humanity of Christ, but there were other factors too. The crusader rulers of Jerusalem had undertaken what Bernard Hamilton defined as 'more ambitious than any comparable piece of religious town planning undertaken in the Middle Ages'.[51] Contemporaries were likewise aware of that gigantic project of rebuilding Jerusalem undertaken by the crusaders. John of Würzburg wrote: '... let us describe the new holy places and the venerable ancient ones which have been newly built and dedicated to the service of religion'.[52] As a result, a ruined Moslem city became capital of a Christian kingdom and one of the most important, if not the most important, centres of pilgrimage in the Christian world.

The crusaders rebuilt many of the Byzantine shrines and built new ones.[53] Most of the crusader shrines were pilgrims' shrines. Rebuilding or building on the holy places, the Franks linked certain traditions to certain sites, aiming to represent certain events of the past. Those places became holy places which could be seen

---

47 Eusebius, *Vita Constantini*, III, XLII.

48 Cynthia Hahn, 'Loca Sancta Souvenirs: Sealing the Pilgrim's Experience', *The Blessings*, p. 90.

49 Wilken, *The Land Called Holy*, p. 117 and see also *ibid.*, pp. 97–9. Following the Council of Chalcedon (451) Sabas and Theodosius, the leaders of the cenobitic communities in Palestine, addressed a petition to Emperor Anastasius describing Jerusalem as 'those holy places in which the mystery of the incarnation of our great God saviour took place'. *Ibid.*, pp. 168–70.

50 Paulinus de Nola, *Epistulae*, no. 49, p. 402; McGinn, '*Iter Sancti Sepulchri*', p. 39.

51 B. Hamilton, 'The Impact of Crusader Jerusalem on Western Christendom', *The Catholic Historical Review*, 80 (1994), p. 699.

52 Johannes Wirziburgensis, pp. 130–31.

53 Hamilton, 'The Impact of Crusader Jerusalem', pp. 696–702; *idem*, 'Rebuilding Zion', pp. 105–11. On crusader art see Bianca Kühnel, *Crusader Art of the Twelfth Century. A Geographical, an Historical or an Art Historian Notion*, Berlin 1994, *passim;* Folda, *The Art of the Crusaders*, *passim*.

and touched; they became witnesses to the truth of the Christian religion; oral traditions or written texts as, for instance, the Holy Scriptures, when linked to certain places, bridged the gulf between past and present. Moreover, embellished as they were with mosaics, frescoes and bas-reliefs, and often accompanied by appropriate texts, the shrines represented a visual expression illustrating sacred traditions. The pilgrims could now, as during the Byzantine period, see, touch and read at the very site where different events described in the Holy Scriptures or elsewhere took place. They could now visit and see most of the sites commemorating various episodes in the life of Christ in Jerusalem as described by the Gospels. Moreover, in the course of these visits they could experience with their own eyes the spiritual aspects, as visual images, produced to illustrate factual matters connected to a particular site, were now more readily available than before 1099. The visual media, it should be remembered, were particularly important for the illiterate. Visual images in the places for worship worked together with non-visual media, as, for example, liturgy, to propagate ideas and shape the piety of such believers. For example, Bernard of Clairvaux argued that: 'Bishops have a duty toward the wise and foolish. They have to make use of material ornamentation to rouse devotion in a carnal people, incapable of spiritual things.'[54] A South German Benedictine monk, writing in 1189–90, claimed that pictures in churches were placed there for the sake of simple-minded men or illiterates (*idiotes*), in contrast to the literate (*litteratus*).[55] Thus, in the Church of the Temple of the Lord (*Templum Domini*) the visitors could see a painting of Christ's presentation in the Temple (Luke 2:22–35) and read its superscription: 'The King of Kings, of virgin mother born, / Was here presented. This is holy ground ...'.[56] On Mount Zion, in a chapel commemorating the judgement of Christ (Matthew 27:1–14), the visitors could see a picture of the trial and read the inscription: 'He whom the saints commend, was by sinners' voice condemned, He for His servants' sake did scourge and buffet take.'[57] Pilgrims coming to Jerusalem from the Christian West found themselves now in a city full of Christian shrines whose beauty either reminded them somewhat of their local churches at home or whose foreign oriental splendour surpassed all their expectations and filled them with wonder. To Bishop Henri Sdyck of Olmütz in Bohemia, who came to Jerusalem twice in 1123 and again in 1137–1138, as for other travellers, Jerusalem represented 'an allegory of Paradise'.[58]

---

54 Bernardus Claraevallensis, *Apologia ad Guillelmum Abbatem*, *Opera*, III, ed. J. Leclercq *et al.*, Rome 1963, pp. 104–7, esp. pp. 104–5.

55 *Liber de poenitentia et tentationibus religiosorum*, *PL*, 213, col. 89; M.R. Miles, *Image as Insight. Visual Understanding in Western Christianity and Secular Culture*, Boston, Mass. 1985, pp. 15–39.

56 Johannes Wirziburgensis, pp. 89–90.

57 *Ibid.*, pp. 116–17.

58 Rorgo Fretellus de Nazareth, *Description de la Terre Sainte*, ed. P.C. Boeren, Amsterdam 1980, p. 6.

Yet another factor contributing to the transformation was pilgrimage. The First and subsequent crusades, as well as the significant growth in the dimensions of pilgrimage to the Holy City, increased and enriched in the West the knowledge of the 'Holy Geography' of the Holy Land, including Jerusalem. For example, Theodorich estimated the number of pilgrims in Jerusalem during his visit (*c.* 1174) at sixty thousand.[59] Duke Henry the Lion of Saxony and Bavaria came in 1172 with a retinue of about twelve hundred men. William of Tyre informs us that in 1182 a single ship could carry as many as fifteen hundred pilgrims.[60] The increase in undertaking pilgrimage to the capital of the new Latin Kingdom also resulted in the growing popularity of the *descriptiones Terrae Sanctae* as a literary genre; some of these were composed by pilgrims, some by Europeans who did not go to the East but instead used pilgrims' oral accounts or reworked existing written, often ancient, descriptions. One reflection of the popularity of this literary genre was that descriptions of the holy places of the Holy Land – sometimes only of Jerusalem – often became incorporated into histories of the crusades and even into European chronicles, such as the *Universal Chronicle* of the 'Anonymous of Laon' and the annals of Hanover.[61]

As a result of the rebuilding of Jerusalem, the ever-growing pilgrimage movement, and an increasing familiarity with the holy places of Jerusalem in Europe, the new map of the holy places of Jerusalem which now gradually emerged was an enlarged one that included the entire life of Christ. It placed particular emphasis on his childhood and adolescence on the one hand, and on his Passion on the other, both illustrating the humanity of Christ. The papal concept of Jerusalem was one of the factors that promoted the diffusion of this new trend in European spirituality. Papal encyclicals, which time and again exhorted the believers to render aid to Jerusalem, reached to every part of Europe during the twelfth century; as they were often read aloud in churches and incorporated into sermons, they became an important factor in the creation of the new concept.

---

59 Theodericus, p. 177.

60 *W.T.*, 22, 15 (14), p. 1026. Eng. trans., II, p. 417; E. Joranson, 'The Palestine Pilgrimage of Henry the Lion', *Medieval and Historiographical Essays in Honor of James Westfall Thompson*, ed. J.E. Cate *et al.*, Port Washington, N.Y. 1938 (repr. 1966), pp. 159–65. For maritime traffic and pilgrims' transport see J.H. Pryor, *Geography, Technology and War. Studies in the Maritime History of the Mediterranean, 649–1571*, Cambridge 1988, *passim.*

61 *Chronicon Universale Anonymi Laudunensis*, ed. A. Cartellieri *et al.*, Leipzig–Paris 1909, pp. 41–5. For the *Descriptiones* as a literary genre see A. Grabois, 'Medieval Pilgrims, the Holy Land and its Image in European Civilization', *Pillars of Smoke and Fire. The Holy Land in History and Thought*, ed. M. Sharon, Johannesburg 1986, pp. 65–79. For the descriptions of Jerusalem in crusade chronicles see *Gesta Francorum*, pp. 98–101; Fulcher Carnotensis, pp. 281–92, 331–4; Albertus Aquensis, p. 480; Radulfus Cadomensis, pp. 686–7; *La Continuation de Guillaume de Tyre dite du Manuscript de Rothelin*, ed. M.R. Morgan, Paris 1982 (*Documents relatifs à l'histoire des croisades*, 14), pp. 490–507. For European chronicles see *Chronicon Universale*, pp. 41–6; William of Malmesbury, pp. 422–3; *Annales Stadenses*, *MGH.SS*, 16, pp. 340–41.

Those encyclical letters presented Jerusalem as the scene of the life of Christ.[62] No less important was the propaganda of, so to speak, official papal propagandists, such as Bernard of Clairvaux on the eve of the Second Crusade, Henry of Albano on the eve of the Third Crusade and Martin of Pairis on the eve of the Fourth Crusade. Though most of them were churchmen and some were theologians, their crusade preaching presented theological ideas in drastically modified and even popularized terms, easily understood by the laity and particularly by the European knighthood, to whom crusade propaganda was mainly addressed. Many of the ideas expressed by papal crusade propaganda and the official crusade preachers found their way into popular songs and poems which, while reflecting the official teaching, translated it into the vernacular and into terms of *topoi* appropriate to its audiences' ideas. The most popular encyclical letters in the twelfth century were those of Bernard of Clairvaux. They inspired crusade preaching for the next few centuries, including that of another abbot of Clairvaux, Henry of Albano, as well as Gregory VIII, Henry of Strasbourg, Innocent III, Jacques of Vitry, not to speak of various, often anonymous, sermons and crusade songs and poems.[63]

In his open letter, known as *Ad peregrinantes Jerusalem* (1146),[64] Bernard of Clairvaux described the Holy Land and Jerusalem as: 'His land ... where He was seen and in which He lived among men for more than thirty years; His land ... embellished by His miracles, consecrated with His blood, where the first flowers of His resurrection appeared.'[65] Writing to the duke and people of Bohemia in 1147, he repeated this description, adding that it was the land 'He honoured by His birth' and 'enriched by His burial'.[66] Three years later, writing to Peter the Venerable, Bernard referred to Jerusalem as the place 'where His feet had trodden' (Psalms 132:7).[67] According to Suger of St Denis, both Pope Eugenius III and Bernard of Clairvaux preached the Second Crusade as aimed at the 'land where our Lord Jesus

62 See above, ch. 4, p. 56.

63 Morris, 'Propaganda for War', pp. 79–101; Spreckelmeyer, *Das Kreuzzugslied*, pp. 91–131, 142–84, 198–214, 264–8; Cole, *The Preaching*, *passim*.

64 For the popularity and the wide circulation of Bernard of Clairvaux's encyclical letter, the so-called 'Ad peregrinantes Jerusalem' (*Opera*, VIII, no. 363, pp. 311–17; also edited by J. Leclercq, 'L'encyclique de Saint Bernard en faveur de la croisade', *Revue Bénédictine*, 81 (1971), pp. 295–300 and L. Grill, 'Die Kreuzzugs-Epistel St Bernhards', *Ad peregrinantes Jerusalem, Studien u. Mitteil. zur Geschichte des Benediktiner-Ordens*, 67 (1956), pp. 250–53) see Leclercq, 'L'encyclique', pp. 289–94. Bernard's influence is also reflected in an anonymous sermon composed no later than the end of the twelfth century for the celebration of the Feast of the Liberation of Jerusalem, probably emanating from the *scriptorium* of St Mary of Ripoll (France, 'An Unknown Account', pp. 773–6). That the Library of Ripoll possessed 'Ad peregrinates Jerusalem' in the twelfth century had been shown by Grill, 'Die Kreuzzugs-Epistel St Bernhards', pp. 237–51. See also Leclercq, 'L'encyclique', p. 291.

65 See above, n. 64.

66 Bernardus Claraevallensis, *Opera*, n. 458, p. 435.

67 *Ibid.*, no. 521, p. 483. For this verse see also *ibid.*, n. 377, p. 341; n. 393, p. 365.

Christ turned into man and for the redemption of human race suffered passion on the Cross'.[68] Writing in 1150 to Suger, Peter the Venerable described Jerusalem's holy places as: '... the places where stood the feet of Him who brought salvation into the midst of the earth ... royal Jerusalem, consecrated by the prophets, the apostles, the very saviour of all ... the very Sepulchre of the Lord, which up to now ... was the glory of the entire world ...'.[69] Henry of Albano in an open letter, preaching the Third Crusade, exhorted the German princes to take the cross 'for the sake of the Holy Land, which the Lord sanctified for the Salvation of humanity by His feet'.[70] Another preacher of the Third Crusade, Henry, bishop of Strasbourg, preaching the crusade at the assembly of Mainz in March 1188, defined its goal as the 'inheritance of Christ', the place 'where His feet had trodden'.[71] Peter of Blois, who preached the Third Crusade in England, in his encyclical letter *De Hierosolymitana Peregrinatione Acceleranda* (1187–1189), urged the believers to depart for the Holy Land, 'where His feet had trodden', the place chosen by God as His dwelling, and which through His presence has become a 'sort of other Heaven' (*quasi alterum coelum*); the place He glorified through His miracles, where He created the sacraments, where He preached and where He left the imprint of His blood, where through the acts of Resurrection and Ascension to Heaven, He not only revealed His miraculous greatness to all the world but won for Himself the entire world.[72]

Chronicles, sermons, poems, charters and descriptions of the Holy Land all show that in the lower levels of Christian society, too, the dominant image of Jerusalem was that of the 'City of the life of Christ'. Twelfth-century poetry, including vernacular poetry aimed at the laity, often described Jerusalem as the place where God was 'working Salvation in the midst of the earth' (Psalms 74:12), 'where His feet had trodden' (*ibid.*, 132:7), where He lived as man.[73] For example,

---

68 Suger, *Vie de Louis le Gros*, ed. A. Molinier, Paris 1887, p. 158.

69 Petrus Venerabilis, *Letters*, no. 1, pp. 399–400; J. Phillips, *Defenders of the Holy Land. Relations between the Latin East and the West, 1119–1187*, Oxford 1996, pp. 106–7.

70 *Historia de Expeditione Friderici Primi*, pp. 11–13. This encyclical reflects the influence of Bernard of Clairvaux's letter to Bohemia. Cf. above, n. 66.

71 *Historia Peregrinorum*, *MGH.SS. rer. Germ.*, 5, pp. 123–4. This sermon of Henry of Strasbourg is clearly inspired by the *Audita tremendi* of Gregory VIII. See below, ch. 9, n. 19.

72 Petrus Blesensis, *De Hierosolymitana Peregrinatione Acceleranda*, *Opuscula*, *PL*, 207, cols 1058–70, esp. cols 1063, 1070. Following the fall of Jerusalem (between the years 1187 and 1189), Peter of Blois composed a treatise, the *Opuscula*, consisting of three parts: the first deals with the martyrdom of Reynauld of Châtillon (*Passio Reginaldi*); the second is an exhortation for a crusade (*De Hierosolymitana Peregrinatione Acceleranda)*; and the third an imaginary dialogue between King Henry II of England and the abbot of Bonneval, a Benedictine abbey near Le Mans. See Petrus Blesensis, *Opuscula*, cols 529–34, 957–88, 1057–70; *idem*, *Dialogus*, pp. 97–112. See also R.W. Southern, 'Peter of Blois and the Third Crusade', *Studies in Medieval History Presented to R.H.C. Davis*, ed. H. Mayr-Harting and R.I. Moore, London 1985, pp. 207–18.

73 G. Wolfram, 'Kreuzpredigt und Kreuzlied', *Zeitschrift für Deutsches Alterthum*, 30 (1886), pp. 89–132, esp. 97–101.

a poem lamenting Saladin's conquest of Jerusalem (1187), 'Oh, in lamentable voice I am forced to describe' (*Heu, voce flebili cogor enarrare*), referred to the Holy Land as the land through which the Saviour walked with His feet, where the creator of all was born from a virgin. It was the land to which Christ gave so many gifts, such as the Salvation of the world; the land also sanctified by the feet of the Virgin Mary and John the Baptist.[74]

Whereas in the early Middle Ages, pilgrims defined the goal of their pilgrimage in terms of the Holy Sepulchre,[75] shortly after the First Crusade a shift in motivation may be observed in both pilgrims and crusaders. The crusader Stephen of Neublens declared on 15 June 1100 that he wished to go to Jerusalem, 'where God was seen as man, to adore His feet in the place where they had trodden'.[76] In 1102, writing to Hugh, abbot of Cluny, the emperor, Henry IV, expressed his desire 'to go to Jerusalem to see the Holy Land, where our Lord was seen in flesh (*in carne*) and conversed with men ..., to adore where he [suffered] for us the slaps of whips (*flagella*), as well as the Cross, the death and the burial'.[77] On the evidence of Suger, King Louis VII of France intended to depart for the Second Crusade to see the 'blessed sites of Jerusalem'. According to Richard of Poitiers, the king wished 'to stay as a pilgrim in order to pray (*causa orationis*) in the Holy City'. And indeed, perceiving his journey to Jerusalem as a pilgrimage, Louis VII adopted conduct conforming to the ascetic practices of *peregrini* – prayers, fasting and vigils.[78] Peter of Blois, writing in 1187–1189, related how, following the fall of Jerusalem, King Henry II had expressed a desire 'to liberate the land where the Lord's feet had trodden, where through His birth and death He redeemed us'.[79] The English chronicler Richard of Devizes described the Jerusalem pilgrimage of Hubert, bishop of Salisbury, and of Henry of Champagne in the year 1192, both participants of the Third Crusade, as aimed at 'adoring in the place where Christ's feet had trodden'.[80]

Pilgrims reveal a similar motivation. John of Würzburg came to Jerusalem to visit 'those venerable places which were sanctified by the presence of Christ together with His mother and the apostles'.[81] Theoderich wrote in the introduction to his *Libellus*

---

74 *Carmina Burana*, no. 50, pp. 99–102; Spreckelmeyer, *Das Kreuzzugslied*, pp. 238–49.

75 For the early Middle Ages see above, n. 36. This was also the aim of Saewulf, who came to Jerusalem around 1102–03; Saewulf, pp. 59, 64. See also *Qualiter sita est Civitas Jerusalem* (*c.* 1093–1105), in ed. Tobler and Molinier, *Itinera*, pp. 347–9.

76 *Recueil de Chartes de l'Abbaye de Cluny*, ed. A. Bernard et A. Bruel, Paris 1904, no. 3737, p. 87.

77 Henry IV, *Briefe*, ed. F.J. Schmale, Darmstadt 1963 (*Ausgewählte Quellen zur deutschen Geschichte des Mittelalters*, 12), no. 31, p. 102.

78 Suger, *Vie de Louis le Gros*, pp. 158, 160; Ricardus Pictaviensis, p. 416; Grabois, 'Louis VII Pèlerin', pp. 7–22.

79 Petrus Blesensis, *Dialogus*, p. 106.

80 Ricardus Divisiensis, *Chronicon de tempore Regis Ricardi Primi*, ed. J.T. Appleby, London 1963, p. 84.

81 Johannes Wirziburgensis, p. 79.

de Locis Sanctis that he planned to describe in his treatise the 'Holy Places where our Saviour revealed Himself corporally'.[82] Though, as mentioned above, most of Jerusalem's pilgrims or crusaders defined the aim of their pilgrimage in terms of the 'sites of the life of Christ' or in reference to the formula of *ubi steterunt*, some still, as in previous centuries, insisted upon the Holy Sepulchre. Thus Stephen of Tournai wrote (in October 1177–1180) to Heraclius, then archbishop of Caesarea, to recommend to him one of his relatives who went 'to visit the most glorious Sepulchre of the Lord'.[83] The second pilgrimage of St Teotonicus (†1166) to Jerusalem aimed 'ad gloriosum sepulchrum finem vitae suae devotus expectaret'.[84] Saewulf revealed that he: 'Jerosolimam pergens causa orandi sepulchrum dominicum'.[85] The pilgrimage of Henry the Lion (1172) was described as: 'ad sanctum sepulchrum'.[86] The Templar Andrew of Montbard wrote in 1149 from the kingdom to Everard of Barres and to the Grand Master of the Templar Order (then in France), referring to the need for the defence of 'the Oriental Church and the Holy Sepulchre'.[87]

In the course of the century, descriptions of Jerusalem, which had previously focused on the Holy Sepulchre, began to identify Jerusalem as the 'City of the Life of Christ'. The holy traditions of the early life of Christ, his childhood and adolescence, were multiplied and enriched. Rorgo Fretellus, a chaplain in the Church of St Mary in Nazareth, composed in 1128–1132 a description of the 'Holy Places in the Promised Land', which was later – in 1137 and 1148 – revised in Europe. His Jerusalem is the 'City of the Life of Christ'. Fretellus describes the Church of the 'Temple of the Lord' as the place where Christ was circumcised on the eighth day after his birth, the place of his presentation in the Temple; the place whence he drove out the traders, saved the woman taken in adultery and taught before the Jews. Similarly, he associates other sites in Jerusalem with New Testament traditions of the life of Christ. On Mount Zion, for example, Christ washed the feet of the apostles, saying: 'You also should do as I have done to you' (John 13:15). On the Mount of Olives, near Gethsemane, Christ prayed, 'My Father, if it be possible, let this cup pass from me ...' (Matthew 26:39), and then said to his disciples: '... See, my betrayer is at hand' (*ibid.*, 26:46).[88] Other pilgrims, too, identified the crusaders' church of the 'Temple of the Lord' with Christ's presentation in the Temple.[89] Not far from there, a structure near the southeast

---

82 Theodericus, p. 143.

83 Etienne de Tournai, *Lettres*, no. 78, p. 92.

84 *Itinera Hierosolymitana*, ed. Sandoli, vol. 2, p. 40.

85 Saewulf, p. 59.

86 Arnoldus Lubecensis, *Chronica Slavorum*, *MGH.SS*, 21, p. 121.

87 *Epistola A. Dapiferi Militiae Templi ad Ebrandum magistrum ejusdem militiae*, *RHGF*, 5, pp. 540–41. On the letter see Phillips, *Defenders of the Holy Land*, pp. 104–6.

88 Rorgo Fretellus, pp. 30–37. For much the same view of Jerusalem see Johannes Wirziburgensis, pp. 89–90, 96–7, 113–17; Theodericus, pp. 158–74.

89 Saewulf, p. 68; Rorgo Fretellus, p. 32; Johannes Wirziburgensis, pp. 89–90; Theodericus, p. 162; *De Situ Urbis Jerusalem*, p. 426; *Chronicon Universale*, p. 43 and see below, ch. 6, p. 99.

corner of the Temple Esplanade became in the twelfth century the 'Church of the Bath' or of the 'Manger of the Lord' (*incunabulum*). John of Würzburg described this church as the dwelling place of Simeon the Just, where Christ had often stayed with Mary and where 'the wooden cradle of Christ is still preserved and is shown with great veneration'.[90] An anonymous description of Jerusalem, composed shortly before Saladin's conquest of the city, referred to the Holy Sepulchre in a most offhand way. The anonymous writer described Jerusalem in the following terms: 'In Jerusalem there are many venerable places, such as the Temple of the Lord (*Templum Domini*), where Jesus the child was presented (Luke 2:21–40) and whence he drove out the traders (*ibid.*, 19:45–46) and whence James, the brother of the Lord, was thrown down; Mount Zion, where he held the Last Supper with his disciples and composed the New Testament (*ibid.*, 22:1–38), where also the Holy Spirit appeared and was seen above the heads of the apostles (Acts 2:1–13), and where the Holy Virgin Mary died; Calvary, where he took, for our salvation, the Cross and died (Luke 23:23–46); the Holy Sepulchre, where his body was laid and whence he rose from the dead (*ibid.*, 23:26–56; 24:1–23); the Mount of Olives, where, riding an ass, he was met with honour by the children (Matthew 21:1–9), and whence he miraculously ascended to Heaven (Acts 1:1–11); Bethany, where he raized Lazarus from the dead (John 11:1–44); Silo, where he restored the sight of a man blind from birth (*ibid.*, 9:1–12); Jehoshaphat, also known as Gethsemane, where he was seized by the Jews (Luke 22:47–62) and where the Holy Virgin was buried ...'.[91]

Yet the most important sites – and as such the highlights of any pilgrimage to Jerusalem – were Calvary and the Holy Sepulchre. Having set out from their homes with crosses in their hands, when the pilgrims finally reached the Church of the Holy Sepulchre, they would place their crosses on Mount Calvary to symbolize the fulfilment of their vow. These crosses were burned every year on Easter Evening.[92] Theoderich, describing Mount Calvary, wrote: 'It remains now to speak of Mount Calvary, which shines in the Church as doth the eye in the head; from whence, by the death of the Son of God and the shedding of His blood, light and eternal life shall be poured forth on us ... The place where stood the Cross on which the Saviour suffered death is on the eastern side, raised on a high step covered on the left with the finest Parian marble, and it displays a round hole almost wide enough to take in a man's head, in which it is known that the Cross itself was fixed; into which hole pilgrims, out of the love and respect they bear Him that was crucified, are wont to plunge their head and face ... One descends fifteen steps into the church and reaches the chapel which is called Golgotha, of

90 Johannes Wirziburgensis, pp. 135–46. For this church, where the pilgrims were also shown Christ's bath and the Virgin's bed, see also J. Wilkinson and W.F. Ryan, *Jerusalem Pilgrimage*, pp. 213–34, 240.

91 *Tractatus de locis*, pp. 152–3.

92 Theodericus, pp. 155–6; *Anonymous II*, *PPTS*, 4, pp. 5–6.

great sanctity, but very dark; at the back of which is a deep recess, which enables the beholders to see the end of the cleft in the rock which came down from Calvary. In that place it is said that the blood of Christ rested after having run down thither through the cleft ...'.[93]

Nevertheless, it is noteworthy that neither Calvary nor the Holy Sepulchre ever became a miracle shrine. It was here that the central miracle of the Christian faith, the Resurrection of Christ, had occurred. The annual miracle of the Holy Fire was connected with this central event, but besides this liturgical miracle no miraculous healing or anything of that nature occurred in Calvary or the Holy Sepulchre. Nor were there relics that could work miracles. It would seem, therefore, that Calvary and the Holy Sepulchre, like other major shrines, such as St Peter's and to some extent St James of Compostella, were not renowned for 'ordinary' miracles. Pilgrims coming to the Holy Land could experience such simple miracles at the sepulchre of St Joseph in Nazareth, 'where white water which drops like holy oil from the wall' was collected to heal the sick, or on the Mount of Transfiguration, whose white stones, made into crosses and then dipped in wine, were known to heal the sick. Among the relics that could work cures, collected by pilgrims in the Holy Land, was the hair of the Virgin. However, in Calvary and the Holy Sepulchre, miracles were not necessary, whether to establish a claim to sanctity or to stimulate devotion.[94]

Veneration of the Holy Sepulchre had its effect during the twelfth century, as in previous centuries, in the construction of churches designed in the shape of the Anastasis Rotunda. It was probably the increased contact between Western Europe and Jerusalem, through pilgrimage and the crusades, that occasioned a sharp rise in the number of such churches, which were intended to remind the believers of the Holy Sepulchre in Jerusalem, namely Milan (1099), Augsburg (1128–1130), Pisa (*c.* 1138), Cambridge (*c.* 1125), Northampton (*c.* 1120), Brindisi (S Giovani del Sepolcro), Kappenberg (*c.* 1122), Osterlar in Denmark, San Sepolcro de Torres del Rio in Navarre and St Jean at Le Liget. Some copies, such as the one in Pisa, functioned as baptisteries. As a rule, the architects did not intend to imitate the actual appearance of the prototype, but to reproduce it *typice* and *figuraliter*, as a memento to a venerated site and simultaneously as a symbol of promised salvation. Indeed, *typice* and *figuraliter* are the terms used by a twelfth-century chronicler describing the structure of Santo Stefano at Bologna (rebuilt after 1141) as a reflection of Jerusalem. There, a whole complex of churches was laid out with the particular aim of reminding the pilgrim of a number of venerated sites in Jerusalem. The complex included the Chapel of the Holy Cross, attached to a church imitating the Anastasis by an open colonnaded court. In the centre of the church was a copy of the Tomb of Christ and in the chapel of the Holy Cross, kept

93 Theodericus, p. 155.

94 Schein, *Miracula*, pp. 291–2; Ward, *Miracles*, pp. 120–26. For the miracle of the Holy Fire see above, ch. 3, pp. 40–42.

at a special place called 'Golgotha', was a relic of the True Cross. Elsewhere in Bologna there were copies of the Mount of Olives and the Church of the Ascension, as well as a Valley of Jehoshaphat, a pool of Siloam and a field of Aceldema. The intention was apparently to create a comprehensive, topographical copy not just of the Holy Sepulchre but also of the city of Jerusalem, a sort of 'Jerusalem' in Bologna.[95] As before the twelfth century, some of the copies were not only associated with the Earthly Jerusalem but with its celestial counterpart – the Heavenly Jerusalem. Hence, the Baptistery of Pisa, whose structure, dated to 1153, repeated many features of the Anastasis Rotunda. Frederico Visconti, the bishop of Pisa, referred, in one of his sermons, to the baptistery as the 'mirror of the city and the Gate of Paradise'; the ideal city, the Heavenly Jerusalem, which the earthly city hoped to imitate. The later addition of the Campo Santo, the cemetery of the complex, paved with dirt brought from Jerusalem, only served to strengthen this association. A citizen of Pisa could not only begin but also end his life in the vision of Jerusalem.[96]

The increasing veneration of Calvary, as the place of Christ's Crucifixion, can be seen as yet another symptom of the devotion to the humanity of Christ in which a particular stress was put on the Passion of Christ. This can be felt in twelfth-century Europe as a growing devotion to the Cross, the Holy Lance and the Five Wounds of Christ. The devotion to the Cross of Christ can be seen in the little bronze and wooden crucifixes that played an increasing role in monastic and, later, lay spirituality. In monastic circles some of the crosses became a subject of criticism as they were made of gold and silver and were decorated with engravings and paintings that provoked the leaders of the monastic orders to denounce them as not befitting monastic simplicity. Peter the Venerable decreed that the crosses presented to sick monks should be made of wood rather than gold and silver, because it was 'more rational and devout' for such crosses to be made of the material on which Christ was crucified, and it would avoid any untruths when the phrase 'Behold the wood of the cross' was repeated. He did not condemn crosses of gold and silver but stated that 'the preciousness of the metal, which raises the minds of the simple people to greater veneration of the Holy Cross, is different from the worthlessness of wood, which moves the souls of the spiritual, to more fervent devotion towards the cross and the crucifix'.[97]

Devotion to the cross, combined with the cult of the True Cross, produced a growing demand in the twelfth century for the most important of Jerusalem's relics

---

95 Krautheimer, *Studies in ... Art*, pp. 128–30; Bresc-Bautier, 'Les imitations du Saint-Sépulchre', pp. 319–42; Ousterhout, 'The Church of Santo Stefano', pp. 310–21; *idem*, 'Loca Sancta', pp. 108–24.

96 Morris, 'Bringing the Holy Sepulchre to the West', pp. 31–59.

97 Ousterhout, 'Loca Sancta', pp. 115–16. Earth from the holy places was often brought to Europe in the early Middle Ages as *eulogiae* together with other cheap souvenirs like water, oil, rocks, and so on. See Y. Hen, 'Labels of the Holy Land Eulogiae in Frankish Gaul', *Cathedra*, 82 (1996), pp. 31, n. 23.

– the relic of the True Cross. The cult of the True Cross had attracted pilgrims to the Holy Sepulchre since the fourth century. Though there was a 'miraculous' supply of particles of the True Cross from Jerusalem even during the Moslem rule in the city, the crusaders' take-over led to a meaningful re-evaluation of Jerusalem as the source of the relic. Jerusalem was now regarded as the chief supplier of the relic of the True Cross, thus replacing Constantinople. The description of the discovery of the True Cross (which the crusaders considered to be the self-same relic recovered by Emperor Heraclius in 628) on 5 August 1099, follows the pattern of the legend of the Invention of the Holy Cross by Helena. The relic was enclosed in a silver casket and, from 1149, was kept in a special chapel on the north side of the Anastasis. During the twelfth century many of its fragments found their way into Europe, either by donation or purchase. Kings of Jerusalem were accustomed to giving or sending the relics of the Holy Cross to various European rulers. Thus Sigur, king of Norway, was granted a fragment by Baldwin I in 1107 or 1110. Count Stephen of Blois also received a piece from the same king before 1102. Fulk of Anjou sent a piece of the Cross to the community of Saint-Laud in Angers. In 1164, on the eve of the battle of Alexandria, King Amalric made a vow to send his pectoral cross with a relic of the True Cross to Clairvaux. Not only did the kings of Jerusalem but also the patriarchs and the canons of the Church of the Holy Sepulchre give away pieces of the Holy Cross from time to time. Thus Patriarch William I of Jerusalem (1130–1145) sent a fragment to Bernard of Clairvaux, and Patriarch Heraclius, while in England in 1185, gave one portion of the True Cross to King Henry II.[98]

It seems that being able to obtain a fragment of the True Cross was one of the aims of European pilgrimage and particularly of the European nobility, as many members of this class brought these relics back home with them. They wished to give them to their home churches and thus enhance their aura of sanctity and popularity. For instance, William IV, lord of Montpellier, brought home (*c.* 1128) with him a fragment of the Cross, which he divided between the Church of the Holy Cross in Montpellier and the Cistercian abbey of Grandselve where he became a monk in 1148. Henry the Lion, duke of Saxony and Bavaria, brought home from his pilgrimage a cross-staurotheca which contained a fragment of the True Cross, and gave it in 1173 to the Church of the Holy Cross in Hildesheim.[99]

---

98 H. Frolow, *La relique de la Vraie Croix*, Paris 1961, pp. 305–49; Kühnel, *Crusader Art*, pp. 125–53. For the importance of the Holy Cross in the life of the crusader kingdom see Y. Katzir, 'The Vicissitudes of the True Cross of the Crusaders', *The Crusaders in their Kingdom, 1099–1291*, ed. B.Z. Kedar, Jerusalem 1987, pp. 243–53; G. Ligato, 'The Political Meanings of the Relic of the Holy Cross among the Crusaders and in the Latin Kingdom of Jerusalem: An example of 1185', *Autour de la Première Croisade*, pp. 315–30; A.V. Murray, '*Mighty Against the Enemies of Christ*: The Relic of the True Cross in the Armies of the Kingdom of Jerusalem', *The Crusades and their Sources*, pp. 217–38.

99 Kühnel, *Crusader Art*, pp. 133–4; Hamilton, 'The Impact of Crusader Jerusalem', p. 711.

The relic, it was believed, as in previous centuries, had healing powers and when in 1191 there were fears for the life of Louis, the young son of Philip II Augustus and the future Louis VIII, the monks of St-Denis placed their hopes in a crown of thorns, a nail from the True Cross and the arm of St Simeon. These relics were used to make the sign of the cross over Louis' abdomen, following which his health was restored. The relic was also used for support at death and burial. Charlemagne, for example, was buried with a particle of the True Cross; while on his death-bed (1153) David I of Scotland commanded the relic of the True Cross to be brought to him 'which he honoured with great reverence'.[100]

Also, other relics that originated in Crusader Jerusalem were mainly connected with Christ. There were stones from the Lord's Sepulchre and Calvary, particles of the Sponge, the Crown, the Manger, and the stone from which the Christ ascended to heaven, as well as relics of the Virgin Mary and the apostles. Thus, Robert II Count of Flanders brought back from the First Crusade particles of the True Cross, of the Sponge, the Crown of Thorns, the Sepulchre and the Manger. Maurice, lord of Craon, brought back, in 1169, a large collection of relics which included fragments of the Holy Cross, a chip of rock from the Calvary, a splinter of stone of the Holy Sepulchre, part of the column of the flagellation of Christ, filings from the hammer and a nail of the Passion, and a piece of Our Lady's dress. One of the Zwiefalten reliquaries, produced in Jerusalem and brought between the years 1099 and 1111 to the Benedictine monastery of Zwiefalten by Bertolf of Sperberseck (Sparwarisegge) contained, besides the relics of the True Cross, a relic of the stone from which Christ ascended to heaven, a relic of Christ's manger, relics of John the Baptist and the apostles Andrew, Mark and James. The second Zwiefalten reliquary, also produced in Jerusalem, contained a portion of the True Cross as well as relics of the Virgin's robe, of John the Baptist's ashes and of the apostles Andrew, Bartholomew, Philip, Peter and Paul. Poor pilgrims, who could not afford relics, brought from the Holy Land *eulogiae* such as water from the Jordan and palm fronds, a souvenir of the entrance of Christ to Jerusalem on Palm Sunday (John 12:12–15). It is possible that they also brought souvenirs similar to the ones brought by pilgrims in the early Middle Ages, namely, earth from various holy places, water from the Jordan, oil and rocks.[101]

Yet another manifestation of the devotion to the humanity of Christ and his Passion was the holy geography of Jerusalem. During the twelfth century, the number of holy places commemorating Christ's Passion increased. On the eve of the crusader conquest of Jerusalem the prison of Christ was shown near Golgotha in one of the chapels built by Constantine Monomachus (on the east side of the court that stretched between them and the Anastasis). Nearby were a

---

[100] Snoek, *Medieval Piety*, pp. 106–15.

[101] Kühnel, *Crusader Art*, pp. 133–7. Daniel received from the keeper of the key of the Holy Sepulchre a small piece of the 'sacred rock'. Daniel, p. 81. See for palm fronds, see Riley-Smith, *The First Crusaders*, p. 144. See also n. 97 above.

few other chapels: the first two commemorating where Christ was crowned and dressed in the purple robe, where he was stripped of his garments and where his clothes were shared out; the third was where he was bound to a column and scourged. Then came some steps, going down to the place where the wood of the Cross was said to have been discovered by Empress Helena, and the last chapel in the corner (on the north side) which had been Christ's prison.[102] Most of the sacred traditions of the Passion of Christ were thus located near the site of the Crucifixion. During the twelfth century most of these traditions were transferred from the vicinity of Calvary, and in fact what emerged was the first step towards the creation of the 'Way of the Cross', corresponding more or less to the *Via Dolorosa* as institutionalized by the Franciscans of the *Custodia Terrae Sanctae* in the fourteenth and fifteenth centuries.[103] The prison of Christ was still displayed in the Church of the Holy Sepulchre but rather as the place where he was guarded while the Cross was being made ready on the Mount of Golgotha.[104] At the same time, however, the prison was also claimed to be on Mount Zion, reported not only by Fretellus[105] but also by other pilgrims such as Theoderich and John of Würzburg. Theoderich mentioned two prisons: one in the Holy Sepulchre and another one in the House of Pilate,[106] which he located near the Church of St Anna and the palace of Antonia. John of Würzburg placed the *praetorium* on Mount Zion and claimed the existence of another prison in the Church of the Holy Sepulchre.[107] Theoderich too placed the flagellation of Christ on Mount Zion. There, he reported: 'peregrini ad eius exemplum flagellari solent'.[108]

In *c.* 1160 or even 1170 a new *locus sanctus* based on a tradition related to the Passion of Christ appeared. This marked the place of the Repose of Christ (John 18:1; Mark 1:32; Matthew 26:36). Built by the Templars at the western end of the northern wall of the Temple Esplanade, it was a small domed structure. Near the chapel was a pool, named by the Templars, according to a guide to Jerusalem pilgrimage of *c.* 1170, as 'the Sheep Pool where the Wood of the Cross remained for a long time'. The Templars also renamed the gate on the West of the Haram the 'Sorrowful Gate' and probably presented it as the gate through which Christ was

102 Wilkinson and Ryan, *Jerusalem Pilgrimage*, pp. 34–5.

103 B. Pixner, 'Where was the original via Dolorosa?', *Christian News from Israel*, 27 (1979), *passim;* S. Schein, 'La Custodia Terrae Sanctae Franciscaine et les Juifs de Jérusalem à la fin du Moyen Age', *Revue des études juives*, 141 (1982), pp. 370–71.

104 Wilkinson and Ryan, *Jerusalem Pilgrimage*, pp. 221, 233, 239.

105 Rorgo Fretellus, p. 61.

106 Theodericus, pp. 153, 171–2. After the construction of the Crusader Church of the Holy Sepulchre, that is, after 1149, the prison was located in the eastern part of the ground level of the north transept, namely, in the same place as before. See Folda, *The Art of the Crusaders*, p. 212.

107 Johannes Wirziburgensis, pp. 115–16, 117.

108 Theodericus, p. 172.

led to Calvary.[109] It thus follows that the Templars, like the Hospitallers in their area of Jerusalem, introduced into their quarter holy traditions previously located elsewhere in order to enhance the holiness of their residence in Jerusalem; what characterizes these traditions is that they all depict the Passion of Christ.[110] Therefore, there is clear evidence of the Way of the Cross before 1187 or even two Ways of the Cross: one beginning at the new *praetorium* on Mount Zion and the second, fabricated or created by the Templars, which started at the 'Church of the Repose'.

A further manifestation of the sanctity of Jerusalem as the city of the New Testament was the appearance of a great number of holy places associated with the life of the Virgin Mary. This resulted from the importance of the cult of the Virgin in the East, as well as Marian devotion in the West. It was also influenced by the piety of the participants of the First Crusade, which was originally scheduled to depart on the Feast of the Assumption of the Virgin in 1096 (15 August).[111] The crusaders frequently experienced, in the course of the First Crusade, visions in which Mary herself appeared. In one such vision in mid-April 1099, a priest named Stephen of Valence saw the deceased papal legate, Adhemar of Le Puy, singing the responsory, 'All rejoice in the Virgin' (*Gaude virgo cunctus*), and countless hundreds of thousands of men, an assembly of saints, joining him.[112] In their versions of Pope Urban's declaration of the First Crusade, chroniclers recalled the fate of the Virgin's sepulchre in the Church of St Mary in the Valley of Jehoshaphat.[113] And indeed, the famous procession which took place on 8 July 1099, organized during the siege of Jerusalem – obviously modelled upon the Israelites encircling Jericho, and devised with the intention of placating God and inducing Him to liberate His Holy City and His Holy Sepulchre from the yoke of the pagans – went from the Church of St Mary on Mount Zion to the Church of Stephen Protomartyr outside the Gate of St Stephen, the present-day Damascus Gate. Barefoot and carrying crosses in their hands, singing psalms and praying that God would renew his miracles of old and make the walls of Jerusalem topple like

---

[109] Wilkinson and Ryan, *Jerusalem Pilgrimage*, pp. 73–6, 240; Folda, *The Art of the Crusaders*, pp. 318–21. The stories about the wood of the Cross taken by Christ from the Temple or from the Pool of Bethesda circulated in Europe already before the crusades. For twelfth-century versions see *The History of the Holy Rood-Tree*, pp. 2–35; Honorius Augustodunesis, *PL*, 172, p. 994.

[110] See Sylvia Schein, 'The *Miracula* of the Hospital of St John and the Carmelite Elianic tradition – Two Medieval Myths of Foundation?', *Cross Cultural Convergences in the Crusader Period. Essays Presented to Aryeh Grabois on his Sixty-Fifth Birthday*, ed. Michael Goodich, Sophia Menache, Sylvia Schein, New York 1995, pp. 287–96.

[111] Riley-Smith, *The First Crusade*, pp. 103–4. For the cult of the Virgin Mary in the West see W. Delius, *Geschichte der Marienverehrung*, Munich 1963, pp. 149–70; M. Warner, *Alone of All Her Sex. The Myth and the Cult of the Virgin Mary*, New York 1983, *passim*.

[112] Raimond d'Aguilers, pp. 127–8; Riley-Smith, *Idea and Reality*, pp. 163–5.

[113] See, for example, Baldricus Dolensis, p. 13.

those of Jericho, the crusaders' procession halted on its way at the Church of St Mary in the Valley of Jehoshaphat 'from where her most holy body was carried off to Heaven'.[114]

Not surprisingly, the crusaders' devotion to the Virgin engendered a process of topographical identification of old and even newly created traditions of her life during their rule in Jerusalem. One such tradition (*c.* 1130) was that of the 'Church of the Bath', 'in which [Simeon the Just] is said to have often received the Blessed Virgin Mary, the mother of our Lord, with hospitality, to have cherished her and given her food'. Here it was believed that Mary had spent the night before her purification, and here the pilgrims were shown her bed, as well as the cradle of the Lord and basin of stone in which he was bathed.[115] Another new tradition, located in the court of the Temple of the Lord (the place of the presentation and purification of St Mary in the Temple), was the school of the Virgin.[116] Additional new traditions identified the Golden Gate as the meeting place of St Joachim and St Anne, the Virgin's parents. Thus, with the nearby Church of St Anne, the site of the Virgin's birth, and the Church of St Mary in the Valley of Jehoshaphat, site of her burial, the whole area around the Temple Mount became the setting for the life of Mary. Its centre was the Church of the Temple of the Lord, where the principal altar was consecrated to her.[117] And indeed, on 2 February, the Feast of the Purification of the Blessed Virgin or Presentation of the Lord in the Temple was celebrated in Crusader Jerusalem not only in the Church of the Holy Sepulchre but also in the Church of the Temple of the Lord. On this feast day a procession set out from the Holy Sepulchre to the Temple of the Lord, encircled the church and stopped near its southern gate, where the patriarch of Jerusalem delivered a sermon.[118] Another Marian feast celebrated in Jerusalem by solemn procession was the Feast of the Assumption of the Virgin (15 August); its procession followed the route of Mary's body as it was carried by the apostles from Mount Zion to St Mary in the Valley of Jehoshaphat. The Franks, when they rebuilt Jerusalem, restored the ruined Byzantine church on the site of the Assumption in the Valley of Jehoshaphat, trying to make it as similar as possible to the church of the other empty tomb, the Holy Sepulchre. Here Melisande was buried in 1161, the only queen regnant of Crusader Jerusalem. All the kings, however, were buried in the Church of the Holy Sepulchre. The house of St Joachim and St Anne, served by Benedictine nuns, was rebuilt as a beautiful Romanesque church. On Mount Zion, the Byzantine church built on the site of the house of St John the Evangelist and thus the scene of the Last Supper, the descent of the Holy Spirit and the Dormition of Our Lady was first repaired and then rebuilt.[119]

---

114 Petrus Tudebodus, pp. 137–8. Riley-Smith, *Idea and Reality*, pp. 164–5.

115 Theodericus, p. 166; Johannes Wirziburgensis, pp. 135–6; *Citez de Jherusalem*, p. 210.

116 Theodericus, pp. 157, 159.

117 Johannes Wirziburgensis, p. 90.

118 *Ibid.*, pp. 140–41; Schein, 'Between Mount Moriah', p. 185.

119 Kohler, 'Un rituel', pp. 409–10; 430; Hamilton, 'Rebuilding Jerusalem', pp. 107–15; Prawer, *The Latin Kingdom of Jerusalem*, pp. 177–8.

The popularity of Marian shrines in Crusader Jerusalem must have been considerable, as the community of the Church of St Mary Latina, which was not a shrine church, tried to convert it into one by telling pilgrims how the blessed Virgin had fainted at the Crucifixion and had been carried to a cave on the site of their church; there she had revived and in her grief, torn out a handful of her hair, which they kept in a reliquary.[120] According to an anonymous description of the Holy Land of *c.* 1145, this was the place where Mary, Mary Magdalen and Mary the wife of Cleopas, stayed, tearing out their hair, during the Crucifixion.[121] As a result, the Church of St Mary Latina became, during the twelfth century, one of the major Marian shrines in Jerusalem. In the letter *Quot et quantum* (1191) Pope Celestine III informed the believers about the merits of the souvenirs of the life of Christ and the Virgin which adorned it, extolling it as a singular sanctuary for the believers. According to the pope the house was, *inter alia*, the place where 'the Mother of God, Immaculate Virgin, remained for three and half years after the Ascension of her beloved son'.[122]

Many pilgrims described yet another Marian holy tradition, namely, the place on Calvary where the Virgin stood at the time of the Crucifixion.[123] It follows thus that the Marian shrines on Temple Mount, and in its vicinity, became a part of a large 'cycle marial' that stretched from Golgotha through the Church of St Mary Latina, the place of the Dormition on Mount Zion, the house of St Anna, the Temple Mount up to the Church of St Mary in the Valley of Jehosaphat. These traditions were a mixture of the old and the newly created. The churches, rebuilt over the places which, through a process of topographical identification became associated with the Virgin, demonstrated both the growing importance of her cult in the West as well as its importance in Crusader Jerusalem where it was always central.

The devotion to the Virgin and the newly created shrines commemorating various events in her life produced a demand for her relics. The relics of the Virgin were particularly difficult to find as her body had disappeared from the earth. The patriarch of Jerusalem, asked in 451 by Pulcheria if the body of the Virgin could be sent to her, replied, in one of the first hints at Mary's Assumption, that her body had never been found and gave Pulcheria the Virgin's grave clothes; Empress Eudocia was sent the Virgin's shroud. Throughout the twelfth century, the flow of Mary's relics increased prodigiously and clerics all over Western Christendom discovered sacred and hitherto unknown relics of the Virgin in their treasuries.[124]

---

120 Theodericus, p. 158; Hamilton, 'Rebuilding Zion', p. 112. This was, it seems, a local tradition as Saewulf reports that he heard it from the local Christians. See Saewulf, p. 67.

121 *Innominatus VII*, ed. Tobler, *Descriptiones*, p. 101.

122 *Cartulaire général de l'ordre des Hospitaliers*, no. 154. See also *ibid.*, no. 2674. For St Mary Latina as the shrine of the Virgin in pilgrims' descriptions see Wilkinson and Ryan, *Jerusalem Pilgrimage*, pp. 87, 103–4, 173, 233, 239. See also Schein, 'The *Miracula*', pp. 287–93.

123 Johannes Wirziburgensis, p. 118; Theodericus, p. 158; Wilkinson and Ryan, *Jerusalem Pilgrimage*, p. 239.

124 Warner, *Alone of All Her Sex*, pp. 285–98.

Henceforth, Jerusalem became the main supplier of those relics, as with those of Christ, in this respect overtaking Constantinople. Already, the participants of the First Crusade were beginning to return home with such relics. Most popular of all was the hair of the Virgin, as it was believed that it could work cures. The hair was mostly brought from the Church of St Mary Latina. Thus, for example, the French knight Igor Bigod, a participant of the First Crusade, who found a ball of the Virgin's hair in Jerusalem, divided it among several cathedrals and monasteries in France. Other participants brought home reliquaries containing the blood of the Virgin and the hair she had torn out when mourning Christ's death.[125] In the course of the century some pilgrims also brought home fragments of the Virgin's tunic.[126]

To sum up: the changes that occurred following the crusader conquest of Jerusalem in the attitude to the Holy City transformed Jerusalem from the 'City of the Holy Sepulchre' into the 'City of the Humanity of Christ'. The Holy Sepulchre remained now, as the Church of the Resurrection, the focus of Crusader Jerusalem, the highpoint of the visits of pilgrims and crusaders alike to the Holy City, the Holy of Holies of the 'Jerusalem of the New Testament'. In the twelfth century the central reason for pilgrimage to Jerusalem was the desire to be where Christ had lived and suffered as a man. It was the newly heightened devotion to Christ's humanity, especially in his Passion, that inspired pilgrims to visit the land and city where he had spent his earthly life. As William of Malmesbury put it: 'What can be greater happiness than for a man, in his lifetime, to see those places where the Lord of Heaven was conversant as a man?'[127] The pilgrims were able to do so more successfully and fully than in the previous centuries as a result of the new 'Holy Geography' of Jerusalem. During the twelfth century it was meaningfully enlarged and embellished so as to depict the entire life of Christ and his Passion, as well as the life of the Virgin Mary whose cult and its new importance engendered a process of topographical identification of both the old and the newly created traditions of her life.

---

125 Riley-Smith, *The First Crusade*, pp. 95, 123.

126 See above, n. 101.

127 William of Malmesbury, p. 398.

Chapter 6

The City of the Old and the New Testaments

Although the sanctity of Jerusalem continued to rest on its special link with Christ, by the first decade of the twelfth century the Holy City had already become for Christendom – and particularly for its inhabitants the Franks – the city of both Old and New Testament. It was a city in which the new temple of Constantine, the centre of the Jerusalem of the New Testament, now faced the temple on the Temple Mount – the 'Temple of the Lord'. Both symbolically and figuratively represented, the Jewish Temple was the centre of the Old Testament's Jerusalem. Consequently, the concept of the Holy City was historically expanded to include its history before Christ, its biblical past and, especially, its past as the city of King David. Thus, Jerusalem emerged in the spirituality of the twelfth century as the embodiment of a double sanctity: the city of the historical Children of Israel as the prefiguration of their successors, the Christians, and the city of the humanity of Christ. From its earliest time Crusader Jerusalem had two temples, the new one built first by Constantine on the place of the Lord's Sepulchre and the old one, now newly reestablished on the Temple Mount – the 'Temple of the Lord'. It was those two temples, the old temple of the Old Testament and the new one, a monument of the New Testament's 'Temple of God', that is, Christ, that became the two foci of Crusader Jerusalem.

This development was in strong contrast to Christian attitudes towards Jerusalem in both the Byzantine and early Moslem periods. During the Byzantine period Jerusalem was often described in biblical terms. Those such as the Bordeaux Pilgrim and Egeria revealed an intense curiosity in the Holy Land as the land of the Bible and visited the places where the great events of biblical history took place, namely, Mount Sinai, Mount Nebo, Mamre, Hebron, Jericho, Bethel and Mount Gerizim. However, whereas outside Jerusalem the sites from the Old Testament and New Testament were combined to create a single sacred geography, Jerusalem's biblical past was either intentionally forgotten and rejected or 'Christianized'. Christian–Byzantine Jerusalem drew its legitimacy almost exclusively from Christ.[1] The sites of Christ's Crucifixion and Resurrection became the New Temple of Constantine's New Jerusalem. Mount Moriah was left in ruins as if to symbolize the victory of Christianity over Judaism, of *Ecclesia* over *Synagoga.*

---

1 Wilken, *The Land Called Holy*, pp. 101–215 *et passim.*

This total rejection of the Temple stemmed from the attitude to the Temple of the New Testament, an attitude which can best be described as ambivalent. The Gospels present the Temple as the house of Christ's Father (Luke 2:49) and Jesus as believing in God's presence in the Temple (Matthew 23:21); Christ declared that 'Jerusalem is the place where one ought to worship' (John 4:20); Christ's act of cleansing the Temple was performed out of concern with the purity of worship in His Father's house: 'Take all this out of here and stop turning my Father's house into a market' (John 2:16). This positive attitude to the Temple is, however, accompanied by other attitudes varying from criticism to total rejection. According to Luke: '... As he [Christ] drew near and saw the city he wept over it, saying, "Would that even today you knew the things that make for peace! But now they are hid from your eyes. For the days shall come upon you, when your enemies will cast up bank about you and surround you, and hem you in on every side, and dash you to the ground, you and your children within you, and they will not leave one stone upon another in you; because you did not now the time of your visitation"' (Luke 19: 41–44; Matthew 24:2). These words of Christ could be understood as a prophecy but also as Christ's curse on Jerusalem as a city 'which kills its prophets and stones those who are sent to you' (Matthew 23:37).

In contrast to the Gospels, both the Book of Revelation and Paul's letters put forward a view rejecting the Temple in most vehement terms. The Revelation to John envisaged a temple that opened in heaven and, in his final vision, a heaven without a temple (Revelation 15:5, 21, 22). In his description of the Heavenly Jerusalem, John stated explicitly that: 'I saw no temple in the city, for its temple is the Lord God the Almighty and the Lamb' (21, 22). No less influential for medieval views of the biblical Temple, and possibly even more so, is the presentation of the Christian believer as an individual, and the Christian Church as a whole as God's temple. Paul argued: 'Do you not know that you are God's temple and that God's Spirit dwells in you? If anyone destroys God's temple, God will destroy him, for God's Temple is holy, and that temple you are' (1 Corinthians 3:16–17).[2] The Temple, in Paul's opinion, prefigured the Church as both the present community and as the glorified Church in heaven; the historical Jewish Temple has outlived its historical purpose and thus needs to exist no longer. The New Testament thus treated the Temple mostly as a symbol, as the symbol of Christ, the Christian community, and of the Christian believer.

The rejection of the historical Jewish Temple and the complete spiritualization of the term 'temple' deeply influenced future generations. In the fourth century, Eusebius, in his *Vita Constantini*, claimed that the New Jerusalem was built 'over against the one so famous of old, which, after the pollution caused by the murder of the Lord, experienced the last extremity of desolation and paid the penalty for the crime of its impious inhabitants'.[3] Early in the fifth century, Jerome could

---

2 See also 2 Corinthians 6:14–18.

3 Eusebius, *Vita Constantini*, III, 33, cols 1094–5.

refer to the Temple Mount as the 'dungheap of the new Christian Jerusalem'[4] that arose around the Holy Sepulchre in the time of Constantine, who was hailed by his contemporaries as 'our most peaceful Solomon who built the Temple'.[5] The Temple Mount visibly expressed the fulfilment of Christ's curse or prophecy: '... there will not be left here one stone upon another, that will not be thrown down' (Matthew 24:2; Luke 19:42–44). As the 'nesting place of owls',[6] it gradually disappeared from the map of the holy places of Jerusalem during the Byzantine period. When the so-called Bordeaux Pilgrim visited Jerusalem around the year 333, he went upon his arrival directly to the Temple Mount and was able to find there the sites of a whole complex of Old and New Testament traditions.[7] Some fifty years later, however, most of those traditions had already been 'shifted' from the Temple Mount to various sites elsewhere in the city, above all to its new centre, the Constantinian buildings on the sites of the Anastasis and Calvary. As a result, the saintly Egeria, who stayed in Jerusalem between 381 and 384, was shown at Calvary not merely the Holy Cross but also King Solomon's rings and the horn with which the kings of Judaea had been anointed, as well as the site of the Presentation in the Temple.[8] Thus, as has recently been noted, 'the Christians took over the meaning of a site without taking over the site itself'.[9] Calvary and Anastasis replaced Mount Moriah not only as the sanctuaries of Christendom, but as the location of traditions of the Jewish Temple, which disappeared from the holy geography of Jerusalem.

The literal physical temple was thus replaced by a spiritual temple, supported by the statement of Isaiah (66:1) that God was not to be contained in any material structure. From the Church's official position that it represented both mystically and universally the Temple, it followed logically that the Messiah's kingdom would supplant the outworn system of the past by raising up a temple of the spirit. The Temple was also transformed into an intellectual one. According to Origen, the rational, intellectual, unpolluted and external nature of divinity resided in such a temple, a higher and purer temple, built of abstract virtues such as simplicity, intellect and trust.[10] In the fourth and fifth century the Fathers of the Church were uncomfortable with the tradition that the Temple would be restored in the Last Days. Jerome, for example, observed with indignation that Jews and those who

---

4 Hieronymus, *Commentariorum in Isaiam*, p. 740.

5 *Ibid.*; Prawer, 'Jerusalem in Christian and Jewish Perspectives', pp. 754–7.

6 Hieronymus, *Commentariorium in Isaiam*, p. 740. For Jerusalem in Christian perspective in the Byzantine period see Prawer, 'Jerusalem in Christian and Jewish Perspectives', pp. 750–74; Linder, 'Jerusalem between Judaism and Christianity', pp. 5–26.

7 *Itinerarium Burdigalense*, p. 15.

8 Egeria, pp. 81, 89; Wilkinson, *Egeria's Travels to the Holy Land*, Jerusalem 1981, pp. 36–3, 298–310.

9 Wilkinson and Ryan, *Jerusalem Pilgrims*, p. 36.

10 H. Nibley, 'Christian Envy of the Temple', *Jewish Quarterly Review*, 50 (1959/1960), pp. 97–109.

were Judaizing, pointed to a plot of ground on the Mount of Olives 'where they say the sanctuary of the Lord, that is, the Temple, is to be built, and where it will stand for ever ... when, as they say, the Lord comes with the heavenly Jerusalem at the end of the world'.[11] In a letter to Presbyter Paulinus, Jerome claimed that as God was not confined to a limited piece of land, one did not need to pray at Jerusalem or Mount Gerizim, because the Spirit was everywhere. 'Those who cry "Temple of God" should remember that *you* are the Temple, inhabited by the Spirit.'[12] For Augustine the *Templum Dei* was Christ and the members of His body in the Church on earth and in heaven. This Temple will be completed when the elect finally rejoin the angels in the Church Triumphant.[13] Moreover, the Temple of Solomon was traditionally perceived, mystically and liturgically, as a prefiguration of the Christian sanctuary. The Temple, erected through divine inspiration, was conceived as an ideal model for a church, since its perfect proportions reflected cosmic harmony.[14]

Following the Moslem conquest of Jerusalem in 638, Patriarch Sophronius, seeing the new rulers of the city taking over the Temple Mount, exclaimed that 'this surely is the abomination of the desolation of the Temple of which David (*sic*) prophesied'.[15] The holy traditions of the Temple Mount remained, however, located elsewhere.[16] Not only were the traditions of the Temple identified largely with the Church of the Holy Sepulchre, but their number significantly diminished. Its 'Holy Geography' being based exclusively on the New Testament, Jerusalem became the city of the Holy Sepulchre.[17] At the same time, Western early medieval writers continued the exegetical tradition of treating the Temple as a *typus* or as a metaphor for things spiritual. One can say that this attitude to the Temple runs parallel to the attitude to 'Jerusalem' which also now became mainly a metaphor, a symbol or a *typos*.[18] For Gregory the Great, the Temple prefigured the two orders of priesthood – the minor and the major; it also was the *fidelis populus*.[19] Gregory also likened the Temple to the

---

11 Hieronymus, *Commentariorum in Hieremiam, CCSL*, 59, pp. 409–13.

12 *Idem*, *Epistulae*, no. 58, pp. 529–32; Prawer, 'Jerusalem in Christian and Jewish Perspectives', pp. 762–3.

13 Augustinus, *Enchiridion*, chs 7, 8, 15, 16.

14 C.H. Krinsky, 'Representations of the Temple of Jerusalem before 1500', *Journal of the Warburg and Courtauld Institutes*, 33 (1970), p. 11. In fact, before the twelfth century, there are scarcely any examples in Western art of the Temple being portrayed as a contemporary church. See *ibid*., pp. 1–19.

15 Nibley, 'Christian Envy of the Temple', pp. 119–20. 'David' is obviously a copyist's mistake for Daniel (11:31). For Sophronius see also Wilkinson and Ryan, *Jerusalem Pilgrims*, pp. 7, 91–2.

16 See, for example, Bede (*c*. 702–703), *De Locis Sanctis*, pp. 254–64; Nibley, 'Christian Envy of the Temple', pp. 97–125, 229–40.

17 See above, ch. 5, pp. 63–5.

18 See below, ch. 7, pp. 111–12.

19 *In Hiezechielem*, no. 10, pp. 287, 381, 382, 394.

soul and the house of God.[20] The Venerable Bede's exegesis concentrated on the biblical–patristic typology of the Temple as the potential Church. For him the Temple of Solomon was the figure of the heavenly *ecclesia*; it was also the eternal rest that awaits the elect.[21] The Carolingian and post-Carolingian commentators presented the biblical Temple as a *typus* of the Church. Carolingian exegesis on the Book of Revelation indicated Heavenly Jerusalem by a set of Old Testament symbols of which one was the Temple and the other, the Ark, Altar, Tabernacle, Jerusalem. Alcuin, in his commentary on Revelation 5:11, wrote: 'Civitas sancta hoc est templum et altare, id est coelestis Hierusalem, quae aedificatur ut civitas.'[22] In a sermon on the dedication of a church, Rabanus Maurus made the Temple of Solomon a prefiguration of the good works and correct liturgy of the Church of Christ.[23] His contemporary, Walafrid Strabo, devoted a treatise to the subject: '*On the Destruction of Jerusalem*'. He summarized the Byzantine and the early medieval attitude to the Temple and to Jewish Jerusalem in general: 'And indeed the city was completely destroyed and not one stone upon another left in it ... This we learn from Josephus [Flavius] and understand it as a proof for the Lord's judgement, namely: "For days shall come upon you, when your enemies will cast up a bank around you and surround you, and hem you in on every side, and dash you to the ground, you and your children within you, and they will not leave one stone upon another in you; because you did not know the time of your visitation" (Luke 19:43–44). And the Lord also said to his disciples [pointing to the Temple]: "Truly, I say to you, there will not be left here one stone upon another, that will not be thrown down" (Matthew 24:2), as well as: "But when you see Jerusalem surrounded by armies, then you know that its desolation has come near" (Luke 21:20). To this very day the place of the city is shown, because where it stood not one stone is left upon another. However, in a later period, outside the city gates, at the site of the Lord's Crucifixion the city was rebuilt by Helius Adrianus and called "Aelia" ...'.[24] Strabo thus describes the ruined, never again rebuilt Jewish Jerusalem and its Temple as being a living eternal proof of the Lord's judgement upon it. Like other early medieval writers, he presented the Temple as a *figura* of the soul (*anima*) of the Christian believer.[25] The Temple of Solomon, erected through divine inspiration, was conceived as an ideal model for a church, since its perfect proportions reflected cosmic harmony. In numerous consecration rites, new churches were presented as the Temple of Solomon.[26]

---

20 *Homiliae*, no. 39, cols 1293–4.

21 Bede, *De tabernaculo*, ed. D. Hurst, *Bedae Venerabilis Opera*, *CCSL*, 119A, p. 42. See also *idem*, *De Templo*, in *Bedae Venerabilis Opera*, p. 147.

22 Alcuinus, *Expositio in Johannis Apocalypsin*, *PL*, 117, col. 1192.

23 Rabanus Maurus, *Homiliae*, *PL*, 110, no. 39, col. 110.

24 Walafridus Strabo, *De Subversione Jerusalem*, *PL*, 114, cols 968–9.

25 *Ibid.*, cols 973–4.

26 Nurith Kenaan-Kedar, 'Symbolic Meaning in Crusader Architecture. The Twelfth-Century Dome of the Holy Sepulcher Church in Jerusalem', *Cahiers Archéologiques*, 34 (1986), p. 113.

Following the crusader conquest of Jerusalem a profound change took place. The new rulers of the city, the Franks, restored the traditions of the Old Testament to their original sites and especially to the Temple Mount. This was now integrated into the urban framework of the city as well as into the *ordo peregrinationis*; in this, the *ordo Templum Domini* played a part second only to the Holy Sepulchre. Jerusalem, whose sanctity now drew from both the Old and the New Testament, was centred on the two foci – the Holy Sepulchre and the Temple of the Lord. This, so to say, 'biblical concept' of Jerusalem's sanctity was characteristic of the period of crusader rule in the city. It disappeared together with the Franks after 1187 and, moreover, it seems that it was symptomatic mostly of the Franks. Many of the Europeans who expressed this concept of Jerusalem in their writings were influenced by the Franks, either through the latter's writings or by learning of biblical traditions from local guides while in Jerusalem, either as pilgrims or as crusaders. Why did the Franks accept the biblical past of Jerusalem and reintroduce it into the holy geography of their capital, rather than rejecting that past, as the previous Christian rulers, the Byzantines, had done? First, the crusaders, unlike the Byzantines, had no reason to feel threatened by Judaism and the biblical past of Jerusalem as part of its Jewish history. Second, contemporaries saw the conquest of Jerusalem both in biblical terms and as a realization of biblical prophecies. The Franks, descendants of the participants in the First Crusade, shared their particular piety and like them identified with the biblical Children of Israel.[27]

Shortly after the First Crusade, the crusader rulers of Jerusalem were presented in the West in relation to the biblical kings of Judaea. Writing (*c.* 1102) to King Baldwin I of Jerusalem, Anselm of Canterbury reminded him that even before the Advent of the Lord, and indeed, because of it, God had chosen Jerusalem from all places in the world as his particular dwelling and wished him 'that Christ [would] rule his heart and deeds, so that he [would] rule for eternity in heaven with King David his predecessor'.[28] Orderic Vitalis, writing between the years 1132 and 1139, described the king of Jerusalem as one who ruled the 'kingdom of David' (*gubernaculum David*) and Jerusalem as the 'capital of the kingdom of David' (*metropolis Davitici regni*).[29] This was also the self-image of the crusader kings of Jerusalem. Godfrey of Bouillon, the first ruler of the crusading kingdom, chose as his palace the al-Aqsa mosque, identified by the crusaders with the palace of King Solomon, and called it the 'Temple of Solomon' (*Templum Salomonis*) or the 'Palace of Solomon' (*Palatium Salomonis*).[30] Even after the kings of Jerusalem

---

27 Prawer, 'Jerusalem in Christian and Jewish Perspectives', pp. 739–82 and see above, ch. 2, pp. 24–5.

28 Anselmus Cantuariensis, *Opera*, no. 324, p. 255; A. Grabois, 'Anselme, l'Ancien Testament et l'Idée de Croisade', *Les Mutations Socio-Culturelles au Tournant des XIe–XIIe siècles. Études Anselmiennes (IVe Session),* Paris 1984, pp. 167–8.

29 Ordericus Vitalis, vol. 5, p. 158.

30 Vogüe, *Les Églises*, p. 281, no. 2; Schein, 'Between Mount Moriah', pp. 179–80.

had moved, probably in 1119, to a new residence near the Tower of David, the banquet traditionally served by the burgesses of Jerusalem, which was part of the coronation ceremonies, was held in the 'Temple of Solomon'.[31] Moreover, the seal of the kings of Jerusalem portrayed three buildings, the landmarks of Crusader Jerusalem and, it seems, the symbols of its kings: the dome of the 'Temple of the Lord' topped by its famous gilded cross, as well as the Tower of David and the open dome of the Lord's Sepulchre.[32] Some of the kings of Jerusalem had chosen to be crowned not in the Church of the Holy Sepulchre but in Bethlehem, clearly not because it was the site of the Nativity, but because it was the birthplace of the House of David. The kings of Jerusalem called their kingdom not only the 'Kingdom of Jerusalemites' or the 'Kingdom of Jerusalem', but also the 'Kingdom of David' (*Regnum David*).[33] Fulcher of Chartres who, as the chaplain to King Baldwin I, was close to the royal house of Jerusalem, reported that the inscription on the tombstone of Baldwin I described him as the 'semblance of Joshua' (*consimilis Josue*).[34] In another version of this inscription (the tombs of the kings of Jerusalem were destroyed in 1808 by a fire), King Baldwin I appeared as another Judas Maccabaeus (*Judas alter Machabeus*).[35] William of Tyre speaks of Godfrey of Bouillon alternately as a new Joshua or Judas Maccabaeus. He compared Baldwin I to King Saul and Fulk of Anjou to *instar David*.[36]

Descriptions of Jerusalem composed in the crusader kingdom expressed the same identification of the Franks with biblical Israel. Fulcher of Chartres' Jerusalem was the city of the kings of Judaea, whose temple stood on Mount Moriah and not on the sites of Christ's Crucifixion and his Sepulchre.[37] For another chaplain of the Jerusalem Church, Rorgo Fretellus of St Mary's in Nazareth, Jerusalem was the 'City of the Life of Christ', but at the same time the glorious capital of Judaea, the royal city of King David, the city where Isaiah had foretold the coming of Christ; it was 'the city of the sanctified sites of the kingdom of David' (*loca sacra regni David*). The crusader rulers of the city were the New Israelites, the new Maccabees, the true successors of King Solomon, those who battled the Amalekites and the Edomites.[38] William of Tyre, describing Jerusalem

---

31 Jean d'Ibelin, *Livre, RHC.Lois*, I, p. 31; Schein, 'Between Mount Moriah', pp. 179–80, 184.

32 Vogüé, *Les Églises*, pp. 453–4.

33 J. Prawer, 'The Roots of Medieval Colonialism', *The Meeting of Two Worlds. Cultural Exchange between East and West during the Period of the Crusades*, ed. V. P. Goss *et al.*, Kalamazoo 1986, p. 24.

34 Fulcher Carnotensis, p. 614.

35 *Corpus Inscriptionum Crucesignatorum Terrae Sanctae*, ed. S. Sandoli, Jerusalem 1974, no. 57, p. 57 and see also the transcription of this inscription by Theodoricus, p. 154.

36 *W.T.*, X, 2, p. 454. Eng. trans. I, p. 416; xiv, 1, p. 631. Eng. trans., II, p. 46; D.W.T.C. Vessey, 'William of Tyre: Apology and Apocalpyse', *Hommages à A. Boutemy*, ed. G. Camlier, Brussels 1976, p. 401.

37 Fulcher Carnotensis, pp. 281–92.

38 Rorgo Fretellus, pp. 6, 7, 31.

in his chronicle written between 1170 and 1184, presented it as the city of both Old and New Testaments. Referring, for example, to the Valley of Jehoshaphat, he mentioned first that it was the place of which it was said: 'For behold, in those days and at that time, when I restore the fortunes of Judah and Jerusalem, I will gather all the nations and bring them down to the valley of Jehoshaphat and I will enter into judgement ...' (Joel 3:1–2); he then mentions the Church of St Mary in the Valley of Jehoshaphat and the Kidron river, the site of the garden where Christ had often met with his disciples (John 18:1–2). Referring to Mount Zion, he noted that 'Zion' means the 'Church Zion' and that the entire city was often called 'Zion' according to Psalms: 'the Lord loves the gates of Zion more than all the dwelling places of Jacob' (Psalms 87:2).[39]

The most striking expression of this double character of the sanctity of Jerusalem was the transformation of the spiritual status of the Temple Mount. And indeed, as Hugh Nibley has observed, 'the Crusaders are a reminder that Christianity was never able to settle for a spiritual Temple or forget the old one'.[40] Whereas previously, in the Byzantine and early Moslem periods, the Temple Mount had disappeared from the holy geography of Jerusalem and its holy traditions were forgotten or removed to other sites, those same traditions were now restored, vindicating the position of the Holy Mount of the Old and New Testaments. Consequently, if in the past the ruins of the Temple on Mount Moriah had been seen in stark contrast to the New Constantinian temple on the sites of Christ's Crucifixion and Sepulchre, the sites of both Old and New Temples were now held to complement one another,[41] representing the two high points in the history of the True Israel.

This transformation took place immediately after the conquest of the city. On 15 July 1099 the victorious crusaders went in procession, barefoot, singing a 'new song to the Lord', to the Holy Sepulchre and from there to the 'Temple of the Lord' on Mount Moriah. On 10 August, after the crusader army had left the city to meet the Egyptian invaders in Ascalon, the clergy, vested and again barefoot, proceeded with crosses from the Holy Sepulchre to the Temple.[42] Similarly, on Easter Day of 1101, when the Holy Fire failed to appear in the lamps of the Holy Sepulchre, the patriarch of Jerusalem ordered a procession, and the clergy, the king with his retinue and most of the people, went barefoot in procession to the 'Temple

---

39 *W.T.*, VIII, 2, 3, pp. 383–4. Eng. trans., I, pp. 340–45.

40 Nibley, 'Christian Envy of the Temple', p. 229.

41 There was also, however, some competition between the Church of the Holy Sepulchre and the Temple of the Lord. See on this subject H. Busse, 'Vom Felsendom zum Templum Domini', *Das heilige Land im Mittelalter: Begegnungsraum zwischen Orient und Okzident*, ed. W. Fischer *et al.*, Neustadt 1982, p. 21; Schein, 'Between Mount Moriah', pp. 181–91; Kenaan-Kedar, 'Symbolic Meaning', pp. 109–17.

42 Raimond d'Aguilers, pp. 54, 137, 144–5, 151, 155; Petrus Tudebodus, pp. 110–11, 137–8, 144–5; Albertus Aquensis, pp. 470–71; Fulcher Carnotensis, p. 305; Riley-Smith, *The First Crusade*, p. 84.

of the Lord'. In Crusader Jerusalem this became typical of both *ordo processionis* and *ordo peregrinationis*,[43] but it was rather to the Temple of the Lord and not the Holy Sepulchre, where the crusaders were accustomed to go to offer their thanks to God following a military victory. This is revealed in the Fulcher of Chartres' description of the reaction in Jerusalem to the news of the conquest of Tyre (7 July 1124): 'When this was heard a most joyful clamour arose. The *Te Deum laudamus* was forthwith sang with exultant voices. Bells were rung, a procession marched to the Temple of God, and flags were raised on the walls and towers. Through all the streets coloured ornaments were displayed, thankful gestures made ... and girls were delightful as they sang in chorus.'[44]

The transformation in the status of the Temple of the Lord resulted in a massive concentration of holy traditions there. As early as 1102–3 the Anglo-Saxon pilgrim Saewulf was able to associate it with a multitude of traditions from both Old and New Testaments. He most probably had genuinely discovered them there, since as a layman he was not sufficiently educated to have invented these traditions. The events he associated with the shrine were therefore already part of a tradition connected with the Temple of the Lord. At that place Saewulf located ancient Bethel, the House of God; there, he believed, Jacob had rested and seen the ladder in his dream (Genesis 28:10–22). It was there, by divine inspiration, that Solomon later built his Temple, and in the middle of that Temple was a rock which was the place of the Holy of Holies. There Solomon had placed the 'Ark of the Covenant with the manna and Aaron's rod that budded and blossomed there and brought forth almonds'; there, too, Solomon had placed the two Tables of the Covenant. There too Christ, 'weary with the contradictions of the Jews, was accustomed to rest'. There was 'the place of confession where his disciples confessed [to be Christ]' (Acts 3:11–16; 5:12–32). There the angel Gabriel had appeared before Zechariah the priest (Luke 1:13), and there, before the Temple and the altar, that same Zechariah son of Berachiah had been slain. There the infant Jesus had been circumcised on the eighth day 'and was called Jesus which means Saviour'. There Christ had been offered on the day of the purification of his mother the Virgin Mary and taken into Simeon's arms (Luke 2:28). There Jesus, aged twelve, had been found sitting among the doctors (Luke 2:46). Thence he later drove out all those who sold and bought in the Temple (Matthew 21:12–13), and there he had said to the Jews: 'Destroy this Temple and in three days I will raise it up' (John 2:19). To that place the woman taken in adultery had been brought before him (John 8:3), and 'there are still visible in the rock the marks of our Lord's feet, when he hid himself and went out of the Temple, as you read in the Gospel, so that the Jews should not cast at him the stones which they took up' (John 8:59).[45] Another

---

43 See above, ch. 3, pp. 41–2 and below, pp. 101–03.

44 Fulcher Carnotensis, p. 465.

45 Saewulf, pp. 67–8; Schein, 'Between Mount Moriah', pp. 186–7. For the sanctity of the holy rock see also Albertus Aquensis, p. 480, who referred to it as *Santa Sanctorum* of the Temple of the Lord. For the tradition of the annunciation to Zechariah see also *De Situ*

pilgrim who came to Jerusalem shortly after Saewulf (*c*. 1106) was the Russian Abbot Daniel. He found additional Old Testament traditions associated with the holy rock. There, 'upon the same stone ... the prophet David saw an angel standing with a drawn sword in his hand smiting the people of Israel, and, entering this grotto, he wept, and addressed his prayer to God: "Lord, it is I that have sinned, but these sheep, what have they done?"' (2 Samuel 24:17). Daniel also remarks that 'on entering the city [through the Gate of David, now the Jaffa Gate], there is a road traversing it, which to the right leads to the Holy of Holies and to the left to the Holy Resurrection of the Holy Sepulchre'.[46] Thus, around 1106, the Temple Mount almost ranked as the Holy of Holies of Crusader Jerusalem.

The importance of the *Templum Domini* increased with the consolidation of crusader rule. Thus Achard of Arrouaise, prior of the Lord's Temple between 1118 and 1131, mentioned it even before the Holy Sepulchre, while expressing his thanks to God who liberated during his lifetime: '*templum suum cum sepulcro*'.[47] Achard, as its prior, certainly sought to publicize the Temple of the Lord. Other writers, however, did not have such vested interests. Chroniclers who wrote a few years after the conquest of Jerusalem, in contrast to those who wrote earlier, such as the anonymous author of *Gesta Francorum* and Raimond of Aguilers, in their accounts of the speech of Pope Urban II in Clermont, presented the recovery of the Temple of the Lord together with that of the Holy Sepulchre as one of the goals of the Crusade. Thus Baldric of Dol, who completed his *Historia Jerosolimitana* in *c*. 1108, mentioned the Temple of Solomon (referring to the crusader Temple of the Lord) as 'a church in which [on the eve of the First Crusade] the barbaric races worshipped their idols, which they had placed there against the law and against religion'.[48] Ekkehard of Aura, who visited Jerusalem in 1101 but composed his *Hierosolymita* after his return to Germany in *c*. 1115, referred in his version of Urban II's speech to the 'famous Temple of the Lord' being desecrated by the Saracens; to him it was a Christian sanctuary.[49] Similarly the chroniclers who

---

Urbis Jerusalem, p. 413. For the holy rock as the place of the Ark of the Covenant, the manna, Aaron's rod, the Dream of Jacob and the death of Zechariah see also *Qualiter Sita est Civitas Jerusalem* (*c*. 1095–1105), p. 348; *Chronicon Universale Anonymi Laudunensis*, pp. 43–4. In Crusader Jerusalem the Moslem Dome of the Rock, identified with the Temple or at least with its site, became a church called the Temple of the Lord (*Templum Domini*) and was served by a community of 'the Augustine Canons of the Temple of the Lord', whose foundation was attributed to Godfrey of Bouillon. During the years 1115–41 the church was restored. Schein, 'Between Mount Moriah', pp. 181–3.

46 Daniel, pp. 11, 20.

47 Acardus de Arroasia, *Tractatus super Templo Salomonis*, ed. P. Lehmann, *Corona Quernea. Festgabe Karl Strecker*, Leipzig 1941 (*Schriften der Monumenta Germaniae Historica*, 6), p. 329, v. 791. See also M. Vogüé, 'Achard d'Arrouaise, Poème sur le Templum Domini', *AOL*, 1 (1881), pp. 562–79.

48 Baldricus Dolensis, p. 11.

49 Ekkhardus, p. 14.

wrote in the kingdom after *c.* 1110 underlined the importance of the Temple of the Lord as one of the goals of the First Crusade. Fulk of Chartres reported that after the conquest the Franks went 'ad sepulchrum Domini et Templum eius gloriosum'.[50] William of Tyre, in his version of Urban II's speech at Clermont, made the pope argue that: 'The Temple of the Lord whence He, in His zeal, drove those who bought and sold that the house of His Father might not become a den of thieves [Matthew 21:12–13], has been made the home of demons. It was this that roused the laudable zeal of Matthias the priest, the great ancestor of the holy Maccabees, as he himself testifies: "Her temple become as a man without honour", and "her glorious vessels are carried away into captivity" (Maccabees 2:8,9)'.[51]

The heightened importance of the Temple Mount in the religious and public life of Crusader Jerusalem is revealed by the part it played in its festivities and feasts. The most important festivity set on the Temple Mount was the coronation ceremony. The coronation itself took place in the Holy Sepulchre. Following that, the royal procession made its way to the Temple of the Lord, where the king placed on the altar the crown that had shortly before been placed on his head, thus symbolically commemorating the presentation of the infant Christ to Simeon (Luke 2:22–39). From the Temple of the Lord the procession crossed the esplanade to the Temple of Solomon, where a banquet was served by the burgesses of Jerusalem.[52]

The great feast days of the Church of Jerusalem were celebrated, like the coronation ceremonies, by solemn processions to the various churches of the city, among which the sanctuaries on Mount Moriah held a position equal in importance to those of the Holy Sepulchre. This is particularly striking in the ritual of one of the great feasts particular to the city of Jerusalem. Celebrated on 15 July, it commemorated two great events in the city's history: its conquest by the hosts of the First Crusade in 1099, and the consecration of the Crusader Church of the Holy Sepulchre in 1149. On the appointed day, the patriarch led a solemn procession from the Holy Sepulchre to the Temple of the Lord. There, at the southern entrance, prayers were recited, and the procession then continued across the esplanade to the area outside the eastern wall (which was the wall of the esplanade as well as of the city), to the graveyard of the crusaders who had died during the city's conquest.[53]

Among the great feast days of Christendom which became linked with the Temple Mount in Crusader Jerusalem were Palm Sunday, the Purification of the Blessed Virgin or Presentation of the Lord in the Temple, and the Exaltation of the

---

50 Fulcher Carnotensis, p. 360.

51 *W.T.* I, 15, p. 132. Eng. trans., I, p. 90.

52 Jean d'Ibelin, *Livre*, *RHC.Lois*, 1, p. 31; Prawer, *The Latin Kingdom*, pp. 111–15; Schein, 'Between Mount Moriah', p. 184.

53 Kohler, 'Un rituel', pp. 427, 436–7; Johannes Wirziburgensis, pp. 190–91; Prawer, *The Latin Kingdom*, pp. 176–7; Schein, 'Between Mount Moriah', p. 185.

Holy Cross. The Palm Sunday procession, attested for Jerusalem as early as the fourth century, was particularly solemn during the time of the crusaders. Before sunrise the Latin clergy of the city set out in a procession from the Holy Sepulchre to Bethany, carrying the Holy Cross. In the meantime another procession, assembled in the precinct of the Temple of the Lord, set out to the Valley of Jehoshaphat, where it met the procession coming from Bethany. Thus united, the celebrants climbed the Temple Mount and entered the Temple area through the Golden Gate. After encircling the Temple of Solomon the procession ended with prayers in the precinct of the Temple of the Lord. The Feast of the Purification of the Blessed Virgin or Presentation of the Lord in the Temple on 2 February was celebrated by a procession which set out from the Holy Sepulchre to the Temple of the Lord, encircled the church, and stopped near its southern gate, where the patriarch delivered a sermon. During the Feast of the Exaltation of the Holy Cross on 14 September, the Golden Gate, closed throughout the year, was opened to admit the procession, as on Palm Sunday. Thus the entrance of the emperor Heraclius through that gate, bearing the Holy Cross which had been recovered from the Persians in 629, was commemorated *in situ*.[54]

A striking feature of these festive processions was that they went from the Church of the Holy Sepulchre to the Temple Mount, thus linking directly the two religious centres of Crusader Jerusalem. The two were linked to each other in more than one way. The *ordo peregrinationis* of Jerusalem during the crusader period presents the same pattern as the thousands of pilgrims who visited the city each year went on to the shrines of the Temple Mount immediately after leaving the Holy Sepulchre.[55] As early as the first decade of crusader rule, these pilgrims found in Jerusalem a whole body of traditions of both Old and New Testaments. Mount Zion, for example, was the site of the Last Supper and of the Dormition of the Virgin, as well as the tombs of King David, King Solomon and other kings of Judaea. It was apparently the crusaders' involvement with the biblical kings of Judaea that perpetuated this last tradition. Mentioned in the anonymous 'Life of Constantine' (*c.* 715–1009), the tradition locating the tombs of kings of Judaea on Mount Zion is referred to by Raimond of Aguilers, Rorgo Fretellus and the so-called 'Innominatus VIII'.[56] On the Temple Mount the pilgrims were shown the 'Temple of Solomon' (*Templum Salomonis*), the al-Aqsa mosque, which was first the royal palace and from the 1120s served as the seat of the Knights Templar, who identified it with the Palace of Solomon.[57] In Siloam they saw the place where

---

54 Kohler, 'Un rituel', pp. 395, 410, 412–14, 424; Johannes Wirziburgensis, pp. 128–9, 190–91; Schein, 'Between Mount Moriah', p. 185.

55 For the ordo see, e.g., *De Situ Urbis Jerusalem (c. 1130)*, pp. 412–13.

56 Wilkinson and Ryan, *Jerusalem Pilgrims*, p. 204; Raimond d'Aguilers, pp. 138–9; Rorgo Fretellus, p. 38; *Innominatus VIII*, p. 194 and see Arce, pp. 105–15; O. Limor, 'King David's Tomb on Mount Zion: The Origins of a Tradition', *Jews, Samaritans and Christians in Byzantine Palestine*, ed. D. Jacoby and Y. Tsafrir, Jerusalem 1988, pp. 11–26.

57 Schein, 'Between Mount Moriah', pp. 179–81.

Christ had restored the sight of a blind man, as well as the tomb of Isaiah; and in the Valley of Jehoshaphat they were shown the tomb of King Jehoshaphat.[58] Thus the entire city was now associated with the biblical kings of Judaea, above all with King David. Pilgrims entered the city through the 'Gate of David,' passed through the citadel, which was called the 'Tower of David' (*Turris Davit*), and made their way to the Holy Sepulchre through the 'Street of David'.[59] The two centres of the Holy City, the Lord's Sepulchre and the Temple of the Lord, were further linked to each other through their architectural form. The 'Dome of the Rock', the crusader *Templum Domini* erected in the seventh century, was influenced by Constantine's Anastasis. Then under the domination of the crusaders, the dome which crowned their new basilica, connecting the Anastasis with Golgotha and the chapels of the Holy Cross and St Helena, was to relate the Rotunda in the same way as the lesser dome of the Temple of Solomon (al-Aqsa) related to the greater dome of the Temple of the Lord. It follows thus that while in the Umayyad period the dome of the Rotunda served as the model for the Dome of the Rock, in the twelfth century the domes of the Temple inspired the disposition of the new crusader church.[60] It is notable that the dome was placed above the site of the *omphalos*, the centre of the world which was originally referred to in the Old Testament as located in the Temple, in its Holy of Holies. In the twelfth century pilgrims identified the centre of the world in both the Church of the Holy Sepulchre and in the *Templum Domini*.[61]

Moreover, it was not only that the two monuments themselves were architecturally related but their 'gates' were very similar too. The double portals of the Church of the Holy Sepulchre were modelled on the double portals of the Golden Gate in the eastern city wall of Jerusalem dating from the end of the sixth or beginning of the seventh century. This stemmed from the Franks' perception of Christ's triumphal entry into Jerusalem through the Golden Gate on Palm Sunday (Matthew 21:7–9; Luke 19:35–36) as a prefiguration of the crusader conquest of the city; each event was symbolized by a double portal edifice.[62]

Moreover, the decorations of the two shrines, those of the crusader Church of the Holy Sepulchre and those of the Temple of the Lord, depicted the specific character of Crusader Jerusalem's sanctity as the City of the Old and the New Testament. The worshipper entering the Church of the Holy Sepulchre saw on the façade, in the western tympanum, a mosaic depicting the Virgin and Child. Below the tympanum were two sculptured lintels. The western lintel had six panels, representing episodes from the life of Christ, all directly connected with Jerusalem

---

58 Petrus Diaconus, *De Locis Sanctis*, p. 178; *Innominatus VII*, p. 104.

59 *Citez de Jherusalem*, pp. 200–201; *L'Estat de la cité de Jherusalem*, *Itinéraires*, pp. 23–8. For the Citadel of Jerusalem as the 'Tower of David' see Raimond d'Aguilers, p. 154.

60 See on the entire subject Kenaan-Kedar, 'Symbolic Meaning', pp. 109–17.

61 See below, ch. 8, pp. 144–5.

62 Kenaan-Kedar, 'Symbolic Meaning', p. 115; *idem*, 'Local Christian Art', pp. 221–2.

and its surroundings: the raising of Lazarus from the dead, Mary and Martha imploring Jesus to perform the miracle, Jesus commissioning St Peter and St John to find a place to celebrate Passover, the finding of the ass to be ridden by Christ on his way to Jerusalem, Christ's entry into the city, and finally the Last Supper.[63]

The interior of the Church of the Holy Sepulchre was lavishly decorated with murals and mosaics. The aedicule of the Holy Sepulchre had a 'picture of mosaic work, that of Joseph and Nicodemus placing our Lord's body in the tomb, with our Lady, His Mother, standing by and the three Marys ... and with the angel also sitting above the sepulchre and rolling away the stone'. The walls of the Rotunda were decorated with mosaics depicting the boy Jesus with his Mother on his left, and on his right the Archangel Gabriel, the twelve apostles and thirteen prophets, 'all of whom have their faces turned towards the beauteous body and reverently address Him, holding in their hands the prophecies with which He inspired them of old'. The murals decorating the arches of the eastern side of the cathedral, beyond the choir, depicted the Virgin Mary, John the Baptist and the Archangel Gabriel; on the ceiling was a depiction of Christ holding his Cross in his left hand and, according to John of Würzburg: 'he was looking royally up towards heaven, with His left foot raised in gigantic stride, His right still resting on the earth, as He enters Heaven while the following are standing around – that is to say, His mother, St John the Baptist and all the apostles'.[64] The largest concentration of figural mosaic work appears to have been in the Calvary Chapel complex. Among other figures and events from the Old Testament were David, Solomon, Isaiah, Habakkuk, Amos, Hosea, Daniel, Elias and the Raven, Elias's Ascension and the Sacrifice of Isaac; the scenes from the New Testament were the Crucifixion, Ascension, Last Supper and Deposition.[65]

Proceeding from the Church of the Holy Sepulchre to the *Templum Domini*, from the New to the Old Temple, one noticed first the gilded cross atop the dome, placed there by the crusaders when the church was consecrated in 1141. Inside the church, paintings and wall inscriptions depicted the history of the Temple. On entering the Temple of the Lord, one came upon a painting of Christ with the inscription, 'My house shall be called a house of prayer' (Isaiah 56:7; Matthew 21:13); a painting of an angel addressing Zechariah: 'Do not be afraid, Zechariah, for your prayer is heard' (Luke 1:13); and paintings of Jacob, David, Solomon, and of Simeon carrying Christ.[66]

---

63 Folda, *The Art of the Crusaders*, pp. 214–27; Kühnel, *Crusader Art*, pp. 42–5. For the lintels see also A. Borg, 'Observations on the Historical Lintels of the Holy Sepulchre, Jerusalem', *Journal of the Warburg and Courtauld Institutes*, 32 (1969), pp. 25–40; Lindner, 'Topography', pp. 81–98.

64 Johannes Wirziburgensis, pp. 122; Kenaan-Kedar, 'Symbolic Meaning', pp. 114–15; Prawer, *The Latin Kingdom*, pp. 452–4.

65 For the plan of the mosaics see Folda, *The Art of the Crusaders*, p. 235.

66 Johannes Wirziburgensis, pp. 89–95; Theodericus, pp. 159–63; Schein, 'Between Mount Moriah', pp. 163–4.

Both the *ordo peregrinationis* and the *ordo processonis* in Crusader Jerusalem led directly from the Church of the Holy Sepulchre, the symbol of the Christian New Testament's Jerusalem, to the Temple of the Lord representing the Temple, the centre and symbol of the Old Testament's Jerusalem. Yet the decorations of the two monuments not only stressed the contrast between these two, the Old and the New Temples, but they visually emphasized the links between them: Christ and Mary were depicted in both the temples' wall paintings as well as prophets, kings of Israel, apostles and so on.[67]

Though pilgrims from the Christian West present Jerusalem in their *itineraria* or *descriptiones Terrae Sanctae* as the City of both the Old and New Testaments, the attitude to the Temple in the twelfth-century West was more complex. The transformation of the status of the Temple Mount produced, it seems, some confusion among twelfth-century writers who, influenced by the exegetical tradition, were used to perceiving the Temple as a metaphor, a symbol or a *typus*. As a result the Temple was somewhat ignored. Most of the writers who did not ignore it were exegetes. Richard of St Victor, in his treatise on the Ark of Moses, referred to the Temple's Holy of Holies as Heavenly Jerusalem.[68] Hugh of Fouilly in his *De Claustro Animae* integrated the 'cloister of the soul' with the Temple in all its dimensions and components.[69] As far as monastic writers were concerned, they relied heavily on the *Glossa Ordinaria* in their interpretations and in fact repeated Bede's exegesis, where the Temple of Solomon stood for, as mentioned above, the heavenly kingdom and the Church.[70] In the second half of the twelfth century, both Adam (1192) and Richard of St Victor (1173) observed with wonder that the Christian Fathers had always gone out of their way to avoid any discussion of the Tabernacle of God, in spite of its great popular interest and its importance in the divine economy. The reason for their strange attitude is the very thing that made the Temple attractive to many Christians. The exciting possibility of a literal and tangible bond between heaven and earth was precisely what most alarmed and embarrassed the churchmen.[71]

The exception among the twelfth-century writers in the West is Bernard of Clairvaux. As argued by Thomas Renna: 'It is only Bernard of Clairvaux among the

---

67 Kühnel, *From the Earthly to the Heavenly Jerusalem*, pp. 164–5. The figure of King David plays a prominent role also in Queen Melisande's Psalter produced in Jerusalem between the years 1131 and 1143. See Kühnel, *Crusader Art*, pp. 67–125. See also Folda, *The Art of the Crusaders*, pp. 137–59.

68 Ricardus de St Victore, *Benjamin major*, *PL*, 196, col. 167; see also his *Mystical Ark*, trans. Zinn, ch. 23.

69 *De Claustro Anima*, cols 1113–30.

70 T. Renna, 'Bernard of Clairvaux and the Temple of Solomon', *Law, Custom and the Social Fabric in Medieval Europe. Essays in Honor of Bryce Lyon*, ed. Bernard S. Bachrach and David Nicholas, Kalamazoo 1990 (*Studies in Medieval Culture*, 28), pp. 74–5.

71 Adam de St Victore, *De tripartito tabernaculo*, *PL*, 198, col. 625; Ricardus de St Victore, *De tabernaculo*, *PL*, 196, cols 211–12.

early post-crusade writers who, without compromising the spiritual interpretation of the Temple as a sign of contemplation, attempts both to combine the dual tradition of the Temple as a soul and Church and to show that the earthly Jerusalem must be retained.'[72] Bernard of Clairvaux's *In Praise of the New Knighthood* (*c.* 1128) was a eulogy not only of the Templars and Crusader Christian Jerusalem, but also of the Temple. Bernard compared the cleansing of the Temple of Solomon by the Templars, 'by installing themselves in this holy house', to Christ's cleansing of the Temple (Matthew 21:12–13; John 2:14–16), claiming that 'it is even more shameful and infinitely more intolerable for a holy place to be polluted by pagans than to be crowded with merchants'. He described Jerusalem as 'the Holy City sanctified by the Most High for his own tabernacle' (Psalms 46:4), as the 'city of the great King' (Psalms 47:3), 'a source of so many joyous and unheard-of marvels', 'mistress of nations and queen of provinces' (Lamentations 1:1), 'heritage of patriarchs' (Genesis 17:8), 'mother of apostles and prophets, source of the faith and glory of the Christian people', 'promised land, source of milk and honey' (Exodus 3:8), '[a place which] now became the source of healing grace and vital sustenance for the whole earth ...'. For Bernard the Temple of the Lord was more than anything else a symbol. As a result of its links with the Templars – the ideal Christian knight (*miles Christi*) was defender of the Church and the Holy Land as well as a permanent candidate for martyrdom – the present Temple of the Lord was the symbol of the 'revenge which Christ contrives against his enemies'; it was a symbol of the triumph of Christianity over Islam (by this he meant the crusader conquest of Jerusalem) and Christian rule in Jerusalem. Little interested in the historical origins of the building, Bernard focused not on the Temple building or the church as the New Israel, but on the fulfilment of prophecies for the new knighthood for which the most perfect model was the Templars.[73]

On a more popular level, a legend that circulated in twelfth-century Europe presented the Dome of the Rock as a replica of the Second Temple or even of the Temple of Solomon. Helena had allegedly had the Dome of the Rock built as a substitute for the destroyed Temple and the building had later been turned into a mosque. The legend evolves, it seems, from the reports brought by pilgrims from Jerusalem. Achard of Arrouaise, who wrote between the years 1118 and 1131 a rhymed history of the Temple, already reported various speculations as to the origins of the Dome of the Rock: 'some think that it was built by Justinian, the emperor of the Romans; some ascribe it to Helena, Constantine's mother, who, when searching for the Cross and the place of Calvary, rebuilt the Temple as it looks now; and some say that it was built by Heraclius at the time he brought back the Cross after killing Chosroes'.[74] John of Würzburg ascribed the Dome of the

72 Renna, 'Bernard of Clairvaux', p. 75.

73 Bernardus Claraevallensis, *De Laude novae militiae*, pp. 222–4. Eng. trans. Grenia, pp. 142–3; Renna, 'Bernard of Clairvaux', p. 80.

74 Acardus de Arroasia, *Tractatus*, p. 329; Busse, 'Vom Felsendom zum Templum Domini', pp. 24–5.

Rock to either Helena, Heraclius or Justinian but added that: 'others [say] that it was built by some Emperor of Memphis in Egypt in honour of *Allah Kebir* ...'.[75] According to Theoderich: 'the Temple which we now behold was built by the Empress Helena and her son the Emperor Constantine, in honour of our Lord Jesus Christ and His holy Mother'.[76]

Yet another hint that there was some change in the West in the attitude to the Temple was the position assigned to it in twelfth-century apocalyptic visions. Jerusalem was, it should be remembered, considered as the location of the final confrontation between Judaism and Christianity. The theme of the Antichrist sitting in the Temple, which he had rebuilt from its ruins in a Jerusalem populated by Jews, was a significant theme in early medieval apocalyptic literature.[77] One of the most popular early medieval apocalyptic visions, that of the Last Emperor, told that a western emperor would come to Jerusalem, lay down his crown and sceptre on the Mount of Olives or Calvary and thus usher in the day of the Last Judgement. At the same time, the Antichrist would come from Babylon and occupy the Temple, which he would rebuild on Mount Moriah. In the twelfth-century version of the vision the Antichrist is no longer linked with either of the buildings on the Temple Mount: the Last Emperor surrenders his insignia not at Calvary or on the Mount of Olives, but before the altar in the Temple. Thus the Temple Mount was portrayed as the place in which the Christian Roman Empire was to reach its appointed end, associated thus with the Last Emperor rather than the Antichrist.[78]

The influence upon the West of the attitude to the Temple in the Latin Kingdom was shortlived. Following the conquest of the Holy City by Saladin, Jerusalem reverted to what it had been in the West prior to the crusader conquest: the City of the Holy Sepulchre. The loss was described in biblical terms;[79] however, Jerusalem's biblical past, including the temples on the Mount of Moriah, disappeared once again from the holy geography of Jerusalem. In the Catholic West the Temple became once again a rhetorical and abstract term, a *typus* of the organizational Church and came also to symbolize the working of grace deep within the soul.[80]

Describing Crusader Jerusalem on the eve of its fall, an anonymous writer, possibly an inhabitant of the Kingdom of Jerusalem, concluded his description of its holy places as follows: 'If we look into the writings of the Old and the New Testaments, there is no mountain, valley, river or pool which has not seen miracles

---

75 Johannes Wirziburgensis, pp. 87–8.

76 Theodericus, p. 164.

77 Linder, 'Jerusalem', p. 11; Nibley, 'Christian Envy of the Temple', pp. 232–8.

78 Alexander, 'Byzantium and the Migration of Literary Works and Motifs', pp. 47–68; *Ludus de Antichristo* in B. McGinn, *Visions of the End*, New York 1979, pp. 117–21 and below, ch. 8, pp. 146–7.

79 See below, ch. 9, pp. 166–8.

80 Renna, 'Bernard of Clairvaux', pp. 84–5.

by the prophets, apostles and Christ himself ...'.[81] This was certainly true where the holy geography of Crusader Jerusalem was concerned. Never before was it so rich and so heterogenous as during the twelfth century. It was the Jerusalem of both King David and Christ, of the Virgin Mary and of both the biblical prophets and the apostles. Centred around its two major shrines, the Church of the Holy Sepulchre and the Temple of the Lord, Crusader Jerusalem embodied the entire history of the Christians under both the Old and the New Testaments.

81 *Tractatus de locis et statu sanctae terrae Jerosolimitane*, p. 153.

Chapter 7

Jerusalem in the Believer's Plan of Salvation

Terrestrial Jerusalem was the city in which one could engage in a physical *imitatio Christi* in a way impossible elsewhere on earth. It was traditionally linked to Jerusalem, the *visio pacis*, the heavenly city whose prime characteristics were freedom (Galatians 4:22–31) and perfect peace (Revelation 21:1–27; 22:1–5). The intimate connection between Heavenly and Earthly Jerusalem affected medieval thought in many ways. As Henri de Lubac has pointed out, in a classic medieval exegesis it had a tropological sense, directly attuned to personal appropriation of the meaning of Jerusalem in the believer's life. The anagogical aspect of the relationship appeared in two quite different manifestations: one, which might be termed 'horizontal anagogy', connected with the Last Things, and conceived in a historical manner; and another, 'vertical anagogy', that sought to achieve the present realization of heaven on earth. It was in the latter sense that the monastic cloister was described as 'Jerusalem', which helps to explain the reservations entertained by some medieval thinkers about the physical Jerusalem.[1] Accordingly, the authoritative exegesis of the early Middle Ages interprets the very term 'Jerusalem' as follows: historically, as the city of the Jews, allegorically, as the Holy Church, anagogically, as the heavenly fatherland and tropologically, as the soul of the Christian believer.[2]

In the first centuries of Christianity the official stand of the Church stressed the importance of Heavenly Jerusalem, almost completely rejecting Earthly Jerusalem. Committed to an allegorical interpretation of the Holy Scriptures, it also insisted upon such interpretation of prophecies, including that of Revelation.[3] Yet it is clear that this attempt failed, and that in the first centuries of Christianity people took literally the New Jerusalem, as described in the book of Revelation. Thus, in 156 in Phrygia, where all Christians were summoned to wait, people expected the New Jerusalem to descend from the heavens.[4] The Church Fathers, whose writings profoundly influenced later medieval writers, bypassed, up to the

---

1 Lubac, *Exégèse Médiévale,* I, pp. 624, 640–42; McGinn, '*Iter Sancti Sepulchri*', pp. 42–3.

2 Lubac, *Exégèse Médiévale,* I, pp. 643–8; Prawer, 'Jerusalem in Christian and Jewish Perspectives', pp. 776–8.

3 *Ibid.*, pp. 773–82.

4 N. Cohn, *The Pursuit of the Millennium*, London 1970, pp. 19–36.

fourth century, the earthly city and spoke only of the Jerusalem above. Irenaeus and Tertullian had cited Galatians 4 to support a belief in a future Jerusalem on earth. Tertullian wrote that in Judaea, a walled city had been seen in the sky every morning for forty days, only to fade away as the day advanced; this was, he said, a sure sign that the Heavenly Jerusalem was about to descend. In Origen's view, however, Paul was speaking not of a model for a city that will one day exist on earth, but of a city in the heavens that will take the place of the Earthly Jerusalem; only this city can be the 'mother of us all'.[5]

It is only following Constantine's construction of the Church of the Resurrection (Anastasis) that the attitude of the Church Fathers towards the earthly city became more positive; yet still it remained what can be best defined as ambivalent. Eusebius seems to have originally believed that worship in a specific place was incompatible with Christian 'spiritual worship'. He paraphrased Christ's teaching as: 'Since I give liberty to all people I teach to look for God not in a corner of the earth, or on hills, or in temples made with human hands, but that each person should worship and adore him at home.'[6] Following the building of the Anastasis, Eusebius' attitude underwent a transformation. Describing Constantine's spectacular edifices in Jerusalem, Eusebius argued that 'this was the New Jerusalem built over against the one so famous of old, which after the pollution caused by the murder of the Lord experienced the last extremity of desolation, and paid the penalty for the crime of its impious inhabitants. Opposite this [old Jerusalem] the Emperor reared, with rich and lavish expenditure, the trophy of the Saviour's victory over death.' He suggests: 'Perhaps this was that strange and New Jerusalem proclaimed in the oracles of the prophets, to which long passages prophesying, by the aid of the Divine Spirit, make countless allusions in song.'[7] It follows thus that the Church Fathers distinguished between the Earthly Jerusalem and two types of Heavenly Jerusalem: one that will unite with the Earthly Jerusalem on earth and one that will remain for ever in Heaven. It was Eusebius who hesitated to say but was, it seems, inclined to believe so that the 'New Jerusalem' of Constantine is that of the prophecy, 'the Holy City, New Jerusalem, coming down from God out of Heaven, prepared as a bride adorned for her husband' (Revelation 21:2).[8]

---

5 Wilken, *The Land Called Holy*, p. 70.

6 Eusebius, *Demonstratio Evangelica*, ed. J.A. Heikel, *Die griechischen christlichen Schriftsteller der ersten drei Jahrhunderte*, Leipzig–Berlin 1913, I, 6, 65; Joan E. Taylor, *Christians and the Holy Places. The Myth of Jewish–Christian Origins*, Oxford 1993, pp. 314–15.

7 Eusebius, *Vita Constantini*, 33, col. 1094; Prawer, 'Jerusalem in Christian and Jewish Perspectives', pp. 754–7. For Eusebius' attitude to Jerusalem see also Walker, *Holy City, Holy Places?*, pp. 93–130 versus Wilken, *The Land Called Holy*, pp. 93–100.

8 Linder, 'Ecclesia and Synagoga', pp. 1030–32; Prawer, 'Jerusalem in Christian and Jewish Perspectives', p. 756.

From the fourth century onwards the attitude towards the merit of Jerusalem pilgrimage became ambivalent. Indeed, some rejected it altogether. Gregory of Nyssa argued that 'when God called on the Blessed to come to the Heavenly Kingdom he did not list the road to Jerusalem among the precepts will lead to this end'. In his opinion God is everywhere, and to judge by what happens in Jerusalem or on the pilgrimage, He is more to be found in Cappadocia than in Jerusalem. Therefore, he says, 'make a pilgrimage, but out of your bodies to the Lord and not from Cappadocia to Palestine'. At the same time, Gregory of Nyssa himself did go on a pilgrimage to Jerusalem.[9] A generation later John Chrysostom went even further, declaring: 'if you are a Christian you have no city on earth. The architect and mason of our city is God; though we may capture the whole world, forever we remained strangers and pilgrims.'[10] To Augustine 'Jerusalem' was only a type of allegory and 'Jerusalem' meant the Heavenly Jerusalem. To him the physical city was significant only because it was a figure of the Celestial City of God. In his opinion, the desire for Heavenly Jerusalem was the true motive for a Christian life: 'If you wish to be armed against temptations in this world, let the desire for the eternal Jerusalem grow and flourish in your hearts.'[11] As to God, he stressed, 'God is in all places and is not contained or enclosed in any one place.'[12] Augustine's contemporary, Jerome, was more positive. While addressing churchmen he argued in the best Pauline tradition that 'from Jerusalem as well as from Britain the *aula* of Heaven is open, because the Kingdom of God is in us', and that 'not to be in Jerusalem but have lived rightly in Jerusalem is praiseworthy'.[13] At the same time, however, Jerome supported and even defended, against Vigilantius, the laity's pilgrimage to Jerusalem, declaring that 'just as Greek history becomes more intelligible to those who have seen Athens, and the third book of Virgil to those who have sailed from Troa by Leucata and Acroceraunia to Sicily and so on to the mouth of the Tiber, so one will get a clearer grasp of the Holy Scripture who has gazed at Judah with his own eyes and has got to know the memorials of its cities and the names, whether they remain the same or have been changed, of the various localities'.[14]

In the centuries following the Moslem conquest of Jerusalem, the official position of the Church, while rejecting Earthly Jerusalem, was to glorify the

---

9 Gregorius Nyssenus, ep. II, cols 1109–1116; Prawer, 'Jerusalem in Christian and Jewish Perspectives', pp. 772–3.

10 Johannes Chrysostomus, *Homiliae, PG*, 49, cols 177, 178; Prawer, 'Jerusalem in Christian and Jewish Perspectives', p. 773.

11 Augustinus, *Enarrationes in Psalmos*, p. 1278.

12 *Idem*, *Contra Faustum*, *PL*, 42, cols 384–6; G. Constable, 'Opposition to Pilgrimage in the Middle Ages', *Studia Gratiana*, 19 (1976), p. 126.

13 Hieronymus, *Epistulae*, no. LVIII, pp. 529, 531; Prawer, 'Jerusalem in Christian and Jewish Perspectives', pp. 762–4.

14 Hieronymus, *Praefatio in librum Paralipomenon*, *PL*, 29, col. 401; Prawer, 'Jerusalem in Christian and Jewish Perspectives', p. 762; Taylor, *Christians and the Holy Places*, *PL*, 29, p. 311.

Heavenly Jerusalem as the symbol of salvation. Hymns such as 'Blessed City called Jerusalem' (*Urbs Beata Jerusalem*), under the influence of John's Revelation, glorified the 'Blessed City', the 'Heavenly Jerusalem', the 'vision of peace and love', 'the halls of Zion', the 'vision of peace', 'heaven of salvation'.[15] This hymn, probably composed at the end of the eighth century, and later a source of inspiration for the many hymns concerning the New Jerusalem, was used for the dedication of churches, as well as at the Baptism on Easter Eve.[16] In the ninth century, Walafrid Strabo ended his treatise on the fall of Jerusalem with a prayer that 'God, who had destroyed Earthly Jerusalem and established the Heavenly one, should eradicate the sins of our body and soul and lead us to the Chambers of his bride, where we shall see unfailingly together with him the felicities of Jerusalem.'[17]

Moreover, since the tenth century, the dedication rite of churches established a relationship between the Christian sanctuary and Heavenly Jerusalem; a church was both liturgically and mystically an image of Heavenly Jerusalem as described by the book of Revelation; the Epistle for the Mass of Consecration was from Revelation 21:2–5. Thus Heavenly Jerusalem became a symbol not only of the Church *per se* but of physical churches. The dedication rite included part of a sermon of Augustine, which emphasized the link between the physical church, made of stones, and the spiritual church, made of 'living stones'; the living stones of earth (which now erect buildings of inert stone) would one day be a part of Heavenly Jerusalem.[18] Both Augustine's sermon and the hymn *Urbs Beata Jerusalem*, also included in the dedication rite, described Heavenly Jerusalem as the prefiguration of Paradise and as the promised destination of true believers after death.[19]

At the same time, however, pilgrimage to Jerusalem was considered, albeit unofficially, as bringing the believer closer to his main goal, namely the salvation of his soul. This became official in the seventh century with the

---

15 *Early Latin Hymns*, ed. A.S. Walpole, Cambridge 1922 (repr. Hildesheim 1966), no. 119, pp. 377–80; Colin Morris, *The Discovery of the Individual 1050–1200*, London 1972, pp. 149–52; H. Ashworth, 'Urbs beata Jerusalem, Scriptural and Patristic Sources', *Ephemerides Liturgicae*, 70 (1956), pp. 238–41. The hymn runs as follows: 'Blessed City of Heavenly Salem / Vision dear of peace and love / Who, of living stones upbuilded / Art the joy of Heaven above / And, with angel cohorts circled / As a bride to earth dost move. / From celestial realms descending / Bridal glory round her shed / By her Lord shall she be led: / All her streets, and all her bulwarks, / Of pure gold are fashioned. / Bright with pearls her portals glitter, / They are open evermore / And by virtue of his merits / Thither faithful souls may soar, / Who for Christ's dear name in this world / Pain and tribulation bore'. Eng. trans. Morris, *The Discovery*, p. 150.

16 See *Early Latin Hymns*, pp. 377–8.

17 Walafridus Strabo, *De Subversione Jerusalem*, col. 974.

18 Augustinus, *Enarrationes in Psalmos* [Ps. 136, 22], *PL*, 37, col. 127 .

19 L.H. Stookey, 'The Gothic Cathedral as the Heavenly Jerusalem: Liturgical and Theological Sources', *Gesta*, 8 (1969), pp. 35–41.

development of the practice of judicial–penitential pilgrimages; the *Poenitentialia* books now recommended pilgrimage as a means of penance, first privately and, from the eighth century on, as a public penance. In the framework of penitential pilgrimages in general, the pilgrimage to Jerusalem came to be considered the most worthy and thus most redeeming, as it was the most difficult.[20] Thus Pope Gregory VII condemned Cencius, who led a revolt against him in 1075, to an enforced journey to Jerusalem; Cencius did not obey, fleeing instead to the protection of Emperor Henry IV, but many others went voluntarily.[21] A rude and violent ruler like Fulk, count of Anjou, nicknamed 'Nerra', undertook the Jerusalem pilgrimage three times, because of the enormity of his sins, and flagellated himself at the site of Calvary.[22] The feeling of the pilgrims was well expressed by Herveus the archdeacon of the Church of the Holy Cross in Orleans: 'I pondered many nights on the frailty of life and so came to live in fear of sudden death. I went to Jerusalem to atone for my sins and to beg for mercy on my knees.'[23] And indeed the last decade of the tenth century and the eleventh century were periods of a meaningful increase in pilgrimages to Jerusalem by laity, clergy and the members of monastic orders including abbots.[24] This can be explained by a number of factors, the main one being the approaching millennium and with it what can best be defined as an obsessive search for the means of redeeming sin.[25] The *peregrinatio religiosa* was considered one of the most efficacious of such means, and among the pilgrimages the Jerusalem pilgrimage, as the hardest one, was seen as the most redeeming. Another factor was the growing devotion in the West to the Holy Sepulchre.[26] Yet another was the change in the nobility, which had undergone a process of evangelization.[27] What may have also influenced the flow of pilgrims to Jerusalem in the eleventh century was the improvement in the physical aspects of the pilgrimage.[28] As a consequence there were, besides individual pilgrimages, also semi-organized group pilgrimages. Hugh of Flavigny, biographer of Richard, abbot of St Vanne's of Verdun, stated that the abbot conducted a pilgrimage of seven hundred pilgrims in 1026. Radulf Glaber reports for 1033, the year believed to be the thousandth anniversary of Christ's Crucifixion, 'an unheard of gathering of people in Jerusalem'. The great German

---

20 C. Vogel, 'Les pélerinages pénitentiels', *Revue des sciences religieuses*, 38 (1964), pp. 113–45.

21 S. Runciman, 'The Pilgrimages to Palestine before 1095', p. 77.

22 Bachrach, 'The Pilgrimages of Fulk Nerra', pp. 205–17.

23 Sumption, *Pilgrimage*, p. 134; Lalanne, 'Des pélerinages en Terre Sainte', p. 28.

24 *Ibid.*, pp. 26–31. Riley-Smith, *The First Crusaders*, pp. 26–33. For pilgrimages by members of the monastic orders see below, pp. 124–6.

25 See below, ch. 8, pp. 146–7.

26 See above, ch. 5, pp. 63–5.

27 See above, ch. 5, p. 69.

28 Runciman, 'The Pilgrimages', pp. 68–78.

pilgrimage of 1064–1065 was said to have involved over ten thousand men and women; in fact it probably numbered seven thousand.[29]

There is, moreover, some evidence that in the eleventh century the notion of Earthly Jerusalem merged with that of Heavenly Jerusalem as the symbol and source of salvation. In view of the belief that 'caelestia exempla sunt terrestrium',[30] Jerusalem came to be considered the gate to the heavenly kingdom, to salvation. Thus pilgrims often described their departure for Earthly Jerusalem in terms of their pursuit of the Heavenly City. St Choloman's journey (*c.* 1012) to Earthly Jerusalem was moved by his love of Heavenly Jerusalem; similarly, Adso of Montier-en-Der intended to 'undertake the road to the Earthly Jerusalem in order to be admitted to our mother the Celestial Zion'.[31] The intimate connection between the two cities and personal salvation is demonstrated by phenomena other than pilgrimage. For example, St Meginwerd (†1034) built a church modelled upon the Church of the Holy Sepulchre in Paderborn, in order to be granted 'Heavenly Jerusalem'.[32]

It seems that in the eleventh century pilgrims felt the need to use the link between the terrestial and celestial to justify their pilgrimage to the Earthly Jerusalem. This changed during the twelfth century. The crusader conquest and Jerusalem's newly acquired position as a Christian city significantly consolidated the link between the earthly and the heavenly cities. Christ's curse (or prophecy) that 'there will not be left here one stone upon another' (Matthew 24:1–2; Luke 19: 42–44) seemed to have been forgotten, as was the image of Jerusalem as the city that 'killed the prophets and stoned those who are sent to you' (Matthew 23:37). The terrestrial city, Earthly Jerusalem, now vindicated its position as the city of the Old and the New Testaments, hence a prefiguration of Heavenly Jerusalem and as such a gate on the believer's road to the celestial city, to one's personal salvation.

This change in the attitude of Christendom to Jerusalem is evident, *inter alia*, in the exegesis of the very term 'Jerusalem'. The authoritative exegesis of the early Middle Ages was dominated by the Pauline historical interpretation of Jerusalem as 'the City of Jews', the city that had slain its prophets and been burned down, first by the Babylonians and then by the Romans. This type of exegesis was typical of such great authorities as Jerome, Hilary of Poitiers, Cassian, Alcuin and Walafrid Strabo, as well as Honorius of Autun, who wrote at the beginning of the

---

29 Radulfus Glaber, p. 109; Hugo Flaviniacensis, p. 393; Joranson, 'The Great German Pilgrimage of 1064–1065', p. 12; Runciman, 'The Pilgrimages', pp. 68–78; F. Micheau, 'Les Itinéraires maritimes et continentaux des Pélerinages vers Jérusalem', *Occident et Orient au Xe siècle. Actes du IXe Congrès de la Société des Historiens Médiévistes de l'Enseignement Supérieur Public (Dijon 2–4 Juin 1978)*, Dijon 1979, pp. 79–111; Riley-Smith, *The First Crusaders*, pp. 26–33.

30 See on this subject Konrad, 'Das himmlische und das irdische Jerusalem', pp. 523–40.

31 *Widrici Vita S. Gerardi, MGH.SS*, 4, p. 488.

32 *Vita Meinwerci*, p. 158.

twelfth century.[33] Following the First Crusade, however, a change took place. Whereas the allegorical, anagogical and tropological interpretations remained unchanged, the association of Earthly Jerusalem with the City of the Jews disappeared from the historical interpretation. Gilbert of Nogent wrote that, according to history, Jerusalem was 'a particular city'.[34] Hugh of St Victor interpreted 'Jerusalem' historically as an earthly city, founded on Mount Zion on the threshing-floor of Ornan the Jebusite, where David built towers, as well as 'a particular city in Judaea'.[35] John Beleth argued that according to history it was the destination of pilgrims.[36] Jerusalem thus became a city in Judaea – not the City of the Jews, of the Second Temple, the killers of Christ – but the City of David, of the biblical Children of Israel, as well as the goal of pilgrimage. In keeping with the traditional exegesis that the historical Jerusalem was the City of the Jews, the site of the Temple,[37] the early twelfth-century writer Honorius of Autun argued significantly that 'Shem [son of Noah], a citizen of heavenly Jerusalem, had founded earthly Jerusalem *in figura* of that which is above'.[38] Thus, although the old exegesis was too authoritative to be easily abandoned, the new concept of Earthly Jerusalem was nevertheless gaining ground.

Another manifestation of the change was the significance of the Sunday of 'Laetare Iherusalem', the fourth Sunday of the forty days' fast (*media quadragesima*), the introit for which was: 'Rejoice with Jerusalem, and be glad for her, all you love her, rejoice her in joy ...' (Isaiah 66:10). The psalm for the day was 'I was glad when they said to me, let us go to the house of the Lord' (Psalms 122:1) and the tractate was Psalms 125:1–2: 'Those who trust in the Lord are like Mount Zion, which cannot be moved, but abides for ever. As the mountains are round Jerusalem so the Lord is round about his people ...'; the psalm for the Eucharist was Psalms 122:23–4: 'Jerusalem, built as a city which is bound firmly together, to which the tribes go up, the tribes of the Lord, as was decreed for Israel, to give

---

33 Hieronymus, *Epistulae*, no. LXVI, p. 335; XVIII, pp. 314–15; *idem*, *Commentatorium in Hiezechielem*, *CCSL*, 75, pp. 160–61; Johannes Cassianus, *Conlationes*, *CSEL*, 13, p. 405; *idem*, *Expositio in Psalmos*, *CCSL*, 13, p. 97; Hilarius Pictaviensis, pp. 854–5; Prosper Aquitanus, *Expositio Psalmorum*, *PL*, 51, col. 322; *Alcuini Epistolae*, *MGH*. *Epistolae Aevi Karolini*, 4, no. 198, p. 327; Walafridus Strabo, *De Subversione Jerusalem*, col. 973; *idem*, *De Universo*, *PL*, 111, col. 379; Honorius Augustodunensis, *De Animae Exsilio et Patria*, *PL*, 172, col. 1245. See also Sedulius Scottus (d. 860) *In Epistolam ad Galatas*, who used Gal. 4:24–7 within his historical interpretation, and thus defined Jerusalem as the city 'which is in bondage with the children'. Lubac, *Exégèse Médiévale*, I, pp. 645–7; Prawer, 'Jerusalem in Christian and Jewish Perspectives', pp. 750–82.

34 Guibertus Novigentis, *Liber*, col. 26.

35 Hugo de S. Victore, *Miscellanea*, *PL*, 177, cols 670, 674; *idem*, *Speculum Ecclesiae*, *PL*, 177, col. 375. See also Hugo de Folieto, *De Claustro animae*, *PL*, 176, col. 1131.

36 Johannes Beleth, pp. 212–13.

37 Honorius Augustodunensis, *De Animae*, col. 1245.

38 *Idem*, *Speculum Ecclesiae*, *PL*, 172, col. 1094.

thanks to the name of the Lord.'[39] In the early Middle Ages the Sunday of 'Laetare Iherusalem' was associated with Heavenly Jerusalem, with Jerusalem, as usual in the Pauline tradition, signifying the Christian Church. In the course of the twelfth century, however, the Sunday of 'Laetare Iherusalem' came to be increasingly associated with Earthly Jerusalem. First, parts of the liturgy of that day were transferred to the liturgy of the new Feast of the Conquest of Jerusalem, celebrated on the anniversary of the day of Jerusalem's conquest, on 15 July.[40] It had also come to be associated with the obligation of Christian knights to aid Earthly Jerusalem. Thus, the Emperor Frederick I convoked the assembly at Mainz, known as the 'assembly of Jesus Christ', *Curia Ihesu Christi*, the purpose of which was to recruit German knighthood for the Third Crusade, on the day of 'laetare', namely 27 March 1188.[41]

Yet another, and perhaps the most important, manifestation of the new attitude to Earthly Jerusalem and its relationship with the Heavenly one was the place assigned to the city in crusade ideology. As early as the eve of the First Crusade, it became accepted that the crusade – which, as a war-pilgrimage on God's behalf, would in any case have ranked as penance for a serious sin – was such a dangerous enterprise that it would purge its participants of their past crimes.[42] Crusade *exitatoria*, like that emanating from Moissac, the so-called encyclical of Sergius IV, claimed that 'Christ gave us His Sepulchre so that only the penitent will reach the Celestial Kingdom'.[43] The participants of the First Crusade did not expect a full remission of sins but rather saw themselves as penitents performing a severe penance, an act of self-sanctification.[44]

The idea of crusading, originally linked almost exclusively with Jerusalem and the Holy Places of what became during the twelfth century the 'Holy Land', was expanded soon after the First Crusade. The ideology of crusades was extended to include goals and aims outside the Holy Land, such as the wars against the Moors in Spain, against the Slavs in the Baltic, and against schismatics, heretics and lay powers in the West.[45] For example, *c.* 1100, Manasses, archbishop of Rheims, writing to the people of Cambrai, argued that fighting the schismatics was the 'path to Celestial Jerusalem'.[46] At the same time, Landulf of Milan claimed that

---

39 For the liturgy of the 'Laetare Iherusalem' Sunday see Gregorius I, *Liber Antiphonarius*, *PL*, 78, col. 667.

40 See above, ch. 2, pp. 20–3.

41 See *Historia de Expeditione Friderici Primi*, *MGH.SS rer. Germ*, 5, pp. 14–15.

42 As early as 1088, Emperor Alexius Comnenus exhorted Robert I of Flanders to liberate the Lord's Sepulchre and thus to be granted a reward in Heaven. *Kreuzzugsbriefe*, no. I, p. 136.

43 Gieysztor, 'The Genesis of the Crusades', p. 33.

44 Riley-Smith, 'Death on the First Crusade', *The End of Strife*, ed. D. Loades, Edinburgh 1984, pp. 14–18; *idem*, *The First Crusaders*, pp. 40–80 esp. pp. 74–6.

45 J. Riley-Smith, *What Were the Crusades?*, London 1977, pp. 18–33.

46 Lambertus Atrebatensis, *Epistolae*, *PL*, 162, col. 655–6.

to free the Church from the yoke of the simoniac ranked with the liberation of the Holy Sepulchre.[47] And Pope Paschal II, caught up in the general excitement surrounding the conquest of Jerusalem, appealed to Robert I of Flanders in 1103 on behalf of the reformers against the German emperor: 'Blessed be the Lord of Israel, who is working the efficacy of His virtue in you, seeing that ... you are striving to attain the Celestial Jerusalem with the deeds of just knighthood ... to arrive at the Heavenly Jerusalem.'[48] In 1124, Archbishop Diego Gelmirez of Compostela justified crusading in Spain on the grounds that the recovery of the peninsula was the key to a better route to the Holy Sepulchre: 'Just as the knights of Christ, the faithful sons of Holy Church, opened the way to Jerusalem with much labour and spilling of blood, so we should become knights of Christ and, after defeating his wicked enemies the Moslems, open the way to the same Sepulchre of the Lord through Spain, which is shorter and much less laborious.'[49] Not only the Spanish *Reconquista* but also the crusades against the pagan Slavs were justified by linking it to Jerusalem. Composing an appeal in 1108 for a crusade against the Wends, an anonymous writer, a Flemish clerk in the circle of the archbishop of Magdeburg, compared the liberation of Jerusalem by the First Crusade with the crusade against the Slavs. The image used was that of the Heavenly Jerusalem. He argued as follows: 'Rise up and come, all lovers of Christ and of the Church, and make preparations just as did the men of Gaul for the liberation of Jerusalem. Our Jerusalem, free from her origin, has been made a slave-girl by the cruelty of the Gentiles. Her walls have fallen because of our sins, but these ruins are in your hands; let all the precious stones of her wall, and the towers of our Jerusalem, be built with Jewels.'[50] The attempts to link the European crusades to Jerusalem show clearly that Jerusalem crusade was much more attractive. It also shows that Jerusalem remained the touchstone against which all other crusades were tested. Though by definition all the crusades and their participants enjoyed equal status, participation in a crusade aimed at the defence of Jerusalem or – after 1187 – its recovery was considered more meritorious than crusading with some other goal in mind.[51] Significantly, the papacy often justified crusades with destinations other than the Holy Land through the cause of Jerusalem. In 1199, for example, Innocent III presented the crusade for the recovery of the papal fief of Sicily as a precondition for the

---

47 Landulfus, *Historia Mediolanensis, MGH.SS*, 8, p. 83; Norman Housley, 'Crusades against Christians: Their origins and early development, c. 1099–1216', *Crusade and Settlement*, p. 20.

48 Pascalis II, *Epistolae*, *PL*, 163, col. 108; Housley, 'Crusades against Christians', p. 20.

49 *Historia Compostellana*, *CCCM*, 70, p. 379; Housley, 'Jerusalem', pp. 35–6; Riley-Smith, *The First Crusaders*, pp. 79–80.

50 Housley, 'Jerusalem', pp. 37–8.

51 Riley-Smith, *What Were the Crusades?*, pp. 11–33; Housley, 'Crusades against Christians', pp. 17–36.

recovery of Jerusalem. Similarly, in 1204, he justified (*post factum*) the conquest of Constantinople by the Fourth Crusade.[52]

Perceiving Earthly Jerusalem as the figure of Heavenly Jerusalem and 'mother of our redemption and faith' (*mater nostrae redemptionis ac fidei*), crusaders leaving for Jerusalem often attributed their departure to their concern for the salvation of their souls. Nivelo, a castellan of Freteval in the diocese of Chartres, upon his departure on the First Crusade (*c.* 1098), announced that he was going on pilgrimage to Jerusalem to expiate his crimes. He confessed to having acted 'in an atrociously tyrannical manner' towards the abbey of Saint-Père de Chartres whenever 'the onset of knightly ferocity stirred' him up. Therefore, 'in order to obtain the pardon for his crimes which God can give' he was going 'on pilgrimage to Jerusalem which until now has been enslaved with her sons'.[53] Godfrey of Bouillon and his brother, Baldwin of Boulogne, were reported to have taken the cross, 'seized by hope of an eternal inheritance and by love, prepared to go to fight for God in Jerusalem'.[54] Such pronouncements reveal not only that the crusade was regarded as a pilgrimage and of the same meritorious value, but also the extent to which European knighthood was influenced by the idea of the Just War, as propagated by the 'Peace of God' (*Pax Dei*) movement.[55]

Already during the First Crusade, Earthly Jerusalem, as the target of the crusade, merged with Heavenly Jerusalem. The goal became both physical and spiritual, with the crusaders believing that as a reward for fighting for the earthly city, they would be granted access to the glory of Heavenly Jerusalem. On the march they sang the song 'Jerusalem the Marvellous' (*Iherusalem mirabilis*), which promised those who died on crusade the 'benefits of Heaven' (*bona caeli*).[56] In September 1098 the leaders of the crusade called upon the pope to join them and to open for them 'the gates of both Jerusalems'.[57] Writing after the conquest of Jerusalem in November or December 1099, Archbishop Manasses of Rheims

---

52 Innocentius III, *Opera*, *PL*, 215, cols 514–16; *idem*, *Register*, II, no. 212, p. 414; Norman Housley, *The Italian Crusades*, Oxford 1982, pp. 62–9.

53 *Cartulaire de ... Saint-Père de Chartres*, II, no. XXXVI, pp. 428–9. Riley-Smith, *Idea and Reality*, pp. 98–100; *idem*, *The First Crusaders*, pp. 62, 71–2, 114–15. Another crusader, Stephen of Neublans, departing for the crusade of 1101, declared that he saw the cross as the sign of salvation and was going on crusade for the redemption of his soul. *Recueil des Chartes de l'Abbaye de Cluny*, V, no. 3737, pp. 89–90; G. Constable, 'Medieval Charters', p. 76.

54 *Triumphus Sancti Lamberti de Castro Bullonio*, *MGH.SS*, 20, pp. 498–9; Riley-Smith, *The First Crusade*, p. 44; H.E. Mayer, *Mélanges sur l'histoire du royaume latin de Jérusalem*, Paris 1984 (*Mémoires de l'Académie des Inscriptions et Belles Lettres*, 5), pp. 43–8.

55 On this subject see Hoffmann, *Gottesfriede und Treuga Dei*, Stuttgart 1964, *passim;* Housley, 'Crusades', pp. 21–6; Bull, *Knightly Piety*, pp. 21–69; Riley-Smith, 'Crusading as an Act of Love', pp. 177–85; *idem*, *The First Crusaders*, pp. 40–52.

56 *AH*, 54b, no. 96, p. 78.

57 *Kreuzzugsbriefe*, no. XVI, p. 164.

described the conquered city as follows: 'This is that Jerusalem which resides on high in gladness and delight, a glorious gift which its God has bestowed upon it in our time. Jerusalem, the city of our redemption and glory, rejoices with new gladness, because it has been delivered from the most cruel bondage of the pagans by the labour and incomparable power of God's children.'[58]

Under the pen of the chroniclers of the First Crusade, the crusaders' journey to Jerusalem became a guarantee to their entry into Heavenly Jerusalem as long as they kept faith with Christ; this was doubtless one of the most compelling ideas of the crusades. Baldric of Dol and Robert the Monk had Pope Urban II at Clermont promising that those who joined the crusade for the remission of their sins 'should be sure of attaining the eternal glory of the Heavenly Kingdom'.[59] Citing those words supposedly addressed to the crusaders at the siege of Jerusalem in July 1099, Baldric of Dol described the crusade as the 'path to life', to eternal life, to salvation; it was a double goal because Jerusalem 'is the cause for all our labours. This Jerusalem is the likeness of Heavenly Jerusalem; this city has the form of that to which we aspire. ... This Jerusalem, which you see, to which you come, at which you are present, prefigures and reflects the heavenly city.'[60] The bond between the two cities was in Baldric's opinion even stronger. What would happen if the Moslems should remain in Jerusalem? 'Obviously', he argued, 'one should fear that the heavenly city should not be closed to us, should not be taken from us, if we in our indolence should be disowned of our house by malicious strangers.'[61] Thus a mystical relationship existed, a magical bond between the two cities. The loss of Earthly Jerusalem might prejudice the right of the crusaders to the Heavenly City.

Not only was the crusaders' goal a double one – the same was true of their reward. Adhémar of Le Puy, the papal legate, in a letter to the bishops in the West dated January 1098 from Antioch, addresses them as 'brothers in Heavenly Jerusalem' (*consortium caelestis Iherusalem*); writing in the name of the clergy, he claimed that the crusaders would receive a double reward: life, that is, salvation, as well as the land of milk and honey.[62] Shortly after the First Crusade, Gilo wrote that in fighting for Jerusalem the crusaders had fought for two kingdoms, for two Jerusalems; the conquest of Earthly Jerusalem was necessary for the possession of Heavenly Jerusalem, where the crusaders would live eternal life.[63] Albert of Aachen and other historians of the First Crusade describe a dream of Peter the Hermit in which Christ appeared to him, promising that the gates of Paradise would be opened to those elect

---

58 *Ibid.*, pp. 175–6; Housley, 'Jerusalem', p. 28.

59 Baldricus Dolensis, p. 15; Robertus Monachus, p. 729.

60 Baldricus Dolensis, p. 15.

61 *Ibid.*, pp. 100–101; Prawer, 'Jerusalem in Christian and Jewish Perspectives', pp. 749–50.

62 *Kreuzzugsbriefe*, no. IX, pp. 146–9, esp. 148.

63 Fulco et Gilo, *Historia Gestarum viae nostri temporis Hierosolymitanae*, *RHC.Hocc*, 5, p. 798.

and devoted men who would purify the holy places of Jerusalem and restore the Christian cult.[64] This was also the message of the Old-French rhymed epic of the First Crusade, the *Chanson d'Antioche*, according to which Christ, while on the Cross, had promised those crusaders who came to avenge him and to recover his sepulchre that they would inherit the 'heavenly paradise'.[65]

Even more important than fighting for Jerusalem was the possibility of dying for its sake. According to Baldric of Dol, Pope Urban II declared that 'it ought to be a beautiful ideal for you to die for Christ in that city where Christ died for you'.[66] William of Malmesbury, in his description of the same occasion, attributed to the pope the statement that 'those who may die [on the crusade] will enter the mansions of heaven; while the living shall behold the Sepulchre of the Lord'.[67] Ralph of Caen described a vision experienced by the crusader Anselm of Ribemont shortly before his death (in February 1099). Anselm saw a splendid palace and innumerable handsome men passing through its door; one of them told him that they were crusaders, now crowned as martyrs, and informed him that he would be the next to join them: 'Those are the Jerusalem pilgrims who from the beginning entered upon the way of God, in which you too are still striving. They have departed this life and wear everlasting crowns. Unless by chance you yet refuse, you also will ascend to us; for you have fought the good fight, and you have completed your course.'[68]

The significance of the idea of martyrdom in the battle for Jerusalem during the First Crusade was that it explained how crusaders, with a view to obtaining remission of their sins, could follow Christ in pilgrimage to the city where He had suffered and died, and at the same time fight a war. The eternal reward for the crusader who failed to reach Jerusalem was that he would die a martyr; he was therefore to be judged fortunate in being admitted so speedily to the Heavenly City. The choice for the crusader in any particular battle was therefore either victory or martyrdom; in the wider context, the choice was between reaching the Earthly City and gaining admission to the Heavenly City. In the latter case he would have reached his ultimate goal as a Christian; in the former he would have done much to ensure his success in finally reaching it.[69]

---

64 Albertus Aquensis, pp. 273, 489; Raedts, 'Jerusalem: Purpose of History', p. 38 n. 9. For the dream of Peter the Hermit and the diffusion of this story in the West see Blake and Morris, 'A Hermit', pp. 79–107. See also Morris, 'Peter the Hermit', pp. 21–34.

65 *Chanson d'Antioche*, pp. 22, 25–7.

66 Baldricus Dolensis, p. 15.

67 William of Malmesbury, *De Gestis*, l. 4, p. 398.

68 Radulfus Cadomensis, pp. 680–81; Riley-Smith, 'Death on the First Crusade', p. 24. For the idea of martyrdom during the First Crusade see *ibid.*, pp. 14–31; H.E.J. Cowdrey, 'Martyrdom and the First Crusade', *Crusade and Settlement*, pp. 46–53. For a different opinion see Jean Flori, 'Mort et martyre des guerriers vers 1109. L'exemple de la première croisade', *Cahiers de Civilisation Médiévale*, 34 (1991), pp. 121–39. For a different version of the same vision see Raimond of Aguilers, pp. 108–9.

69 Cowdrey, 'Martyrdom and the First Crusade', p. 53; Green, *Millstätter Exodus*, pp. 267–71.

Some three decades after the First Crusade, the theme of dying in battle for Jerusalem, as a guarantee for martyrdom and access to the glory of Heavenly Jerusalem, was developed and expanded by Bernard of Clairvaux. His treatise *In Praise of the New Knighthood* (*De Laude novae militiae*), composed for the new military order then founded in Jerusalem, the 'order of the Poor Knights of the Lord and the Temple of Solomon' (the Templars), is in fact an evocation of the glories of martyrdom. The Order of the Temple is presented as a collective of knights dedicated to martyrdom. It is not only a community of potential martyrs but a sort of 'nursery of martyrs'. The supreme virtue and essential quality of the Templars is their vocation for martyrdom; in fact, they are permanent candidates for martyrdom. They fight the Lord's battle fearing neither sin nor death, 'because to slay for Christ is not sin but a first claim for glory', and moreover there is no more just cause for fighting than for Zion, 'a strong city that sets up salvation' (Isaiah 26:1). The Earthly City is the prefiguration of Heavenly Jerusalem: 'The glory of the Earthly City does not deny the glory of its celestial figure but builds it up, as long as we remember that the one is the figure of the other, which is above and which is our mother' (Galatians 4:26).[70]

Preaching the Second Crusade, Bernard attempted to resolve the active pursuit of Earthly Jerusalem with the contemplative pursuit of Heavenly Jerusalem. For him the crusade was, like pilgrimage, a personal and liturgical gesture towards salvation, an exercise in personal penance; an opportunity granted believers by the grace of God for the redemption of their souls.[71] In his encyclical letter *Ad peregrinantes Iherusalem* (1146), he wrote: 'Has the hand of the Lord been shortened or has it been made powerless to save, just because he has called poor little worms to guard his inheritance and restore it to him? Can he not send "more than twelve legions of angels" (Matthew 26:53) or simply say the word and the land will be set free? It is completely within his power to do so when he wishes it. But I say that the Lord is putting you to the test ... For the Lord has pity on his people and is providing a salving remedy for the gravely fallen. Consider how much skill he uses to save you and be astonished. ... He does not desire your death, but that you should turn from your way and live, because this is his way of offering you a favourable opportunity, not of destruction but of salvation ... Now, strong soldier and man of war, you have a battle you can fight without danger; in which to win will be glorious and to die will be gain ... Take the sign of the Cross. ... If it is worn on a faithful shoulder it is certain to be worth the Kingdom of God.'[72] For Bernard

---

70 Bernardus Claraevallensis, *De Laude Novae Militiae*, pp. 215, 217, 224.

71 For Bernard's concept of the crusade see E. Williams, 'Cîteaux et la seconde croisade', *Revue d'histoire ecclésiastique*, 49 (1954), pp. 116–51; Y.M.J. Congar, 'L'église et Cité de Dieu chez quelques auteurs cisterciens à l'époque des croisades', *Mélanges offerts à Etienne Gilson*, Paris 1959, pp. 173–81; J. Prawer, *Histoire du Royaume Latin de Jérusalem*, 2 vols, Paris 1969–1970, vol. 1, pp. 347–53; Cole, *The Preaching*, pp. 42–61.

72 Bernardus Claraevallensis, *Opera*, VIII, no. 363, pp. 313–15. Eng. trans. Riley-Smith, *Idea and Reality*, pp. 95–6. For this letter see above, ch. 5, p. 77, n. 64.

of Clairvaux, therefore, the fighting for Jerusalem, unlike other kinds of warfare, which were evil, was a path to the Kingdom of Heaven. It was an opportunity for the knighthood to enter Heavenly Jerusalem, not by abandoning their profession but by applying it to the right cause, Earthly Jerusalem: 'You should not desist from your earlier militarism, but you should desist from your malice, with which you are accustomed to kill one another, destroying one another in such a way that your are bringing one another to ruin ...'.[73]

Bernard's crusade propaganda deeply influenced the participants of the Second Crusade. They considered the crusade a pilgrimage to Jerusalem, its reward being salvation. Thus, for example, Arnold of Seschaves took the cross in 1147, out of concern for the salvation of his soul and the souls of his parents.[74] Other participants of the Second Crusade, such as Conrad III of Germany, Henry Jasomirgott of Austria, and Welf of Bavaria, saw themselves as aiming directly at the Celestial Jerusalem.[75] A poem composed on the eve of the Second Crusade, 'Knights, now is this a time of grace' (*Chevalier, mult estes guariz*), translated Bernard's ideas into terms clearly accessible to knights: 'Whoever goes with Louis now, / Need never fear the devil's horde, / His soul will go to Paradise / With the angels of the Lord.'[76] In an undated poem, 'For the perfect joy in Paradise' (*Por joie avoiz perfite en Paradis*), a crusader lamented leaving his love in order 'to have perfect joy in Paradise'.[77] Two poems, one composed on the eve of the Second Crusade and the other on the eve of the Third, repeated almost *verbatim* some of Bernard's phrases in *Ad peregrinantes Iherusalem*. That composed on the eve of the Second Crusade, 'Faith with idolatry' (*Fides cum Idolatria*), described Jerusalem as a market in a camp of liberty set up by the King of Heaven; those prudent merchants who wished to acquire life, that is, 'salvation', should hurry to depart for the Jerusalem crusade.[78] Another poem, dated to *c.* 1189, 'You who

---

73 Bernardus Claraevallensis, *Opera*, VIII, no. 363, p. 315. For this idea see also *idem*, *De Laude Novae Militiae*, pp. 217–19.

74 Constable, 'Medieval Charters', p. 76. For the crusade as pilgrimage to the Holy Sepulchre see *Historia Comitum Ghisnenium et Ardensium*, *MGH.SS*, 24, p. 633; *Chronica regia Coloniensis et continuationes*, *MGH.SS rer.Germ.*, 11, pp. 82–3; J. Von Hormayer, *Die Bayern im Morgenlande*, Munich 1832, p. 43; *Annales Herbipolenses*, *MGH.SS*, 16, p. 3; G. Constable, 'The Second Crusade as seen by Contemporaries', *Traditio*, 9 (1953), p. 242, no. 161.

75 *Monumentorum Boicorum Collectio Nova*, v. 28, no. XV, p. 227; Röhricht, *Die Deutschen im Heiligen Lande 650–1291*, Innsbruck 1984, pp. 28, 36.

76 Schöber, *Die altfranzösische Kreuzzugslyrik*, pp. 72–5. English trans. Morris, 'Propaganda for War', p. 95; M. Routledge, 'Songs', *The Oxford Illustrated History of the Crusades*, ed. J. Riley-Smith, Oxford 1995, pp. 97–8.

77 *Chansons de croisade*, no. XXII, pp. 283–6.

78 *Carmina Burana*, no. 44, p. 90. This poem also used another of Bernard's phrases, declaring that the Holy Sepulchre, glorified by prophets, should be kept from dogs (that is, the Moslems), to whom what is holy should not be given, like pearls to swine (Matthew 7:6). See Bernardus Claraevallensis, *Opera*, no. 363, p. 313.

love with true love' (*Vos qui ameis de vraie amour*), includes the following verses: 'I have heard it said proverbially / A sensible merchant spends money from his purse / And he has a very fickle heart / Who sees what is good and chooses what is evil. / Do you know what God has promised / to those who wish to take the cross? / God help me, a very fair wage: / Paradise, by firm promise.'[79]

Warfare for Jerusalem as the gateway to the Heavenly Kingdom, described in terms of Heavenly Jerusalem, became the central theme in the preaching of the Third Crusade. Pope Gregory VIII, proclaiming the crusade in his encyclical *Audita tremendi*, of October–November 1187, followed Bernard of Clairvaux in arguing that the crusade was an act of God whose purpose was to grant believers an opportunity to achieve eternal life: 'The Lord indeed could save it [the Holy Land] by His will alone, but it is not for us to ask Him why He has acted thus. For perhaps the Lord has wished to find out and bring to the notice of others whether there is anyone who has knowledge of Him or is seeking Him and might joyfully embrace the chance of penitence offered to Him and, in laying down his life for his brothers, may be killed in a brief moment and gain eternal life ...'. Moreover, the pope declared that 'those who were making their way to the Heavenly Jerusalem ... may not reach it except by the exercise of good works and through many temptations'.[80] The crusade should therefore be combined with repentance, penance and works of piety.

Such was also the opinion of one of the preachers of the Third Crusade, the Cistercian Henry of Albano. He argued that, spiritually speaking, the journey to Earthly Jerusalem was far less important than the journey to Heavenly Jerusalem. Earthly Jerusalem would not be saved unless Christians returned in their hearts to Heavenly Jerusalem. The return to Heavenly Jerusalem was therefore a precondition for the restoration of Earthly Jerusalem to Christianity, its recovery in the Third Crusade. Henry of Albano, like Bernard of Clairvaux, saw Heavenly Jerusalem as superior to Earthly Jerusalem in the believer's path to salvation and again, like Bernard of Clairvaux in preaching of the crusade, he described the journey to Earthly Jerusalem as a means to enter Heavenly Jerusalem.[81]

Peter of Blois, who preached the Third Crusade in England, was evidently less hesitant regarding the nature of the relations between the two, Earthly and

---

79 *Chansons de croisade*, no. II, pp. 20–22; Riley-Smith, *Idea and Reality*, pp. 89–90. Cf. Bernardus Claraevallensis, *Opera*, no. 363, p. 315: 'If you [strong soldier and man of war] are a prudent merchant, if you are a man fond of acquiring this world's goods, I am showing you certain great markets ...'.

80 Gregorius VIII, *Audita tremendi*, p. 8; Riley-Smith, *Idea and Reality*, pp. 65–6; Cole, *The Preaching*, pp. 63–5 and below, ch. 9, p. 164, n. 19.

81 Henry d'Albano, *De Peregrinante Civitate Dei*, *PL*, 204, cols 350–2; *idem*, *Ex quo vox*, pp. 11–13; Y.M.J. Congar, 'Henri de Marcy, Abbé de Clairvaux, Cardinal-Evêque d'Albano et légat pontifical', *Studia Anselmiana*, 43 (1958), pp. 56–90. For a similar opinion see Radulfus Niger, *De re militari et triplici via pererinationis Ierosolimitane*, ed. L. Schmugge, Berlin 1977, pp. 92–3, 147, 179, 182–3, 187–8, 194, 218–19.

Heavenly Jerusalem; he declared that all who worked 'for the recovery of the inheritance of the Lord', thus denying themselves 'an inheritance in this world', would win 'celestial inheritance' (*celis haereditas*) – they would inherit Heaven. Jerusalem as His 'sanctuary' (*Domini sanctum*) (Psalms 20:2) is the gateway to life, to Heavenly Jerusalem. In his treatise *Passio Reginaldi* (The Passion of Reynald of Châtillon), Peter portrayed his hero, as well as others who died during the fall of the Kingdom of Jerusalem, as martyrs for the faith, as elected ones who had been transported from Earthly to Heavenly Jerusalem.[82]

By 1187 this idea was deeply rooted in all levels of European society. A chronicler from Pegau in Saxony, recording the fall of the Kingdom of Jerusalem, announced that all who were killed in battle or taken as prisoners of war were collected as citizens by Heavenly Jerusalem.[83] The author of the 'History of the Crusade of Frederick I' (*Historia de expeditione Frederici I*) describes the German Crusaders departing for the Third Crusade as aspiring to that Jerusalem which is the figure of the Heavenly Jerusalem.[84] Magnus of Reichersberg, in Austria, claimed that the emperor himself had gone on crusade for the redemption of his soul, in order to be accepted after death to Heavenly Jerusalem.[85] While Europe was mobilizing its forces for the Third Crusade, poets and troubadours also presented participation in the crusade as the gate to the other Jerusalem. Their message to the faithful was to abandon Babylon and the fight for the 'celestial region', 'the water of life'; through aiding Jerusalem, to gain access to Heavenly Jerusalem, to paradise.[86] One of the best loved of crusading songs, 'Alas, O love' (*Ahi amors*), composed by Conon of Béthune on the eve of his departure on the Third Crusade, included the following verses: 'So let the great and the little all take note / They should take the field in chivalry / Where paradise and honour can be won ...'.[87]

Nevertheless, the idea that Earthly Jerusalem, whether as the goal of a crusade or of a pilgrimage, was a gateway to Heavenly Jerusalem, was not universally or wholeheartedly accepted during the twelfth century. The sources reveal a variety of opinions, ranging from complete negation of the salvational merit of Earthly Jerusalem to its negation in regard to certain groups in European society. This would be within the wider context of opposition to the participation of these

---

82 Petrus Blesensis, *Passio Reginaldi, Opuscula, PL*, 207, cols. 957–76. For this treatise see above, ch. 5, n. 72. For Reynard of Châtillon as 'beatus' see also Radulfus Niger, *Chronica universalis*, pp. 336–7.

83 *Annales Pegavienses*, *MGH.SS*, p. 266.

84 *Historia de Expeditione Frederici I*, p. 59.

85 Magnus Reicherspergensis, *Chronicon*, *MGH.SS*, 17, p. 509.

86 'Crucifigat omnes', *Carmina Burana*, no. 47, p. 92; 'Miror cur tepeat', *AH*, 33, p. 320; 'Jerusalem civitas inclita', in H. Hagenmeyer, 'Deux poésies latines relatives à la IIe croisade', *AOL*, 1, pp. 582–4; and see below, ch. 9, pp. 177–9.

87 *Chansons de croisade*, no. III, pp. 32–5. English trans. Riley-Smith, *Idea and Reality*, pp. 157–8.

groups in crusades or pilgrimages. Moreover, even those who preached Earthly Jerusalem as the gateway to the Heavenly City did not do so unreservedly. On the whole, the twelfth century should be noted for the great popularity of the idea of the meritorious value of Jerusalem as a goal of either pilgrimage or crusade among and for the laity, especially for the European nobility. At the same time, however, there was a sharp and significant increase in opposition to Earthly Jerusalem on the part of the spiritual leaders of European monasticism.

In the early Middle Ages, pilgrimage was considered as one step towards the ascetic way of life, close to monastic circles. The terms 'pilgrimage' and 'monasticism' were often considered equivalent; they represented the same way of life, a form of permanent asceticism, of permanent voluntary exile, a way of imitating Christ in his poverty; an authentic evangelic life in accordance with the precept: 'If any man will come after me, let him deny himself and take up his Cross and follow me' (Luke 9:23; Matthew 16:24) as well as: 'One must suffer for the sake of my name' (Acts 9:16). By the eleventh century, pilgrimage had come to be regarded by monks, even more than previously, as the renunciation of this world for the sake of Christ, an imitation as it were of Abraham, who was told: 'Go from your country and your kindred and your father's house ...' (Genesis 12:1; Hebrews 11:8–9). Consequently, pilgrimage to Jerusalem became popular among the members of the monastic orders, including abbots. Thus, St John of Parma (*c.* 982), following the example of Abraham, left his parents and his homeland and became a monk in Jerusalem; according to his *Vita* he visited Jerusalem six times. Emiraldus, a member of the community of Anzy-le-Duc, lived in Jerusalem for seven years before returning to France in about the year 1000. Another John, the future Abbot John III and Liutius, both of Monte Cassino, stayed in the Holy City at that time. St Simeon of Trier went on pilgrimage to the Sepulchre of the Lord (*c.* 1017), moved by his desire to undertake a pilgrimage for the sake of Christ and to suffer a poverty like that of the Saviour. Similarly, St Gerard of Csanád travelled to Jerusalem to see the Holy Sepulchre, in order to imitate Christ's poverty. Richard, abbot of St Vanne of Verdun, came to Jerusalem in 1026, accompanied by a large group, and in 1048 Thierry of St Evroul in Normandy renounced his position as abbot to venerate the Holy Sepulchre. Between 1020 and 1080, too, such respectable monks and abbots as Gervin of Saint-Riquier, Ulrich of Cluny, Leonius of Saint-Bertin, Poppo of Stavelot, John of Fécamp (between 1056 and 1076), Gerald of La Sauve-Majeure of Corbie (†1095), Raoul, abbot of Mont-Saint-Michel (in 1058), Nicholas, abbot and founder of Sauve Majeure in the diocese of Bordeaux (before 1095) – all made pilgrimages to the Holy Sepulchre.[88]

---

[88] J. Leclercq, *Aux Sources de la Spiritualité Occidentale. Étapes et Constantes*, Paris 1964, pp. 35–77; Lalanne, 'Des pèlerinages en Terre Sainte', pp. 30–31; B. Hamilton and P.A. McNulty, 'Orientale Lumen et Magistra Latinitas: Greek Influences on Western Monasticism (900–1100)', *Le Millénaire du Mont Athos, 963–1963. Études et Mélanges,* vol. 1, Chevetogne 1963; repr. Bernard Hamilton, *Monastic Reform, Catharism and the*

However, some representatives of monastic circles, stressing the insistence of the Benedictine rule of obedience and 'stability', criticized pilgrimages as a serious threat to the monastic vocation. Typical of this sort of attitude to pilgrimage on the eve of the First Crusade is that of Odilo of Cluny. On the one hand, he induced many of his lay friends to go to Jerusalem while maintaining, on the other, that Cluny was the Promised Land. What need was there to go and see Christ the King in the Holy Land? Cluny was itself Jerusalem, a Jerusalem illuminated with glory.[89] As the twelfth century progressed, opposition to pilgrimages by the members of monastic orders increased. The position of the leaders of the monastic world now emphasized in the life of a monk the importance of the ideal of 'stability' (*stabilitas*) and obedience (*obedientia*), as well as that of inner rather than outer behaviour. Whereas early monastic groups had adopted the ideal of 'stability in pilgrimage' (*stabilitas in peregrinatione*), this idea was replaced by a new one: 'pilgrimage in stability' (*peregrinatio in stabilitate*).[90] The goal thus became a spiritual one: the ideal of pilgrimage was superseded by a spiritual, internal pilgrimage, an ideal that completely denounced actual pilgrimage, even to Jerusalem, as contrary to the monastic ideal of stability.[91] This was, as recently demonstrated by Giles Constable, a symptom of the general tendency in the twelfth century to interiorize and spiritualize all aspects of monastic life and morality.[92]

This was evidently an opinion shared by Pope Urban II. Originally a Benedictine at Cluny, Urban II extended the existing provision forbidding monks to go on pilgrimage without the permission of their abbots;[93] writing on 7 October 1196 to the congregation of Vallombrosa, he expressed his dismay 'that those who have abandoned the world and have vowed themselves to spiritual warfare should either bear arms or go on this journey'.[94] Other more recent monastic orders forbade pilgrimage and crusading altogether in the course of the twelfth century. Amongst those who did so were the Order of Grandmont, the later followers of Stephen of Muret and the Cistercian Chapter-General of 1157.[95] Monastic writers

Crusades (900–1300), London 1979, no. V, 187, 192; Constable, *The Reformation*, p. 33; A. Jotischky, *The Perfection of Solitude*, p. 2.

89 E. Delaruelle, 'The Crusading Idea in Cluniac Literature of the Eleventh Century', *Cluniac Monasticism in the Central Middle* Ages, ed. N. Hunt, London 1971, pp. 210–11, and see, on the entire problem, Jestice, *Wayward Monks*, pp. 57–8, 198–200 *et passim*.

90 Constable, 'Opposition to Pilgrimage', pp. 125–32.

91 See on this subject J. Leclercq, 'Monachisme et pérégrinations du IXe et XIIe siècles', *Studia Monastica*, 3 (1961), pp. 3–35; G. Constable, 'Monachisme et pèlerinage au moyen âge', *Revue historique*, 258 (1977), pp. 3–27; *idem*, 'Opposition to Pilgrimage', pp. 125–46.

92 *Idem*, *The Reformation of the Twelfth Century*, pp. 150, 266–2.

93 *Kreuzzugsbriefe*, no. III, pp. 137–8; Riley-Smith, *Idea and Reality*, pp. 39–40.

94 Wiederhold, 'Papsturkunden in Florenz', pp. 313–14; Riley-Smith, *Idea and Reality*, pp. 39–40.

95 *Statuta Capitulorum Generalium Ordinis Cisterciensis*, ed. J. Canivez, Louvain 1933–1935, I, no. 53, 66; Stephanus Grandimontensis, *Sententiarum Liber*, col. 1114. See

and apologists repeatedly pointed to the overriding superiority of the monastic vocation; the monk was superior to the crusader or pilgrim because he aspired to the heavenly rather than the earthly city. Geoffrey of Vendôme argued, during the First Crusade, that monks who carried the Cross of the Lord and followed him in their monasteries had no reason to seek His Sepulchre; he claimed, moreover, that 'Not all those who have seen Earthly Jerusalem deserve to receive the Jerusalem, the one which is in Heaven.' This attitude found its way into canon law.[96] Gratian, treating the subject within the framework of monastic obedience, remarked that no vow, including a vow of pilgrimage, taken by a monk without the consent of his abbot, was valid. He argued that monks, as well as clerics, 'avoiding regular discipline were not permitted to take the vows of pilgrimage, lest in this way they returned to a secular way of life.'[97]

On the whole, leaders of the monastic world supported the participation of laymen in the crusades to Jerusalem, considering it as spiritually meritorious; but they were opposed, often fiercely, to both crusading and participating in pilgrimages by members of the monastic orders. An example of how the *iter Hierosolimitanum* and the foundation of a crusading kingdom in the Holy Land altered the attitude of the monastic orders to Jerusalem, and in general to warfare, is provided by Anselm of Canterbury. A decade before the First Crusade (*c.* 1086), Anselm, then an abbot of Bec, urging a young Norman to become a monk at Bec instead of following his elder brother to the East in the military service of Byzantium, wrote as follows: 'Be not ashamed to call yourself a poor man of Christ, for yours will be the Kingdom of Heaven. Fear not to make yourself the soldier of such a king, since this king will be with you in every danger. Do not wait in this life to follow a better way than that you have decided upon, lest you put off receiving the blessed crown in the next life. I warn, advise, beg, beseech and command you, as one most dear, to renounce Jerusalem, which is now a vision not of peace but of tribulation, and the treasures of Constantinople and Babylon, which must be seized with bloodstained hands. Embark instead upon the way to the Heavenly Jerusalem, which is the vision of peace (*visio pacis*), where you will find real treasures which may be had only by those who despise the other [treasures].'[98] Between 1086 and 1096, however, Anselm's attitude changed, and he supported Urban II's crusade plan, describing it as *iter Hierosolimitanum*.[99] At the same time, he maintained his opposition to the

---

also Bernard Claraevallensis, *Opera*, VIII, no. 544, pp. 511–12; J. Leclercq, 'La Lettre de Gilbert Crispin sur la vie monastique', *Studia Anselmiana*, 31 (1953), pp. 119, no. 3; Constable, 'Opposition to Pilgrimage', pp. 135, 140–41 and n.59.

96 Goffridus Vindocinensis, *Epistolae*, no. XXI, cols 162–3; no. XXIX, col. 127.

97 Gratianus, *Decretum*, c. XX, 9.4, p. 852.

98 Anselmus Cantuariensis, *Opera Omnia*, III, no. 117, p. 254; Grabois, 'Anselme', p. 165; Constable, 'Opposition to Pilgrimage', pp. 132–3.

99 Anselmus Cantuariensis, *Opera Omnia*, IV, no. 195, pp. 85–6; Cowdrey, 'Pope Urban II's Preaching', pp. 183–5; Grabois, 'Anselme', pp. 166–8.

departure of monks to Jerusalem. In late autumn 1096, Anselm, by this time archbishop of Canterbury, wrote to Bishop Osmund of Salisbury, instructing him to prevent Abbot Hamo and his monks of Cerne from going to Jerusalem and to forbid any monk, under pain of anathema, to join the crusade; he went even further by demanding the correction of Hamo, who had apparently encouraged young monks to join the crusade and even planned to do so himself.[100]

This rejection of Earthly Jerusalem as a means of salvation for monks, in contrast to laymen, alongside active involvement with the crusading movement, became the policy of two leaders of the monastic world, Bernard of Clairvaux and Peter the Venerable of Cluny. Bernard preached the Second Crusade on behalf of Pope Eugenius III and was an ardent supporter of the Templars. At the same time he completely rejected monastic pilgrimages to Jerusalem. In 1124 he wrote to Arnold of Morimond begging him to consider the 'great ruin' he would cause by carrying out his proposed pilgrimage. In his opinion, by going to Jerusalem without permission, Arnold was disobeying the statutes of the Order and was threatening the principle of unity enshrined in the *carta caritatis.*[101] Describing Arnold's behaviour in a letter to Bruno of Cologne (1124), Bernard wrote of it as 'a grave scandal for the whole order'.[102] Writing to one of Arnold's companions, the monk Adam, Bernard reminded him of his oath to remain in the cloister and that the virtue of stability is a permanent and inviolable rule for all monks.[103]

Moreover, as he wrote to Pope Calixtus II between December 1124 and January 1125, Bernard thought that Jerusalem was in much more need of soldiers who could fight rather than of monks who 'sing and weep'.[104] 'For the object of monks', he wrote, 'is to seek out not the earthly but the Heavenly Jerusalem, and this not by proceeding with the feet but by progressing with feelings.'[105] In 1147 Bernard had written to all Cistercian abbots, reminding them not to permit their monks to go to the Holy Land and recommending that monks who violated the prohibition should be excommunicated. This was included in the statutes of the Cistercian Order at the Chapter-General of 1157.[106]

Around 1129 Bernard of Clairvaux wrote to Alexander, bishop of Lincoln, on behalf of the latter's young clerk named Philip who, on his way to Jerusalem had stopped at Clairvaux and found his vocation there. He described Philip's journey as follows: 'He has already crossed the great and wide sea, has reached the desired shores and has landed here in the heaven of salvation. And his feet are standing in Jerusalem, and he adores Him whose voice was heard in Ephrata ... , but this

100 Anselmus Cantuariensis, *Opera Omnia*, IV, no. 195, p. 85; Constable, 'Opposition', p. 134; Riley-Smith, *The First Crusaders*, p. 90.

101 Bernardus Claraevallensis, *Opera*, vol. VII, no. 4, pp. 24–7.

102 *Ibid.*, no. 6, p. 30.

103 *Ibid.*, no. 7, pp. 31–43.

104 *Ibid.*, vol. VIII, no. 359, p. 305.

105 *Ibid.*, no. 399, pp. 379–80; Constable, *The Reformation*, p. 267.

106 *Ibid.*, no. 544, pp. 511–12; *Statuta*, vol. I, p. 53.

Jerusalem is not that near Mount Sinai in Arabia which is in slavery with the children, but the Jerusalem above, which is free and she is our mother (Galatians, 4:25–26). And that is, if you would know, Clairvaux.'[107] It is this famous statement together with his rejection of monastic pilgrimages on the one hand and his preaching to laity of the crusades to Jerusalem as the gateway to the Heavenly Jerusalem on the other, that made Bernard's attitude to Jerusalem a subject of controversy. In the best Cistercian and thus Pauline tradition, his letter to the bishop of Lincoln declaring Clairvaux as Heavenly Jerusalem has been interpreted as a total rejection of Earthly Jerusalem.[108] Additionally it was argued that Bernard, who always adapted his message to his hearers, while preaching to the laity, presented the crusade for the Earthly Jerusalem in terms of the reward of heaven. This was said to be for the reason that he regarded laymen as people whom he could interest in their own salvation by using only the crudest of devices, and by offering them gain. Moreover, he felt that he could not expect them to comprehend as fully as monks that the true Jerusalem was spiritual and invisible.[109]

There is no doubt that Bernard's preference was given to Heavenly Jerusalem. This, however, does not mean that he acknowledged the place of Earthly Jerusalem in the plan of salvation of laymen just because it was a useful slogan in his propaganda for the Second Crusade. As a matter of fact I tend to agree with Peter Raedts[110] on two points: first of all, Bernard always spoke, whether addressing learned monks or simple laymen, about the Heavenly City; secondly, Bernard sincerely cared about the Earthly Jerusalem. For Bernard the true Jerusalem was in heaven. Its original citizens were the angels. Even the devil and his companions were once sons of Jerusalem. And it was their fall that made God create man so that man might fill the empty places and restore the ruins of the heavenly city. From then on, the true Jerusalem, although still only one, is partly in heaven, partly on earth. In heaven, it is firm like a house, here ramshackle like a tent; there it is a house of praise, here of prayer. Here on earth, Jerusalem is the fortress of our strength; in heaven it is the city of our rest. As long as man cannot join the angels in seeing God (and only the angels had the perfect and direct knowledge of God

---

107 Bernardus Claraevallensis, *Opera*, vol. VII, no. 63, pp. 157–8; Prawer, 'Jerusalem in Christian and Jewish Perspectives', pp. 781–2.

108 See A. Bredero, 'Jérusalem dans l'Occident médiéval', *Mélanges offerts à René Crozet*, ed. P. Gallais *et al.*, 1, Poitiers 1966, pp. 169–71: 'Pour l'abbé de Clairvaux, Jérusalem était le symbole de la béatitude éternelle, où l'on entre par sa conversion spirituelle et monastique ...'. See also Prawer, 'Jerusalem in Christian and Jewish Perspectives', p. 782.

109 See P. Raedts, 'St Bernard of Clairvaux and Jerusalem', *Prophecy and Eschatology*, ed. M. Wilks, Oxford 1994 (*Studies in Church History*, *Subsidia*, 10), pp. 171–4. For an opposing view see M. Diers, *Bernhard von Clairvaux. Elitäre Frömmigkeit und begnadetes Wirken*, Münster 1991 (*Beiträge zur Geschichte der Philosophie und Theologie des Mittelalters*, 34), pp. 349–69, 397–8.

110 Raedts, 'St Bernard', pp. 169–82; Diers, *Bernhard von Clairvaux*, pp. 349–69, 397–8 and Jotischky, *The Perfection*, pp. 10–16.

because they were the original and the true citizens of Jerusalem) he must attempt to conform to them through his way of life. Of all men on earth none are closer to the heavenly Jerusalem than the monks; they were the first in the Church; they are closer to angels because of their chastity and the ardour of their love.[111] While speaking to monks Bernard did not once mention the Earthly Jerusalem, while when addressing the laymen he always spoke about both Earthly and Heavenly Jerusalem. In his opinion, for monks, the Earthly Jerusalem was an obstacle as monks did not need visible signs. For the laymen, however, such visible signs as Earthly Jerusalem were necessary. Therefore, when Bernard addressed the laity he described this Jerusalem as the prefiguration of the Heavenly City.[112] Thus he claimed that all Christians are on the road to the true Jerusalem; but only the road of laymen leads through the Earthly Jerusalem, as the gateway to the Heavenly one.

According to Bernard, the Earthly Jerusalem, now a capital of a Christian, Latin Kingdom, served a concrete purpose. The reconquest of Jerusalem in 1099, foretold by prophets, as in Isaiah 52:9: '... for the Lord hath comforted his people, he hath redeemed Jerusalem', offered Christian knights an awareness of their heavenly reward as the earthly city was the image of that which is in heaven. Bernard claimed that this land of promise, that once gave only milk and honey to its inhabitants, offered in his day the remedies of salvation and the food of life to all the world.[113] Seeing Jerusalem as the image and the gateway of the heavenly city as far as laity was concerned, Bernard was deeply concerned with the fate of Crusader Jerusalem. In May or June 1150 he wrote a gentle but firm rebuke to Peter the Venerable for not attending the Council of Chartres, held in April 1150 to discuss future plans for the Holy Land Crusade, and begged Peter to attend such councils in the future, stressing the threats which at present face the kingdom.[114] In 1153, writing to his uncle Andrew, Master of the Temple, Bernard once again expressed his concern regarding the fate of the Latin kingdom; he blamed the present troubles of the kingdom on its rulers 'whose labours will fail because their eyes are on the Earthly Jerusalem, not the Heavenly one'.[115] For Bernard, Jerusalem was a sacrament, a visible sign referring to an invisible reality. Thus, when Bernard called himself 'a monk and a Jerusalemite',[116] this definition certainly summed up his entire attitude to Jerusalem, complex but not ambivalent, as has often been argued.[117]

---

111 Raedts, 'St Bernard', pp. 172–5.

112 See above, nn. 71, 72, 73.

113 Bernardus Claraevallensis, *De Laude Novae Militiae*, pp. 218–19, 225–6; Raedts, 'St Bernard', pp. 180–81.

114 Bernardus Claraevallensis, *Opera*, VIII, no. 521, p. 483.

115 *Ibid.*, no. 288, p. 203.

116 Bernardus Claraevallensis, *Sermones super Cantica Canticorum*, 55, *Opera*, II, p. 112; Raedts, 'St Bernard', p. 172.

117 See above, nn. 108, 109, 110.

Peter the Venerable exhorted knights to go to Jerusalem, whose holy places he eulogized in his sermon *In Laudem Sepulchri Domini*,[118] but he forbade monks to go and worship there. In his *On Miracles* (*De Miraculis*) he praised knights whose pilgrimage to Jerusalem was a kind of prologue to taking the monastic vows. Thus, he told the story of a knight who gave his horses, his good clothes and all his money to Cluny on the eve of his departure as a pauper to Jerusalem; upon his return from the Holy Land he took monastic vows and spent the rest of his life at Cluny.[119] At the same time, Peter was opposed to the pilgrimages of monks who had already taken vows. Writing in 1146 or 1147 to the knight, Hugh Catula of Châlons, who wanted to go to Jerusalem after joining Cluny, Peter tried to dissuade him from making the pilgrimage and reminded him that 'serving God perpetually in humility and poverty is greater than making a journey to Jerusalem with pride and luxury. To visit Jerusalem, where His feet trod, is good, but to look towards heaven, where He is seen face to face, is far better. He who promised the better cannot replace it by the worse.'[120] Writing to a patriarch of Jerusalem (probably William I, 1130–45), Peter cites 'the prohibition of the monastic order' as an explanation for his inability to come to Jerusalem 'to see the super-celestial places of our salvation, to pour out our tears physically and worship in the place where the Lord's feet have trodden'. At the same time he asks the patriarch to send him some relics from Jerusalem, of the Holy Sepulchre and of the Virgin Mary.[121]

Yet Peter the Venerable reveals a surprisingly sympathetic attitude towards abbots going on pilgrimage. When Theobald, abbot of St Colomba, took the cross at Vézelay and announced his intention to go to the Holy Land, Peter declared (in 1146) 'that pilgrimage to the distant Holy Sepulchre will please God as much as the pilgrimage of holy women to the Sepulchre of Christ on Easter Sunday'. He praised the abbot's piety and devotion but warned him to shun levity, curiosity and the love of vainglory; for he would only derive spiritual benefit from the pilgrimage if his intentions were sincere and his conduct befitted a monk rather than a knight.[122] And indeed it seems that, on the whole, participation of abbots in the crusade or their undertaking pilgrimages aroused less opposition than in the case of ordinary monks; they were allowed to do so with the consent of the pope or, in the case of Cistercian houses, of the abbot of Cîteaux and the Chapter-General of the Order.[123] Moreover, as shown above, some abbots openly disregarded the opposition to monks' pilgrimages. Hamo, abbot of Cerne, and Pontius of Cluny not only went to Jerusalem themselves but openly encouraged their monks to follow their

---

118 See above, ch. 5, p. 71. For Peter the Venerable's ideas see also J.P. Torrell and D. Bouthillier, *Pierre le Vénérable et sa vision du monde*, Louvain 1986, pp. 78–85.

119 Petrus Venerabilis, *De Miraculis*, p. 55.

120 *Idem*, *Letters*, no. 51, p. 152.

121 *Ibid.*, no. 83, p. 220.

122 *Ibid.*, no. 144, pp. 358–9; Siberry, *Criticism of Crusading*, pp. 42, 96.

123 *Ibid.*, pp. 43–4.

example. One of those monks, Gelduin of Le Puiset, a relative of King Baldwin II of Jerusalem, became abbot of the monastery of St Mary in the Valley of Jehoshaphat in 1120; in 1119 he brought to Cluny a relic of St Stephen which Pontius placed over the high altar a year later.[124] In 1122 Pontius was forced to abdicate and left for Jerusalem, where he planned to spend the rest of his life; he returned after a year and a half, was unable to regain his former position and finally died in prison in Rome on 28 December 1128.[125] Among other abbots who went to Jerusalem were Geoffrey, founder and prior of Chalard, and Norbert of Xanten or Magdeburg, the founder of the Premontratensian Order.[126]

Though rejecting Earthly Jerusalem as the gateway to salvation of monks save for abbots, Peter the Venerable was absolutely convinced that it provided just such a gateway for the laity. This is reflected in his letter of April 1150 to Bernard of Clairvaux: 'Who will not be concerned if the Holy Land, recently liberated from the yoke of the wicked through the zeal of our fathers, through the spilling of so much Christian blood, is again enslaved by the wicked and sacrilegious. Who will not be concerned that the so salvational path of pilgrimage for repentant sinners, which for the last fifty years has saved from hell innumerable thousands and restored them to paradise, is closed [to pilgrims] because the Saracens will stand on their way as a barrier?'[127] This is almost a repetition of Baldric of Dol's statement that if Jerusalem remained in Moslem hands the heavenly city would remain closed to the crusaders.[128] Peter's letter of 1146 or 1147 to Suger of St Denis is a eulogy of Earthly Jerusalem: 'Is not the most important matter of all to care that the Holy Land should not pass to the dogs? That the places trodden by the feet of Him who from old has worked salvation in the midst of the earth (Psalms 74:12) should not be trampled under the feet of the wicked? That majestic Jerusalem, sanctified by prophets, apostles and particularly by the Saviour Himself ... should not be enslaved by wicked and pagan people? That the Cross of our Salvation ... should not be captured? That the Holy Sepulchre itself, hitherto ... the glory of the entire world, should not perhaps be totally destroyed, as they [the Saracens] threaten to do?'[129]

---

124 H.E.J. Cowdrey, 'Cluny and the First Crusade', *Revue Bénédictine*, 83 (1973), pp. 309–10 and here above, n. 93. See also *Chartes de Terre Sainte provenant de l'Abbaye de N.-D. de Josaphat*, ed. H.F. Delaborde, Paris 1880, no. VIII, p. 33; no. XII, p. 37; no. XVII, p. 43; no. XVIII, p. 45.

125 H.E.J. Cowdrey, 'Abbot Pontius of Cluny', *Studi Gregoriani*, 11 (1978), pp. 233–43.

126 Bernardus Claraevalensis, *Opera*, VII, no. 56, p. 148; Constable, 'Opposition to Pilgrimage', p. 141. It is also known that Suger of St Denis planned to go to Jerusalem with the Second Crusade. Brown and Cothren, 'The Twelfth-Century Crusading Window', p. 22 and n. 96.

127 Petrus Venerabilis, *Letters*, no. 164, p. 397; Berry, 'Peter the Venerable', p. 160.

128 See above, n. 61.

129 Petrus Venerabilis, *Letters*, no. 166, pp. 399–400; Berry, p. 160. See also Peter's letter to King Louis VII of France. Petrus Venerabilis, *Letters*, no. 130, pp. 327–30; Berry, p. 148. The letters to Bernard of Clairvaux and to Suger share common topics with Eugenius III's 'Quantum praedecessores'. See Eugenius III, 'Kreuzzugsbulle', pp. 302–5.

A comparison between the attitudes of Bernard of Clairvaux and Peter the Venerable to the Earthly Jerusalem shows that both were opposed to monastic pilgrimages, though Peter less vehemently so than Bernard. Both believed in the importance of the Earthly Jerusalem in the laity's plan for salvation. Both fully recognized the sacred quality of the Earthly Jerusalem. Peter the Venerable, it seems, unlike Bernard of Clairvaux, was scarcely prepared to acknowledge the earthly city as the image of the celestial one. He was aware of the sanctity of Jerusalem and especially of the Holy Sepulchre. However, whereas for Bernard of Clairvaux, Jerusalem was a sacrament, for Peter of Cluny it was one of the holy places, and even the most holy of all by virtue of the presence in it of the Holy Sepulchre.[130]

Not only the leaders but also the members of the monastic orders themselves considered the undertaking of the Jerusalem pilgrimage by fellow monks as the temptation of the devil. Writing in 1122–1124 to Odo, abbot of Marmoutier, who wished to go for his second pilgrimage to Jerusalem, Geoffrey of Vendôme claimed it as an act of transgression against the monastic profession. He argued that it was recommended to laity but forbidden to monks by the Apostolic See and that Pope Urban II had encouraged the laity to go to Jerusalem but had forbidden it to the monks. When Benedict of Nursia referred to 'monk–pilgrims' (*peregrini monachi*) he meant those who renounced the world and devoted themselves to the service of the Lord. He spoke of those whose whole profession was in a fixed place, believing they should carry the Cross of the Lord, following him in their profession and not going on pilgrimage to seek out strange sepulchres. In Geoffrey's opinion the pilgrimage is an act of Satan, whereas the true Promised Land is to be found within the monastic profession.[131] The Carthusian Guido I, fifth prior of La Grande Chartreuse, urged monks to 'let others go to Jerusalem, and you [go] to humility and patience. For you thus go outside the world; he goes within it.'[132] It follows that within the general tendency to interiorize and spiritualize all aspects of monastic life, the *peregrinatio* became a spiritual rather than physical ideal; the true pilgrimage of the monk was one inside his soul. A case in point is a vision experienced in 1133 by Gunthelm, a novice in a Cistercian monastery. Gunthelm, who was famed for his physical strength, longed to utilize it in God's service by fighting His enemies in the Holy Land. He asked his abbot for guidance, but the latter advised him that if he wished to visit Jerusalem, he should approach the Heavenly City, where he would be more than just a temporal citizen. Gunthelm heeded the abbot's advice and took his vows, but the desire to go to Jerusalem – the 'devil's temptations' – continued to torment him. Finally, Gunthelm was 'saved' by

130 See above, n. 118.

131 Geoffroy de Vendôme, *Epistulae* in *Oeuvres*, ed. and trans. G. Giordanengo, Turnhout 1996, no. 196, pp. 480–83.

132 Guido de la Chartreuse, *Meditationes*, ed. A. Wilmart, Paris 1936 (*Études de philosophie médiévale*, 22), no. 262, pp. 111, 232; Constable, *The Reformation*, p. 267.

a vision: the Virgin Mary appeared to him and asked him if he was a constant dweller in her house; Gunthelm answered that he would never forsake her service.[133]

It is as the act of the devil that pilgrimage to Jerusalem appears in the writings of the Abbess Hildegard of Bingen. Writing to an unknown abbess, Hildegard argued that the pilgrimage which this abbess wished to undertake was 'the act of the Devil's deceit'.[134] Another twelfth-century Benedictine and visionary, Elizabeth of Schönau, expressed a similar view while writing to a monk from Odenheim, who wrote to her about his intention to go on pilgrimage to Jerusalem. Elizabeth's reply is unequivocal. She contrasted evildoers who were required to travel to Jerusalem to seek penance with monks who are sons of light, who need only to seek God in their hearts. She warned the monk against this dangerous seduction and then proceeded to remind him that the daily life in the abbey, not the exotic distraction of the pilgrimage, should be the focus of his concerns.[135] It follows thus that for both Hildegard of Bingen and Elizabeth of Schönau there was only one Jerusalem as far as the members of the monastic orders were concerned – the Heavenly one. Hildegard of Bingen accordingly described nuns as being permanently 'on their journey to the Heavenly Jerusalem'; the Heavenly Jerusalem is for her 'the glory and honour of those who have shed their blood for the love of virginity or in radiant humility preserved their virginity for the sake of Christ'. One of her most beautiful poems is 'O Jerusalem', inspired by Revelation 21:11, 12, 19 and composed for the dedication of the abbey of Rupertsberg's church in 1152. It is an invocation of the Heavenly Jerusalem's other worldly splendour accompanied by a vision of St Rupert as its citizen, as one of its 'living stones'.[136]

Pilgrimage to Jerusalem is described in similar terms by an anonymous Benedictine from southern Germany in his 'Book on the penitence and temptations of monks' (*Liber de penitentia and tentationibus religiosorum*). Writing during the Third Crusade in 1189–90, the author compared pilgrimage to Jerusalem with painting in churches. Both pilgrimage and pictures were, he claimed, designed for simple-minded men (*idiotis*) but not for the literate (*litteratis*). There was no need to go on pilgrimage in order to pray in Jerusalem. The true Jerusalem was not the Earthly City but the city above, the 'vision of peace' (*visio pacis*). A monk did not have to go to Jerusalem – because he was already in Jerusalem: the monastery was his Holy City; the altar was his

---

133 G. Constable, 'The Vision of Gunthelm and other visions attributed to Peter the Venerable', *Revue Bénédictine*, 66 (1956), pp. 102–3, 106.

134 Hildegard von Bingen, *Epistolae*, *PL*, 197, cols 329–30. Eng. trans. Dronke, *Women Writers of the Middle Ages*, Cambridge 1984, pp. 186–7.

135 Elizabeth von Schönau, *Die Visionen der Elisabeth und die Aebte Ekbert und Emecho von Schönau*, ed. F.W.E. Roth, Brünn 1884, p. 143; Clark, *Elizabeth of Schönau*, p. 127.

136 Hildegard von Bingen, *Scivias*, Eng. trans. C. Hart and J. Bishop, New York 1990, II, V, 6, p. 205.

Bethlehem; his cell was his Bethany; fear of the Last Judgement was his Valley of Jehoshaphat; his misericord in the choir was his Mount of Olives, and his baptistery was his Jordan River. Thus, for this author, the entire significance of Jerusalem's holy places existed inside the monastery. Its residents, therefore, had no need of visiting the holy places of Jerusalem across the sea.[137] The *mot juste* of the entire argumentation is a line from Horace's Epistles: 'those who run across the sea exchange climates but not their souls'.[138] The same anonymous author rejected not only monastic pilgrimage but even Earthly Jerusalem in general, as the gateway of salvation for knighthood. Citing Christ's words to Peter: 'Put your sword into its sheath' (John 18:11), and also those attributed to St Martin: 'I am a knight of Christ, not permitted to fight', he argued that a knight of Christ should, if he chooses to fight, do this like Moses – by praying: 'it is not fighting with a sword for Jerusalem that grants salvation, but being at peace in Jerusalem which is the vision of peace'.[139]

The preference for the celestial Jerusalem produced, in monastic circles, a number of treatises eulogizing the glory of Heavenly Jerusalem, drawing upon the Book of Revelation and John Fécamp (d. 1078). Peter of Celle described monks as the citizens of Heavenly Jerusalem *par excellence* and saw monastic life as an anticipation of life in the celestial city. Similarly, Hugh of Fouilloy, in his 'On the Cloister of the Soul' (*De Claustro animae*), described, in the middle years of the twelfth century, the Heavenly City through an analogy with monastic life.[140] In that case the author modelled the monastic life upon the court of heaven. The man-made cloister resembles the God-made 'cloister' in Jerusalem. God the Father is the great abbot and Christ is the great prior. This probably evolved from a tendency, which developed in the eleventh, but became popular in the twelfth century, to concentrate the paradisiacal image on the cloister rather than on the

---

137 *Liber de poenitentia*, cols 891–4. For the treatise see Schmugge in Ralph Niger, *De re militari*, pp. 71–2. For a similar opinion see Thibaut d'Etampes, *Improperium cuiusdam in monachos*, ed. R. Foreville et J. Leclercq, *Studia Anselmiana*, 41 (1957), pp. 92–111; *Annales Corbeienses, MGH.SS*, 3, pp. 13–14. For the importance of pictures for *idiotes* see above, ch. 5, pp. 74–5.

138 *Liber de poenitentia*, col. 891. This phrase was popular among the critics of monastic pilgrimages and crusades. See Petrus Venerabilis, *Letters*, no. 80, p. 216; *Annales Stadenses*, p. 343. Master Isenbold of St Paul at Halberstadt, writing during the Second Crusade to the monk Elvingus of Corbie, praised the latter's desire to visit Jerusalem, but at the same time reminded him that his vocation was Celestial Jerusalem, 'which is our mother and which is built in heavens from living stones', *Annales Corbeienses*, pp. 13–14.

139 *Liber de poenitentia*, cols 893–4.

140 Hugo de Folieto, cols 1167–72; E. Lamirande, 'Jérusalem céleste', *Dictionnaire de Spiritualité*, 1, cols 950–55; G. Constable, 'Renewal and Reform in Religious Life: Concepts and Realities', *Renaissance and Renewal in the Twelfth Century*, ed. R.L. Benson *et al.*, Oxford 1977, pp. 48–51 ; T. Renna, 'The Idea of Jerusalem: Monastic to Scholastic', *From Cloister to Classroom. Monastic and Scholastic Approaches to Truth*, ed. E. Rozanne Elder, Kalamazoo 1989 (*Cistercian Studies Series*, 90), pp. 99–100.

Church as a whole, and to see the monastery as the *hortus conclusus* of the Scriptures. Now Heavenly Jerusalem was often compared to a cloister. The cloister is Heavenly Jerusalem; it is Paradise. Thus Gaucherius, the founder and the first prior of Aureil, strove, according to his *Vita*, 'to build the Heavenly Jerusalem'.[141] Bernard of Clairvaux compared Clairvaux to the Heavenly Jerusalem as the port of salvation.[142] Anselm of Canterbury wrote to Turold, a monk of Bec, on his decision to leave the world claiming that he was directed by divine mercy to 'paradise'.[143] Peter the Venerable tried to attract Hatto, bishop of Trier, to retirement in Cluny by comparing the monastery to Paradise;[144] at the same time Vézelay was called 'the other Jerusalem'.[145]

The Heavenly Jerusalem was often described in the same terms as those applied to Earthly Jerusalem. Bernard of Cluny, for example, wrote in his *De Contemptu mundi*: 'City of Zion, fair city, country of harmony and light, to your joys you are always drawing the pious heart. Blessed Jerusalem, our home ... Golden City of Zion, country of milk, beautiful in your people ...'.[146] An anonymous Benedictine monk praised Heavenly Jerusalem as the 'land which surpasses all other lands, glorious land blessed by God (I Chronicles 13:14), land of milk and honey'. This anonymous author, possibly a disciple of John de Fécamp, described the Heavenly Jerusalem using biblical verses depicting the Earthly Jerusalem.[147] At the same time, however, the Earthly Jerusalem was described in terms traditionally used depicting the Heavenly one. Thus, Orderic Vitalis described the Earthly Jerusalem as the mother of the believers, an image traditionally used (following Paul in Galatians 4:25–26) for the Heavenly one. Orderic used the image while describing the siege of Jerusalem by the armies of the First Crusade: 'The Christians invested Jerusalem on 6 June, not as stepsons would a stepmother, but as sons would a mother. As friends and sons they surrounded her ... not to deprive her of freedom, but to free her from captivity.'[148] Making use of Galatians 4:25–26, William of Tyre presented Jerusalem on the eve of its conquest by the First Crusade as 'the mother of salvation, ... now forcibly held by a people without God, the son of the Egyptian handmaiden...'.[149] Bernard of Clairvaux also used a very literal interpretation of Jerusalem's image as mother. In his sixth sermon on Christmas Eve he argued that although somehow the Earthly Jerusalem was the mother of

[141] *Monumenta Vizeliacensia*, p. 540.

[142] Bernardus Claraevallensis, *Opera*, VII, no. 64, pp. 157–8; Constable, *The Reformation*, pp. 33, 70.

[143] Anselmus Cantuariensis, *Opera Omnia*, V, no. 418, pp. 363–4.

[144] Petrus Venerabilis, *Letters*, no. 86, p. 227.

[145] J. Leclercq, 'Une élévation sur les gloires de Jérusalem', *Mélanges J. Lebreton*, 2 vols, Paris 1952 (*Recherches de science réligieuse*, 40), 2, p. 334.

[146] Bernardus Clunensis, *De contemptu mundi*, pp. 11, 12–16.

[147] Leclercq, 'Une élévation', pp. 330–34.

[148] Orderic Vitalis, vol. 5, p. 157.

[149] *W.T.*, I, 15, p. 132. Eng. trans. I, p. 89.

Christ, she had not treated him with a mother's affection, but thrown him out of the city.[150]

Similar opposition, and consequently prohibitions, made in the course of the twelfth century was directed against the clergy. On the eve of the First Crusade, Urban II urged the clergy to seek their bishops' consent before taking the cross. Gratian ruled in his *Decretum* that a bishop should seek papal permission before accompanying a military expedition. Time and time again canonists and theologians repeated the canonical ruling that clergy could not bear arms.[151] However, this opposition was explained not in religious–spiritual terms but rather through practical necessities. The main reason was the concern for the clergy's parochial duties and the perils to which priests exposed their congregations by going on pilgrimage or a crusade. The clergy were often accused of deserting their congregations in order to obtain remission of their sins, which they could acquire in any case in their own churches through acts of penitence. Adam, abbot of Perseigne, argued (*c.* 1204–1215): 'Christ did not shed his blood for the acquisition of Jerusalem, but rather to win and save souls. And it seems to me that those who ask to acquire His land and in doing so neglect the salvation of souls cloak the business of their own perdition with an appearance of piety.'[152] Still, it is well known that clergy, including bishops, took more than an active part in the crusades, especially in the Third Crusade.[153] A comparison between the context of opposition to clerical pilgrimages and crusading and to similar undertakings by members of the monastic orders shows that in the latter case the main grounds were spiritual, such as the superiority of the monastic vocation and the importance of the concept of stability. Critics of clerical pilgrimages were concerned with the duties of the clergy and the fate of

150 Bernardus Claraevallensis, *Sermo in vigilia Nativitatis Domini*, *Opera*, IV, pp. 243–4.

151 Ivo Carnotensis, *Decretum*, *PL*, 161, cols. 234–5, 877; *idem*, *Panormia*, *PL*, 161, cols 1144–5; Gratianus, *Decretum*, p. II, c. XX, q. IV; dist. V, c. XXXVII; Rolandus Bandinelli, *Summa*, c. XX, q. IV, pp. 73–4; *The Summa Parisiensis*, c. XX, q. IV, p. 198; Innocent III, *Register*, I, no. 69, p. 102; Siberry, *Criticism of Crusading*, pp. 32–3.

152 Adam de Perseigne, 'Correspondance', ed. J. Bouvet, *Archives Historiques du Maine*, 13 (1951), pp. 39–40; Radulfus Niger, pp. 195, 211, 224–5; Siberry, *Criticism of Crusading*, pp. 28–35.

153 For the First Crusade see Riley-Smith, *The First Crusade*, pp. 79–88. For the Second Crusade see Suger, pp. 158–9. For the Third Crusade see *Cronica S. Petri Erfordensis*, *MGH.SS rer.Germ.*, 42, 1880, pp. 195–6; *Historia de Expeditione Friderici Primi*, pp. 14–15, 18–24; Rigord, *Gesta Philippi Augusti*, *RHGF*, 17, p. 25; *Grandes Chroniques de France*, ed. J. Viard, Paris 1920–1953, vol. 6, pp. 168–9; Ambroise, *The Crusade of Richard Lion-Heart*, Eng. trans., M.J. Hubert and J.L. La Monte, New York 1941, p. 36; William of Newburgh, *Historia rerum Anglicarum*, *RS*, 82/1, p. 272; Gerald of Wales [Giraldus Cambrensis], *Expugnatio Hibernica*, ed. A.B. Scott and F.X. Martin, Dublin 1978, p. 207. For numerous examples of both clergymen and monks undertaking crusade or pilgrimage see Constable, 'Medieval Charters', pp. 73–93; Röhricht, *Die Deutschen im Heiligen Lande*, *passim*.

their congregations, rather than with the compatibility of the clerical vocation in participating in pilgrimage and crusading.

A total rejection of the Earthly Jerusalem in favour of the Celestial one, without any specification, was relatively rare in the course of the twelfth century. Among its protagonists were Gerhoh of Reichersberg, Ralph Niger and Henry of Albano. Gerhoh of Reichersberg, who wrote in the 1130s and the early 1140s; described Jerusalem as 'the City of God, Holy City'; in his commentary upon Psalms he described the First Crusade as the fulfilment of Psalms 33 (34): the conquest of Jerusalem is the victory of the standard of Christ and the liberation of the Holy Sepulchre is the greatest victory known in all times, one that should be celebrated with the singing of *Laetare Hierusalem* according to Isaiah 66.[154] However, some twenty years later, in his *Libellus de investigatione Antichristi* (composed in 1161–1162), Gerhoh included an *Inclamatio contra Jerusalem terrenam.* He argued that Jerusalem, which in the past killed its prophets and stoned those who were sent to it (Matthew 23:37), now kills the Christians who are sent there and is thus full of Christian blood. He compares the avarice of the inhabitants of Jerusalem with that of Judas Iscariot. The latter sold the Lord for thirty pieces of silver whereas the inhabitants of Jerusalem caused the death of thousands of Christians; because of their avarice they called upon the Christians of the West to come to the East. The true Jerusalem is the Heavenly Jerusalem; the true Temple of the Lord is not the one on Mount Moriah cursed by Christ – it is the Church.[155]

A similar attitude appears within the reaction to the fall of Jerusalem in 1187. Cardinal Henry of Albano, until 1179 abbot of Clairvaux, preached the Third Crusade on behalf of the papacy in France and Germany. Nevertheless, in the last chapter of his *De Peregrinante Civitate Dei*, composed probably in 1188, he argued, in the Pauline tradition, that the sanctity of the Earthly Jerusalem was one of an inferior status. Only those who do not have the spiritual strength to elevate themselves directly to the heavenly city use the Earthly Jerusalem to be able to perceive, with its aid, the Heavenly Jerusalem which is the true *sancta sanctorum.* Heavenly Jerusalem precedes the Earthly Jerusalem: the latter had fallen because of the loss of the Heavenly Jerusalem, namely, the forsaking of God by the Christians. Only the renewal of faith, the return to God, namely to Heavenly Jerusalem, can bring about the liberation of the earthly city of Jerusalem. Only those who truly and sincerely aim for the Heavenly Jerusalem and those chosen for the Kingdom of Heaven will be able to hold the Earthly Jerusalem.[156]

---

154 Gerhoh Reichersbergensis, *Liber de Simoniaciis, MGH.Ltd*, 3, p. 243; *idem*, *Libellus de Ordine donorum Sancti Spiritus*, *MGH.Ldt*, 3, p. 280; *idem*, *Commentarius in Psalmos (PS XXXIV)*, pp. 429–31. Gerhoh also argued that a person could be healed by looking on a citizen of Heavenly Jerusalem, from whose example he is refreshed as from a fountain. *Idem*, *De aedificio Dei*, *PL*, 194, col. 1305.

155 *Idem*, *Libellus de investigatione Antichristi*, pp. 314–20, 330, 374–8.

156 Henry d'Albano, *De Peregrinante Civitate Dei*, col. 351 and below, ch. 9, p. 178. For the influence of Bernard of Clairvaux on Henry d'Albano see Congar, 'Henri de Marcy', pp. 57–90.

Ralph Niger, who wrote his *De re militari* at the turn of 1187–1188, goes even further in his rejection of the salutary role of Earthly Jerusalem. He argues that a spiritual pilgrimage to the Heavenly Jerusalem was much more important than a physical pilgrimage to the earthly city and should be its precondition; it was more meritorious to follow the way of faith, grace and justice than to undertake any bodily pilgrimage. One who failed to build Jerusalem, the *visio pacis*, within himself would derive no benefit from his journey to its earthly counterpart.[157]

To sum up: the crusader conquest of Jerusalem resulted in the decline in the opposition to Earthly Jerusalem which had dominated early Christianity and the early Middle Ages. This resulted in part from the fact that Jerusalem was now the capital of a Christian kingdom, whose very existence depended on a constant and regular influx of warriors from the West. Clearly, then, one could not preach a crusade to aid Jerusalem and at the same time deny that city's value for personal salvation. In fact, the very opposite occurred. Earthly Jerusalem as the gateway to its celestial counterpart, Heavenly Jerusalem, became the central theme in the crusade propaganda of the period. Even those who, in their theological writings and within the monastic world, opposed Earthly Jerusalem preached in its favour and for its sake, urging the European knighthood to take the cross. These developments, demonstrating as they do the influence of practical political considerations on the spiritual status of Earthly Jerusalem, represent the continuity in the interdependence between the political status of the city and the Christian attitude, so convincingly argued by Joshua Prawer for the early Middle Ages. Whereas the Moslem conquest of Jerusalem (638) had caused a significant change in Christian attitudes, namely, relegation of the earthly city to a secondary position, its occupation during the First Crusade restored to Jerusalem its status as an *Urbild* of the heavenly, cosmic, eternal Jerusalem and, moreover, as a gateway to that celestial city. In other words, Jerusalem enjoyed this status only when ruled by Christians.[158] Not only did the Franks[159] try to rebuild the Earthly Jerusalem as the likeness of that heavenly one, but the former was often described in terms previously used exclusively for its celestial counterpart.

The conquest of Jerusalem intensified the link between the Earthly Jerusalem and the European nobility. Jerusalem was never so greatly extolled as the gateway to its celestial counterpart to salvation as it was during the twelfth century. The nobility, too, were increasingly preoccupied with the pursuit of personal salvation, as if the individual's destiny was the centre of attention, rather than the general

---

[157] Radulfus Niger, *De re militari*, pp. 92–3, 147, 167–70, 179, 182–3, 187–8, 194, 218–19; Siberry, *Criticism of Crusading*, pp. 96–7; G.B. Flahiff, 'Deus non vult: A Critic of the Third Crusade', *Medieval Studies*, 9 (1947), pp. 162–88.

[158] Prawer, 'Jerusalem in Christian and Jewish Perspectives', pp. 739–812; Kühnel, *From the Earthly to the Heavenly Jerusalem*, pp. 113–67 has successfully applied this thesis to the portrayal of Jerusalem in early medieval art. For Jerusalem after 1187 see below, ch. 9, pp. 177–8.

[159] See above, ch. 5, pp. 174–6.

eschatological expectations of the Last Judgement. Not that the wider idea was denied; but the heart of the matter lay in personal deliverance from sin.[160] During the twelfth century the pursuit became a general one, or at least the centre of attention of the European nobility, which focused its personal longing for salvation on Jerusalem.

In the course of the century, one observes an increase in the opposition of the monastic orders both to pilgrimages to Jerusalem, and to the meritorious position of Earthly Jerusalem as far as the members of the monastic orders were concerned. This was a result of a greater insistence upon the ideals of stability and obedience, as well as a shift in monastic moral teaching from external behaviour to internal attitudes. This emphasis on stability and obedience may well have been a reaction to the actual increase in monks participating in pilgrimage following the First Crusade. As a result the monasteries, which before 1099 used to tolerate those of their members who would not refrain from undertaking the journey to Jerusalem, now reacted sharply and resolutely against the ideal, rejecting it as totally unsuitable for monks. In fact it seems as if the rejection of the Earthly Jerusalem in favour of the Heavenly now became one of the self-made characteristics of the monastic orders as a group in face of the laity.

The increasing belief, sometimes verging on obsession, in the salutary value of Earthly Jerusalem as the goal of both pilgrimages and crusades by the laity on one hand, and its total negation by the monastic orders on the other, points to a profound change that occurred during the twelfth century in European society. Now the name 'Jerusalem' no longer held the same meaning for all. It meant one thing to the nobility and another to the monastic orders and to some extent the clergy. Whereas for the nobility the road to salvation led through Earthly Jerusalem, the monastic orders (and to some extent also the clergy) rejected this road altogether. Thus only the final aim remained the same for all – the Heavenly Jerusalem.[161]

---

[160] Morris, *The Discovery*, pp. 148–77.

[161] During the twelfth century, as in the previous centuries, grants given to various ecclesiastical institutions included the 'so-to-say' *formula* of 'ut dominus noster Jesus Christus michi et meis in celesti Jerusalem eterna gaudia retribuat' or: 'ut Dominus noster Jesus Christus nos et nostros in celesti Jerusalem collocare dignetur'. See *The Cartulary of the Knights of St John of Jerusalem in England. Secunda Camera, Essex*, ed. M. Gervers, Oxford 1982 (*Records of Social and Economic History. New Series*, 6), n. 285, p. 175 (*c.* 1165–80); no. 343; p. 201 (before 1180). See also *ibid.* no. 353, pp. 205–6 (for 1153).

Chapter 8

Jerusalem – Centre of the World and Scene of the Last Days

'City of the Old and New Testament', 'Inheritance of Christ', 'Gateway to the Heavenly City' – Jerusalem was also considered, in the twelfth century, to be the geographical centre of the world and the future site of the events of the Last Days. Both ideas, which expressed the centrality of Jerusalem in Christian spirituality, were already current in the early Middle Ages. In the twelfth century, however, the theological concept of Jerusalem as the centre of the world became almost universal and was translated into real, geographical and cartographical terms. One may say that it was in the twelfth century that Jerusalem became known as the centre in both *sensus allegoricus* as well as *sensus literalis*.[1] Similarly, Jerusalem, as the scene of the End of the World, became a regular and constant feature of twelfth-century eschatology.

The Christian idea of Jerusalem as the centre of the world derives from the Old Testament (Ezekiel 38:12; Psalms 74:12) and a Midrashic interpretation of Ezekiel: Eretz Israel (Palestine) is the navel of the world, Jerusalem is the centre of Eretz Israel, the Temple is the centre of Jerusalem and the centre of the Temple is the foundation rock, from which the world was created.[2] As early as the third century, however, Christianity, while adopting the centrality of Jerusalem from Judaism, transferred the centre of Jerusalem from the Temple Mount to Calvary. According to Victorinus of Pettau (d. *c.* 303), 'Calvary is the place which is thought of as the centre of the world (*medius terrae*)'.[3] In the fourth century, Cyril of Jerusalem also argued that Calvary was the centre of the world: 'This is not my idea but the prophet's: "He is working salvation in the midst of the earth" (Psalms 74:12).'[4] Jerome described Calvary as the 'navel of the world' (*umbilicus terrae*), associating two passages in Ezekiel: 'Thus says the Lord God: This is Jerusalem; I have set her in the centre of the nations, with countries round her' (Ezekiel 5:5) and '... those who dwell at the centre of earth' (*ibid.*, 38:12) with Psalms 74:12. Therefore, he declared that Jerusalem was situated at the centre of the world

---

1 D.R. French, 'Journeys to the Center of the Earth: Medieval and Renaissance Pilgrimages to Mount Calvary', *Journeys Towards God: Pilgrimage and Crusade*, ed. Barbara N. Sargent-Baur, Kalamazoo 1992, pp. 45–81.

2 *Midrash tanhuma*, Kedoshim 10, p. 78; Friedman, 'The City of the Kings', p. 207.

3 *Victorinus Petavionesis*, in *PL*, 5, col. 294.

4 Cyrillus in *Enchiridion Locorum Sanctorum*, p. 624.

(*medium mundi*) and is, according to Ezekiel, the 'navel of the world'. It is the place where, according to Psalms 74:12, God worked salvation in the midst of the earth (*in medium terrae*).[5] Isidore of Seville, in his influential *Etymologiae*, also had Jerusalem placed, if not explicitly as the centre of the world, at least implicitly as the centre of Judaea.[6]

In the fourth century, Christianity, taking over the biblical–Jewish image of the world, transferred the centre of the world from the old Jewish centre, the Temple upon Mount Moriah, to its own temple, namely the Constantinian buildings on the sites of Calvary and the Lord's Sepulchre. This was done as part of a 'Christianization' of Old Testament traditions of the Temple, many of which were transferred, in the fourth and fifth centuries, from Old Jerusalem to the 'New Jerusalem' of Constantine.[7] Around the middle of the sixth century, the so-called *Breviarius de Hierosolyma* provided a full, well-elaborated catalogue of such 'Christianized' Jewish traditions. Visitors to Calvary were shown the horn from which King David and Solomon were anointed, the ring with which King Solomon sealed the demons, the place where Adam was created, the site where Abraham offered his son Isaac in sacrifice – and the centre of the world.[8] In the famous Madaba mosaic (*c.* 565) Jerusalem was placed in the centre of Palestine. This map emphasized the centrality of Constantine's buildings, locating them in the heart of the city, exactly in the middle of the *cardo maximus*.[9]

For two centuries, the inhabitants of Jerusalem, such as its patriarch, Sophronius, and its pilgrims, saw Calvary as the centre of the world.[10] Between *c.* 670 and 870, however, the tradition 'migrated' once again. The pilgrim Bernard the Monk, who stayed in Jerusalem in *c.* 870, placed the centre of the world not at Calvary but in the unroofed court between the four churches – Calvary, the Church of the Resurrection, the Church of the Invention of the Holy Cross and the Chapel of St Mary.[11]

In the twelfth century, the idea of Jerusalem as the centre of the world became almost universal. It was now held not only by theologians and self-styled experts on the 'Holy Geography' of Palestine, such as the authors of descriptions of the

---

5 Hieronymus, *Commentatorium in Hiezechielem*, pp. 55–6. It should be pointed out that Jerusalem was not shown as the centre of the world in the two maps attributed to Jerome; moreover, neither map marks the centre of Jerusalem; the city was represented by a circle without centre. A.D. van den Brinken, 'Mappa mundi und Chronographia. Studien zur *imago mundi* des abendländischen Mittelalters', *DA*, 24 (1968), pp. 137–9, 173.

6 Isidorus Hispalensis, l. XIV, III, 21.

7 See above, ch. 6, pp. 91–5.

8 *Breviarius de Hierosolyma*, *CCSL*, 175, p. 110; Prawer, 'Jerusalem in Christian and Jewish Perspectives', pp. 767–8.

9 M. Avi-Yonah, *The Madaba Map*, Jerusalem 1954, *passim*; Kühnel, *From the Earthly to the Heavenly Jerusalem*, pp. 89–93.

10 Sophronius in *Enchiridion Locorum Sanctorum*, p. 640; Arculfus, pp. 194–9 and see also Beda, *De Locis Sanctis*, p. 258; Eucherius, p. 243.

11 Bernardus Monachus, p. 93.

Holy Land, but by Europeans in general. This may have been a result of the literary output of the First Crusade. For example, Robert the Monk's *Historia Iherosolimitana* described Jerusalem as 'the navel of the world ... a royal city, placed at the centre of the world'.[12] In the course of the century similar sentiments were voiced by theologians like Peter Abelard,[13] leaders of the monastic world such as Peter the Venerable,[14] chroniclers, poets and pilgrims.[15] According to Abelard: 'The soul of the world is found at the middle of the world; consequently Jerusalem, from whence comes salvation, is found at the centre of the world.'[16]

Moreover, in the twelfth century this historical–spiritual concept of Jerusalem as the centre of the world acquired a new permanence, being translated into geographical–cartographical terms. The known world maps (*mappae mundi*) of the twelfth century all placed Jerusalem in the exact 'literal centre of the world'. Beginning with the Byzantine Oxford T-O map, brought back to England or Ireland in 1100, Jerusalem was marked as the centre of the world on *mappae mundi*. This became a kind of law in medieval cartography, as shown, for example by the 'Psalter Map' (*c.* 1250), the Ebstorf map (*c.* 1250) and the Hereford map (*c.* 1283).[17] Jerusalem appeared now for the first time in the Christian West in the centre of the *mappae mundi*. It took the form of a circle (in the Psalter map) or was shaped like a sepulchre-reliquary (as on Ranulf Higden's map of the fourteenth century); on some other maps Jerusalem appeared in the form of a square. This square or circular representation was also typical to the representation of Heavenly Jerusalem in the illustrated Apocalypses of the early Middle Ages, for example, the Beatus Apocalypse and the Apocalypses of Valenciennes and Paris.[18] The fusion between apocalyptic visions of Jerusalem and the topographical

12 Robertus Monachus, p. 729.

13 Petrus Abelardus, *Theologia Christiana*, *PL*, 178, col. 1150. Jerusalem is 'in medium terrae', constituted by God as his *Caput regni* according to Psalms 74:12 and to Isaiah 2:3: 'For out of Zion shall go forth the law and the word of the Lord from Jerusalem.'

14 Petrus Venerabilis, *Sermo in Laudem Sepulchri Domini*, pp. 238–9 and see below, n. 20.

15 Hugo Floriacensis, p. 394; *Continuatio Zwettlensis Altera*, *MGH.SS*, 9, p. 543; Hugo de S. Maria, p. 364; *AASS*, February, vol. III. p. 114 (=*Vita S. Theotonici*); *Monachus Anonymus Scaphusensis*, p. 335; *Innominatus I* in *PPTS*, VI, p. 2; Rorgo Fretellus, p. 31; *Innominatus VI*, p. 426; *Tractatus de locis et statu sancte terre ierosolimitane*, p. 144; *Jerusalem, luge, medio dolor orbis in orbe*, *PL*, 155, cols 1095–8.

16 See above, n.13; French, 'Journeys', p. 62.

17 C.R. Beazley, *The Dawn of Modern Geography*, vol. 2, London 1901, pp. 556; French, 'Journeys', pp. 62–8; Kühnel, 'Geography and Geometry of Jerusalem', pp. 311–16.

18 Kühnel, *From the Earthly to the Heavenly Jerusalem*, pp. 112–65. For the maps of Jerusalem see M. Levy-Rubin, 'Medieval Maps of Jerusalem', *The History of Jerusalem. Crusaders and Ayyubids (1099–1250)*, ed. Joshua Prawer and Haggi Ben-Shammai, Jerusalem 1991, pp. 418–62; M. Levy-Rubin and R. Rubin, 'The Image of the Holy City in Maps and Mapping', *City of the Great King. Jerusalem from David to the Present*, ed. Nitza Rosovsky, Cambridge, Mass. 1996, pp. 352–62.

presentations of Earthly Jerusalem depicted once again the closeness that the Heavenly and the Earthly City enjoyed in the twelfth century.

Another concept that became universal in the course of the twelfth century was that of the Holy Sepulchre as the navel of Jerusalem and, thus, the centre of the world. In the numerous maps of Jerusalem of the twelfth and thirteenth centuries, the centre of the city is not marked.[19] Nevertheless it is obvious from other sources that this was a common conception. Peter the Venerable, in his *Sermo in Laudem Sepulchri Domini*, argued as follows: 'It is obvious that there [at the site of the Sepulchre] is the centre of this world, the place called by the Saviour the heart of the world. Surely the Sepulchre of the Saviour is the heart of the world; it is called the place of the centre of the world, like the city of Jerusalem itself, and the entire land, about which it was said that it is the Promised Land ... Just as the human or animal heart is the centre of their bodies, so is this land, where our fathers received the Christian religion, situated in the middle of the world ... All the land of our Holy Fathers, but particularly the city of Jerusalem ... where is the Holy of Holies and even more so the Lord's Sepulchre, should rightly be called the heart of the land, because there rested the Holiest among the Holy.'[20]

Similarly, most of the Jerusalem pilgrims, with the exception of Theoderich, located the 'navel of the world' close to the Holy Sepulchre. Two anonymous descriptions of Jerusalem placed it in the new crusader cathedral (*Catholicon*) east of the Sepulchre, at the centre of the choir, which they identified as the place where Christ was laid after being taken down from the Cross.[21] According to Daniel and some other pilgrims, the place was now covered by a small building over the vault, on which Christ was represented in mosaic with this description: 'The sole of thy foot serves as a measure for the heaven and for the earth'.[22] John of Würzburg described it as a monument 'formed into the shape of an altar ... beneath which are certain circles on the ground, according to the text: "His salvation hath He wrought in the midst of the earth".'[23] Theoderich located the site of Christ's descent from the Cross in the *Catholicon*, but not that of the centre of the world. Although he did not say so explicitly, he apparently located it in its original place, on Mount Moriah, in the Church of the 'Temple of the Lord' near the Chapel of St James.[24] Thus, though the tradition locating the centre of the world in the *Catholicon* near the Holy Sepulchre gained currency during the twelfth century, there were some,

---

19 See above, n.18.

20 Petrus Venerabilis, *Sermo in Laudem Sepulchri Domini*, pp. 238–9.

21 *Innominatus VI*, p. 101; *Citez de Iherusalem*, p. 204.

22 Daniel, p. 13. See also Saewulf, p. 39; W. Simpson, 'The Middle of the World in the Holy Sepulchre', *Palestine Exploration Fund Quarterly*, 20 (1888), pp. 260–63.

23 Johannes Wirziburgensis, pp. 119–20; *Innominatus VIII*, p. 194. See also the Icelandic pilgrim, Abbot Nicholas, who came to Jerusalem in *c.* August 1133 in Kedar and Westergard-Nielsen, 'Icelanders in the Crusader Kingdom', pp. 204–5.

24 Theodericus, pp. 151, 160–61 and see also Theodericus, *Libellus de Locis Sanctis*, ed. M.L. and W. Bulst, Heidelberg 1976, p. 65, n. 3.

still loyal to Old Testament tradition, who placed it in the old Jewish Temple rather than the new one, the Church of the Holy Sepulchre.

The physical–geographical centrality of Jerusalem derived from its identification as the place of salvation, of the humanity of God made man. But Jerusalem was also identified as the place where the events of the End of the World would take place. Like the idea of Jerusalem as centre of the world, the belief in the special role of Jerusalem in the 'Last Days' originated in the Old Testament and Jewish writings of the Second Temple period. The books of the Old Testament, full of prophecies, reconciliation, the Day of Judgement and the End of Time, associated the Day of Judgement with the Valley of Jehoshaphat. The Hebrew name 'Yehoshafat' namely 'Jehovah judged', identified it topographically as the Valley of Kidron: 'I will gather all the nations and bring them down to the Valley of Jehoshaphat and I will enter into judgment with them there ... Let the nations bestir themselves and come up to the Valley of Jehoshaphat, for there I will sit to judge all the nations round about' (Joel 3:2–12). The 'Mount of Olives which is before Jerusalem on the east' (Zechariah 14:4) was, in Zechariah's vision of the Lord's coming, identified with the Mount of Olives. This mount was already linked by the Jews with the divine presence (*shekhinah*) that was exiled with the nation from the city after the fall of Jerusalem (AD 70). The Mount of Olives came to be connected with the place of the Ascension and was identified as the place of the Second Coming. The nearby Mount Moriah was associated with the belief produced by the New Testament, that the Last Days would include a period in which the Antichrist would rule the world from the Temple in Jerusalem. According to Paul, 'and the man of lawlessness is revealed, the son of perdition who opposes and exalts himself against every so-called god or object of worship so that he takes his seat in the temple of God, proclaiming himself to be God' (2 Thessalonians 2:3–4).[25]

Since the fourth century, the Church had attempted to discredit millenarian eschatology, and advocated an allegorical interpretation of the Scriptures. In the fifth century, Augustine argued that the Revelation to John should be understood as a spiritual allegory: the Millennium had already begun with the birth of Christianity and was fully realized in the Church.[26] His contemporary and friend Jerome stated, in his commentary on the book of Isaiah, that 'to understand the Book of Revelation literally would be "judaising"; yet "to set" it out spiritually, as it is written, it seems that we shall be contradicting the opinions of many ancients. From among the Latins: Tertullian, Victorinus, Lactantius; from among the

25 Prawer, 'Jerusalem in Christian and Jewish Perspectives', pp. 779–80. For Earthly Jerusalem as the seat of the Antichrist see Victorinus Petavionensis, *Commentarii in Apocalypsim*, pp. 128–9; Hieronymus, *De Antichristo*, pp. 914–44; Sulpicius Severus, *Vita Sancti Martini*, *CSEL*, 1, p. 133; Beda, *Sibilinorum verborum interpretatio*, cols 1185–6; Honorius Augustodunensis, *Elucidarium*, *PL*, 172, col. 1163. See also Bousset, *Der Antichrist*, pp. 104–59. Augustine was, however, sceptical. See Konrad, p. 538 and n. 73.

26 Bredero, 'Jérusalem dans l'Occident', pp. 260–62.

Greeks, bypassing others, I mention Irenaeus, bishop of Lyons, against whom the most eloquent Dionysius, bishop of Alexandria, wrote an elegant treatise, ridiculing the millennial fable and the golden and bejewelled restoration of the Temple in the lands of Jerusalem ...'. In Jerome's opinion, the Old Testament prophecies were already fulfilled with the coming of Christ.[27]

Such opinions were to dominate the official position of the Church in centuries to come. Yet the message of the Holy Scriptures, understood literally, was too attractive to be abandoned. The ideas held such enormous fascination that no official condemnation could suppress them, and they appear again and again – and by no means exclusively in the circles 'of the unprivileged, the oppressed, the disoriented and the unbalanced'.[28] People in the early Middle Ages lived in constant consciousness of the End of Time, and the phrase 'the Lord is at hand' was almost a reality, especially at certain times. There is evidence that, especially in certain periods of high eschatological tension – the eleventh century, for example – such beliefs as the Millennium, the Last Days as taking place in Jerusalem, including the figure of the Last Emperor and Antichrist, loomed large among the clergy, including its highest ranks;[29] at any rate, our written evidence came from churchmen. Building on the foundations of Scriptures, in the face of the opposition of the Fathers of the Church, an elaborate Christian picture of the End of the World emerged; and its climax – the cosmic struggle between Good (Christ) and Evil (Antichrist) as well as the Last Judgement – was located in Jerusalem.

Most influential in the formation of this picture was Pseudo-Methodius (*c.* 660–680).[30] The so-called *Pseudo-Methodius* prophecy, attributed to Methodius of Patara, a bishop martyred in the early fourth century, was composed in its Latin version in *c.* 660–680, either in Palestine or Mesopotamia.[31] After the Book of Daniel and the Revelation to John, the *Revelation of Pseudo-Methodius* became one of the most widespread of medieval apocalyptic texts; in its Latin version it has survived in some one hundred and ninety manuscripts, twenty-one of these pre-dating the twelfth century.[32] The *Revelation*, the earliest surviving witness to the legend of the Last World Emperor, was probably responsible for creating the abdication scene, which became most deeply rooted in Christian eschatology,

---

27 Hieronymus, *Commentatorium in Isaiam*, pp. 740–41; Prawer, 'Jerusalem', pp. 757–8.

28 For this opinion see Cohn, *Pursuit of the Millennium*, p. 30 and *passim* who obviously follows Alphandéry and Dupront, *La chrétienté*, I, pp. 57–79.

29 For a view discrediting the eleventh century as an age of high apocalyptic expectations see McGinn, *Visions of the End*, p. 88. For a different opinion see H. Focillon, *L'An Mil*, Paris 1952, pp. 41–64.

30 E. Sackur, *Sibyllinische Texte und Forschungen*, Halle 1898, pp. 117–87, esp. pp. 185–6; McGinn, *Visions of the End*, pp. 43–50.

31 See on this subject Alexander, 'Byzantium', pp. 53–61; McGinn, *Visions of the End*, p. 70.

32 See *ibid.*, p. 70 and n. 1.

locating it at Golgotha. Its vision of a coming emperor who would defeat the Moslems, the enemies of Christ, and restore Roman glory, incorporated the rise of Islam into the Christian apocalyptic scheme of history. According to the *Revelation*: 'When the Son of Perdition has arisen, the King of the Romans will ascend Golgotha, upon which the wood of the Holy Cross is fixed, in the place where the Lord underwent death for us. The king will take the crown from his head and place it on the cross and, stretching out his hands to heaven, will hand over the kingdom of the Christian to God the Father. The cross and the crown of the king will be taken up together to heaven ... When the Cross has been lifted up on high to heaven, the King of the Romans will directly give up his spirit. Then every principality and power will be destroyed so that the Son of Perdition may be manifest ...'[33]

In the tenth century, the figure of the Last Emperor became associated with Otto III, who had wished to go to Jerusalem and to become a monk there.[34] The Antichrist, too, cut a no less tangible figure. In 991, the Synod of Rheims declared that Pope John XV was the Antichrist, 'who sits in the Temple of God pretending to be God'.[35] Prominent in contemporary prophecies of the Last Days were the words 'and his dwellings shall be glorious',[36] which constantly reminded believers of the Last Judgement. Pilgrims came to the Holy City identifying it as the site of that event, often believing that the End was at hand. Expectations of the End, including the Second Coming and the Last Judgement, though always present, gained a new ascendancy at the turn of the millennium, in the year 1000. In the eleventh century, Jerusalem became a focus for apocalyptic expectations, which in turn became a powerful impetus for pilgrimage to the Holy City. In 1033, as reported by Ralph Glaber, an unprecedented number of people of all classes, from all over the world, gathered in Jerusalem, expecting the advent of the Antichrist.[37] About thirty years later, in 1064, a large company of pilgrims departed from Germany, bound for the Holy Sepulchre. Among the participants in this pilgrimage were several prominent German bishops and a large number of clergymen. It was inspired by the then popular opinion that the Day of Judgement would take place at Easter 1065. This opinion was based on the fact that the Easter of that year fell on 27 March, the date ascribed in medieval calendars to the Resurrection of Christ.[38]

The period around 1000 also produced treatises that embellished the end-drama with a brief moment of present-day triumph. The most influential one was the treatise 'On the Antichrist' of Adso of Montier-en-Der, disseminated at the

---

33 Sackur, pp. 89–94; McGinn, *Visions of the End*, pp. 75–6.

34 Konrad, 'Das himmlische ... Jerusalem', pp. 537–8.

35 *Gerberti acta concilii Remensis*, *MGH.SS*, 3, p. 672.

36 The Vulgate version of Isaiah 11:10 is: 'et erit sepulchrum eius gloriosum'.

37 Radulfus Glaber, p. 109.

38 *Vita Altmanni*, p. 230; Joranson, 'The Great German Pilgrimage', p. 12.

beginning of the eleventh century by 'Albuin' of Gonze and later by Benzo of Alba.[39] Adso, abbot of Montier-en-Der, died in 992 on a pilgrimage to Jerusalem believing, like many at the time,[40] that pilgrimage to terrestrial Jerusalem would gain him access to 'our mother, the celestial Zion'.[41] Some thirty years earlier (*c.* 950), he had written his own version of the Last Days, actually a version of Pseudo-Methodius' *Revelation*, dedicating it to Gerberga, sister of Otto I and wife of Louis of Outremer.[42] Included in this work, besides a description of the Antichrist's rule in Jerusalem and its climax – the final battle between Christ and Antichrist on the Mount of Olives – was the story of the Last Emperor of the West who would come to Jerusalem, ascend the Mount of Olives and there deposit his sceptre and crown. In addition, it described a period following the slaying of the Antichrist and before the Day of Judgement, granted to the elect to do penance.[43]

Other apocalyptic works of the time clearly reveal their authors' interest in contemporary events. According to the text of the Cumaean Sibyl, composed *c.* 1090 by a person involved in the Investiture Contest, the Last Emperor, before coming to Jerusalem, would destroy Rome, that is, Babylon, and fight the Saracens; his victory would be followed by a period of peace until the time of the Antichrist. Then the Last Emperor would come to Jerusalem and place his crown on the Cross at Calvary, and subsequently Christ would come to the Last Judgement.[44]

Benzo of Alba, also involved in the Investiture Contest on behalf of Henry IV, exhorted (*c.* 1085) the emperor to go to the Holy Sepulchre in Jerusalem and be crowned there again. Familiar with Pseudo-Methodius, Benzo claimed that, after conquering Apulia and Calabria, the emperor would lead an expedition to Jerusalem and, having rescued the Sepulchre and the other sanctuaries of the Lord, be crowned, to the praise and glory of the One who lives for ever and ever. 'Babylon in amazement will come to Zion desiring to lick the dust of His feet. Then will be fulfilled what is written: "And his sepulchre will be glorious" (Isaiah 11:10) ...'[45] Thus Benzo, like Pseudo-Methodius (but unlike Adso), linked the Last

---

39 *De Antichristo. Quomodo Nasci Debeat auctore Albuini*, pp. 68–74; Benzo Albensis, pp. 605–41; Erdmann, 'Endkaiserglauben', pp. 403–8; Riley-Smith, *The First Crusaders*, pp. 26, 29–30.

40 See above, ch. 7, p. 114.

41 *Widrici Vita S. Gerardi*, p. 488.

42 Alexander, 'Byzantium', pp. 52–3, 61.

43 Adso Dervensis, pp. 20–30. For the popularity of Adso's treatise see Konrad, 'Das himmlische ... Jerusalem', p. 537 and n. 69. According D. Verhelst, 'Les texts eschatologiques dans *Le Liber Floridus*', in *The Use and Abuse of Eschatology in the Middle Ages*, ed. Werner Verbeke *et al.*, Leuven 1988 (in his introduction to Adso, pp. 3–18), the treatise has survived in 171 manuscripts. Adso was also influenced by Haymo of Halberstadt's Commentary on the Second Epistle to the Thessalonians. See Haymo Halberstatensis, *Exposito in Epistolas S. Pauli*, *PL*, 117, cols 777–84.

44 See the text in Erdmann, 'Endkaiserglauben', pp. 396–8.

45 Benzo Albensis, p. 605 but see the version of Erdmann, 'Endkaiserglauben', pp. 403–8; McGinn, *Visions of the End*, pp. 90–91.

Emperor's journey to Jerusalem with the Holy Sepulchre. It follows thus that on the eve of the First Crusade, the liberation of the Holy Sepulchre from the Moslems was clearly seen as part of the Last Days, which many believed to be imminent. For example, in 1088, exhorting Robert I of Flanders to liberate the Holy Sepulchre, Emperor Alexius Comnenus referred to the coming of the Antichrist and the Last Judgement.[46] Moreover, in the period immediately preceding the First Crusade an increasing number of people identified an expedition for the recovery of the Holy Sepulchre with the Last Emperor's journey to Jerusalem.[47]

The First Crusade showed the importance of the place that Jerusalem held in the contemporary vision of the Last Days. The crusade moved in a climate of messianic expectations. Moreover, though the apocalyptic factor was by no means the prime mover behind the crusaders' conquest of Jerusalem, apocalypticism cannot be excluded from the variety of motives entertained by the participants of the First Crusade; after all, Jerusalem was the apocalyptic city *par excellence*.[48] Some participants of the crusade in fact considered the conquest a necessary precondition for the beginning of the Events of the End.[49] On the eve of the crusade, as reported by Ekkehard of Aura, popular preachers recounted what he called 'fables', namely that Charlemagne would rise again from the dead and lead the crusade; Charlemagne was thus identified with the Last Emperor.[50] Influenced by the Last Emperor's prophecy that the Jews would be converted to Christianity on the eve of the Second Coming, the crusaders persecuted the Jews of the Rhineland in the early summer of 1096, forcing them to convert in fulfilment of the prophecy.[51] During the crusade the Moslem enemy was regarded as attendants of the Antichrist.[52]

These apocalyptic expectations characterized the so-called armies of the poor in the crusade. These included peasants, petty knights and all sorts of servants, referred to by chroniclers of the crusade as 'paupers' or 'paupers of God' (*pauperes; plebs pauperum Dei*). They believed that the time of the Second Coming was close at hand, that St Peter had liberated them from their poverty and

---

46 See n. 45 above.

47 P. Munz, *Frederick Barbarossa*, London 1969, p. 376. For plans for the recovery of the Holy Sepulchre before the First Crusade see above, ch. 1, pp. 9–11.

48 Alphandéry and Dupront, *La Chrétienté*, *passim*, present the crusades as motivated by strong apocalyptic expectations. This, as other studies have shown, is an exaggeration. See Erdmann, 'Endkaiserglauben', pp. 382–414; McGinn, *Visions of the End*, pp. 88–9 and more recently Raedts, 'Jerusalem: Purpose of History', pp. 31–40.

49 See above, ch. 2, p. 24.

50 Ekkehardus, *Hierosolymita*, p. 19. For this identification see also R. Folz, *L'idée d'Empire en Occident du Ve au XVe siècle*, Paris 1953, pp. 128–32.

51 Riley-Smith, *The First Crusade*, pp. 34–6. For the persecutions of 1096, known in Jewish history as 'Gezerat Tatnu', see *ibid.*, pp. 49–57.

52 Robertus Monachus, pp. 828, 874, 876; Radulfus Cadomensis, p. 695.

would lead them to the land of milk and honey. They knew that the road would be hard; that the enemy they were about to fight was the Antichrist and his legions; their goal was 'Jerusalem the Golden'. The atmosphere among these crusaders was a mixture of popular enthusiasm, a quest for a better life in the East and expectations of the Last Days and the Second Coming.[53] They were also strongly influenced by the evangelical idea that the poor were the elected and that they alone would inherit the Kingdom of God. In the course of the crusade a unique ideology of poverty emerged, portraying the poor as the true and real heroes of the drama of the Last Days. A major bearer of this ideology was a well-organized body of poor who, calling themselves 'Tafurs', were ruled by a king alleged to have been a Norman knight who had no lord and had been reduced to the ranks. Motivated by their hope for the establishment of a millennial kingdom in Jerusalem, they believed that they should carry only defensive arms, as it had already been decreed by providence which of the crusaders would be killed and ascend to Heaven, and who among them would reach the Earthly Jerusalem and thus be among the first to enter the Heavenly Kingdom after the Last Judgement.[54]

The apocalyptic state of mind was not, however, confined solely to the poor. It was also to be found among the clergy who had joined the crusade – witness, for example, the letter of Patriarch Daimbert, Godfrey of Bouillon and Raimond of St Gilles to the pope in September 1099. They wrote 'that after we break the power of the Saracens and that of the Devil, the Kingdom of Christ and of the Church will stretch from sea to sea'.[55] Indeed, apocalyptic expectations reached their apex following the conquest of Jerusalem. Some among the poor were opposed to the election of a ruler for the new kingdom claiming that the Kingdom of Heaven was close at hand. The Provençal Raimond of Aguilers, who was close to the poor during the crusade, describes their arguments as follows: 'One must not choose a king where the Lord suffered and was crowned. Because he [the King] will say in his heart: "I sit on the throne of David and hold his Kingdom", he will actually alienate himself from the faith and virtue of David. [And therefore] perhaps God may ruin him and will be made angry with the land and the people. This is the reason the prophet said: "When the Holy of Holies comes the unction will cease"; and it is clear to all nations that he has come.'[56]

---

53 Baldricus Dolensis, p. 16; Guibertus Novigentis, pp. 149–50; Runciman, *A History*, vol. I, pp. 113–15.

54 Raimond d'Aguilers, pp. 99–104; Petrus Tudebode, pp. 114–15. The meaning of the world 'Tafur' is not clear; some interpreted it as 'poor'; others as 'king' or 'gypsy'. See L.A.M. Sumberg, 'The "Tafurs" and the First Crusade', *Medieval Studies*, 2 (1959), pp. 224–46. On the poor in the First Crusade see Alphandéry and Dupront, *La chrétienté*, pp. 64–135; Schein, 'Die Kreuzzüge', pp. 126–30.

55 *Kreuzzugsbriefe*, n. XVIII, pp. 171–2.

56 Raimond d'Aguilers, p. 143. The verse cited by Raimond 'when the Holy of Holies comes, the unction will cease' is based on Daniel 9:24: 'Seventy weeks of years are decreed concerning your people and your holy city, to finish the transgression, to put an end to sin,

Europeans describing the conquest of Jerusalem regarded it as the realization of scriptural prophecies,[57] as well as part of the events of the Last Days and a precondition for the Second Coming. According to Guibert of Nogent, who was evidently inspired by Adso of Montier-en-Der, Pope Urban II argued at Clermont that 'the end of the world is already near, although the gentiles have not been converted to the Lord: according to the apostle Paul there must be a revolt from the faith (2 Thessalonians 2:3). But first, before the coming of Antichrist, a renewal of the Christian empire in those regions is necessary according to the prophecies, either through you or by means of those whom God chooses, so that the head of all the evil ones, who will sit on the throne of the Kingdom in that place, should discover some support of the faith against which he may fight ... Think, I beseech you, of the hearts giving birth to such rejoicing when we see the Holy City revived by your assistance and the prophetic, nay, rather, divine, predications fulfilled in our own times. May what the same Lord said to the Church stir your memory. "I will bring", he said, "thy seed from the East and gather these from the West" (Isaiah 43:5).' But the Antichrist could hardly appear in Jerusalem unless it was occupied by Christians, so that he could attack them. Guibert therefore presented the conquest of Jerusalem as a prelude to Doomsday, particularly since the liberation of Jerusalem from the Gentiles was associated by Luke with the Last Days: 'Jerusalem will be trodden down by the Gentiles, until the times of the Gentiles are fulfilled' (Luke 21:24). Guibert concluded his description of the conquest of Jerusalem with a declaration that it was God's deed, aimed 'against the fury of the procreation of Antichrist': 'We have no doubt that God did not undertake this conquest merely to liberate one city, but to scatter the seeds that will grow long and far against the madness of the Antichrist.'[58]

Such ideas were certainly in the air, as Ekkehard of Aura, too, was echoing them: 'In the time of the Emperor Henry IV ... according to the prophecies in the Gospels, everywhere nation arose against nation and Kingdom against Kingdom, and there were great earthquakes in diverse places, and pestilences and famines and terrors from heaven and great signs. And because already in all nations the evangelical trumpet was sounding the coming of the Just Judges, the universal Church beheld throughout the whole world the portents in prophetic signs.' Among those 'signs' was one seen by those 'who kept watch feeding horses, namely, the likeness of a city in the air ... and various crowds hurrying to it from different cities, both by horseback and on foot.'[59] This city in the air was obviously the

---

and to atone for iniquity, to bring in everlasting righteousness, to seal both vision and prophet, and to anoint a most holy place.' In the Vulgate this verse has a different meaning. See Prawer, 'Jerusalem in Christian and Jewish Perspectives', p. 747 and n. 14.

57 See above, ch. 2, pp. 24–7.

58 Guibertus Novigentis, pp. 138–9, 193. Eng. trans. Levine, p. 43, 143. See also Riley-Smith, *The First Crusade*, p. 143 and n. 30. This verse of Luke, inspired by Isaiah 63:18 and Daniel 8:13, is reworked in the Epistle to the Romans 11:25 and Revelation 11:2.

59 Ekkehardus, *Hierosolymita*, pp. 12, 18–19; Riley-Smith, *The First Crusade*, p. 33.

Heavenly Jerusalem.[60] Baldric of Dol, Sigebert of Gembloux and Orderic Vitalis, who relied on Baldric's history of the First Crusade for their own accounts, all reported similar signs.[61]

During the period of crusader rule, Jerusalem retained its position as the place of the Last Days, though the high eschatological tension characterizing the period prior to the conquest seems to have subsided. This is attested by the pilgrims' itineraries, by the apocalyptic literature of the period and above all by the reaction of Christendom to the loss of Jerusalem in 1187. Some of the numerous pilgrims who visited Jerusalem came to the city because they believed that the Antichrist was about to appear there and thus to usher in the Last Days. For example, this was the motivation of Norbert of Magdeburg, founder of the Premonstratensian order, who in 1124 'declared himself quite certain that it would be during this present generation that he [the Antichrist] would be revealed'.[62] Pilgrims sometimes mentioned the valley of Jehoshaphat as the site of the Last Judgement. Theoderich declared that the Last Judgement would take place in the 'valley of Ennon' and that simple-minded pilgrims (*simplices peregrini*) used to put up piles of stones marking where they would sit at Doomsday.[63]

At the same time, some intellectuals of the period began to reject such beliefs completely, whereas others supported them. At the beginning of the century, Honorius of Autun, discussing the location of the Last Judgement and whether it was to be in the Valley of Jehoshaphat, declared that 'this valley is this world'; the Last Judgement would take place in the entire world and not in the Valley of Kidron, in Earthly Jerusalem. None the less, he elaborated upon the current descriptions of the Antichrist's rule in Jerusalem: he would rebuild ancient Jerusalem and Jews from all over the world would flock to him and recognize him as God.[64] Bishop Otto of Freising called the identification of the Valley of Jehoshaphat with the Last Judgement 'a pious error'. His eschatology, dominated

---

60 This is already found in a number of ancient authors, among them Tertullianus, pp. 419–22.

61 Baldricus Dolensis, p. 16; Sigebertus Gemblacensis, p. 367; Ordericus Vitalis, vol. 5, p. 8.

62 This information about Norbert derives from a letter of Bernard of Clairvaux (of 1124 or 1128). Bernardus Claraevallensis, *Opera*, VII, n. 56, p. 148. Commenting on Norbert's expectations of the Antichrist, Bernard was sceptical. He thought that the times of the Antichrist would be preceded by an age in which the Church would be persecuted, and this age, in his opinion, had not started yet. For Bernard's eschatology see his *Opera*, I, pp. 243–4; see also *ibid.*, IV, pp. 404–9; B. McGinn, 'Saint Bernard and Eschatology', *Bernard of Clairvaux. Studies Presented to Dom Jean Leclercq*, Washington, DC 1973 (*Cistercian Studies Series*, 23), pp. 161–85.

63 Theodericus, p. 179 and see also *Innominatus I* in *PPTS*, VI, p. 2. This custom persisted through the next centuries. For the late thirteenth century see Ricoldo da Monte Croce, *Liber Peregrinationis*, ed. J.C.M. Laurent, *Peregrinatores Medii Aevi Quatuor*, Leipzig 1864, p. 111.

64 Honorius Augustodunensis, *Elucidarium*, cols 1163–6.

by a confident belief in the near approach of the end of the world, was concerned with the three major topics: Antichrist, the Resurrection and the end of the two cities. Regarding the location of the Last Judgement, Otto argued that: 'some, in accordance with the prophecy which says "In the Valley of Jehoshaphat will I judge them" (Joel 3:2, 12), which they interpret literally, think that the judgement will take place not far from the Holy City in the Valley of Jehoshaphat; that the Lord, when he comes to judge, may come to the place in which he was judged. Hence certain simple-minded persons (*simpliciores*), on visiting the Lord's sepulchre for the purpose of praying, set up in that valley stones or some other mark, as though they were selecting places for themselves against that hour.' This he believed to be 'a pious error', as the Judgement would actually be held in 'whatever place He appoints'.[65]

Others, urging that the Scriptures be studied in their literal and historical sense before attempting an allegorical interpretation, supported the identification of Jerusalem as the place of the Last Events. Thus, in his celebrated and widely diffused work, the *Historia Scholastica*, Peter Comestor, chancellor of the University of Paris (1168–1178/9), fulfilling the wish of his teacher, Hugh of St Victor, to provide a constructive historical, geographical and chronological basis for reading the Scriptures, interpreted Daniel in the most literal way. In his exegesis of Daniel 8:23–26 he stated that the Antichrist will dwell in Jerusalem, between the Dead Sea and the Mediterranean Sea. The 'Holy Mountain' (*ibid.* 9:16) is the Mount of Olives, where the last and decisive battle between Christ and Antichrist will take place.[66] Bernard of Clairvaux seems to have held a similar view. His 'Praise of the New Knighthood' contains a fairly long paragraph about the Valley of Jehoshaphat as a place intended to remind one that in contemplation of 'the richness of divine mercy, one should not forget the fear of judgement'.[67]

As far as apocalyptic visions are concerned, Earthly Jerusalem appeared in the twelfth century mainly in the Sybilline prophecies. Alongside developing forms of apocalyptic literature centring on the life of the Church,[68] the twelfth century witnessed the continuing popularity of the Last Emperor myth. It was circulating in France in 1146, when Louis VII was preparing to depart on the Second Crusade, and again on the eve of the departure of Philip Augustus on the Third Crusade.[69] And, indeed, it seems that the Second and Third Crusades were, like the First, times of

---

65 Otto Frisingensis, *Chronica*, pp. 413–18.

66 Petrus Comestor, *Historia Scholastica*, *PL*, 198, cols 1463–6; Luscombe, 'Peter Comestor', pp. 109–29.

67 Bernardus Claraevallensis, *De Laude*, pp. 227–8.

68 McGinn, *Visions of the End*, pp. 108–16.

69 *Annales S. Iacobi Leodiensis*, *MGH.SS*, 16, p. 641; Otto Frisingensis, *Gesta Friderici Imperatori, Ottonis et Rahewini Gesta Friderici I Imp.*, *MGH.Scr. Rer.Germ.*, 46, pp. 9–11; McGinn, *Visions*, pp. 117–19. For Philip Augustus see E.A.R. Brown, 'La notion de la légitimité et la prophétie à la cour de Philippe Auguste', *La France de Philippe Auguste – Le Temps des Mutations*, ed. R.H. Bautier, Paris 1982, pp. 85–96, 104–10.

intensified eschatological expectations. Thus chroniclers of the Second Crusade included the prophecy of the Last Emperor in their accounts, identifying him with either the king of France or Emperor Conrad III of Germany.[70] In Germany, the Last Emperor, identified first with Henry IV,[71] was associated from the 1160s with Frederick I Barbarossa. The *Play of Antichrist* (*Ludus de Antichristo*), composed in about 1160 at Frederick's court, was the earliest and the most impressive of the Antichrist plays that were to proliferate during the following centuries. Heavily dependent on Pseudo-Methodius, Adso, as well as upon Lambert of Saint-Omer's *Liber Floridus* (of *c.* 1120),[72] the play is an example of the 'imperial apocalypse', presenting a conservative pro-imperial account of the End. Particularly interesting is its presentation of Jerusalem. In the earliest versions of the prophecy of the Last Emperor, he placed his crown on Mount Zion or Calvary; the Antichrist dwelt in the Temple, which he rebuilt from its ruins. In the *Play of Antichrist*, however, the Last Emperor placed his insignia not at Calvary or on Mount Zion, but on the altar in the Temple. The link between the Antichrist and the Temple disappeared, and instead of being the seat of the Antichrist, it became the site of the Emperor's abdication. This change obviously reflected the transformation, following the crusading conquest of Jerusalem, in the Christian attitude to the Temple Mount. Moreover, the Emperor, rather than the king of Jerusalem, was presented as Jerusalem's defender. According to the play, the Emperor defeated the king of France and turned the king of the Greeks and the ruler of Jerusalem into his subjects. The king of Babylon attacked Jerusalem and was defeated by the Emperor, who then fulfilled the traditional role of the Last Emperor by giving up his crown. Unlike earlier versions of the Last Emperor prophecy, the Emperor did not die but resumed his role as king of the Germans. The Antichrist, whose appearance in Jerusalem was triggered by the emperor's abdication, triumphed over the leaders of the Church; the kings of Jerusalem and of the Franks also agreed to become his subjects. Finally, after the Jews, first converted by the Antichrist's preaching and then convinced by Enoch and Elijah of their error, were slain by the Antichrist, the latter being destroyed by Christ. In the play, when the Last Emperor lays down his crown, an Angel of God appears and sings: 'Judaea and Jerusalem have no fear, / Knowing that tomorrow you shall see God's help! / For your brethren are at hand who will free you / And will overcome your enemies in power.' Then comes the chorus: 'Judaea and Jerusalem have no fear; tomorrow we shall go forth and the Lord will be with you. Alleluia!'[73]

---

[70] *Chronica S. Petri Erfordensis*, p. 176; Otto Frisingensis, *Chronicon*, pp. 365–7; Constable, 'The Second Crusade', pp. 271–2.

[71] Folz, *L'idée d'Empire*, pp. 128–32, 213–14. For twelfth-century versions of the Last Emperor prophecy see also *Liber Anselmi de Antichristo*, *CCCM*, 45, pp. 161–6; *Descriptio Cuiusdam Sapientis de Antichristi nativitate*, *CCCM*, 45, pp. 52–3.

[72] See Verhelst, 'Les texts eschatologiques', pp. 299–305.

[73] K. Langosch, *Geistliche Spiele*, Basel–Stuttgart 1957, pp. 181–239, 267–84; McGinn, *Visions of the End*, pp. 117–21; Young, *The Drama of the Medieval Church*, vol. 2, pp. 369–96. For the Temple Mount see above, ch. 6, pp. 96–102.

One of the specific political intentions of the play was thus to portray the Emperor as the defender and even protector of Jerusalem. As a result, one is tempted to see its reflection in the meeting held in Mainz at Easter 1188, the so-called 'Court of Christ', *Curia Ihesu Christi*, at which the Emperor met his nobles to take up the cross for the Third Crusade. Here he made a striking gesture: he declined to occupy the throne, saying that it belonged to Christ alone, the true leader of the crusade.[74]

Indeed, the fall of Jerusalem on 20 October 1187, and the subsequent renewal of the struggle between Christendom and Islam over the Holy City, produced as its by-product intensified eschatological expectations. It also generated resurgence in the position of the Earthly Jerusalem in the current eschatological theories of history, including apocalyptic speculations over the ages of the Church. The very loss of Jerusalem was seen as part and parcel of the events of the End; to some, the armies of Saladin were *praecursores Antichristi*.[75] The event was incorporated in prophecies that often foretold *vaticinium ex eventu* not only the fall, but the liberation of pessimism. One such prophecy appears in the *Itinerarium peregrinorum*, an account of the Third Crusade composed by an anonymous author after the fall of Jerusalem, thus introducing optimism at a time of crisis. It was a vision allegedly seen by King Guy of Lusignan's chamberlain on the eve of the fall of Jerusalem. In his vision, the chamberlain saw the standard of the crusader army flying above the city, seven ballisters with seven missiles, marching and crying out together in terrible voices: 'Woe to you, O Jerusalem' (Jeremiah 13:27). According to the author, this vision refers to a passage in the book of Psalms: '... [God] has bent and strung his bow; he has prepared his deadly weapons ...' (Psalms 7:12–13). The seven missiles symbolized the 'Seven Deadly Sins' on account of which the crusading army was defeated at the battle of Hattin (4 June 1187).[76] Another vision predicting *vaticinium ex eventu* the fall of Jerusalem was ascribed to William of Tyre by the continuators of his chronicle; one of them, writing several years after 1187, described the patriarchal election of 1180 in Jerusalem and William of Tyre's intervention against the election of Heraclius, then archbishop of Caesarea. William allegedly argued as follows: 'Sirs, I have found it written that a Heraclius won a victory in Persia and brought back the Holy Cross to Jerusalem, and that a Heraclius will remove it from Jerusalem and in his time it will be lost. Therefore I pray for the sake of God that you do not elect him [Heraclius] for patriarch, because if you do, I know [it] well that the city [of Jerusalem] will be lost ...'[77] In another version of the same

---

[74] Young, vol. 2, pp. 390–93.

[75] Berter d'Orléans, *Iuxta threnos Jeremiae*, *The Oxford Book of Medieval Verse*, no. 201, pp. 297–300.

[76] *Itinerarium peregrinorum*, p. 257. For the 'Seven Deadly Sins' see M.W. Bloomfield, *The Seven Deadly Sins*, East Lansing, Michigan 1952 (new edn. 1967), pp. 69–86.

[77] *La Chronique d'Ernoul et de Bernard le Trésorier*, ed. L. de Mas Latrie, Paris 1871, pp. 82–4; P.W. Edbury and J.G. Rowe, 'William of Tyre and the Patriarchal Election of 1180', *EHR*, 93 (1978), pp. 4–6.

prophecy, William of Tyre argued that he 'found it written in a book that Heraclius brought back the Holy Cross from Persia and left it in Jerusalem and [a] Heraclius will take it away from Jerusalem, and in his time the city will fall'.[78]

The power of prophecies is also demonstrated by the famous meeting between King Richard I of England and Joachim of Fiore in 1190 at Messina. Richard, on his way to take part in the siege of Acre, approached the prophet from Calabria, asking him about the fate of his crusade. The meeting was recorded by Roger of Hoveden in two versions. According to his account in the *Gesta*, Joachim identified the seven heads of the beast of the Apocalypse as the seven persecutors of the Church – Herod, Nero, Constantius, Mohammed, Melsemutus, Saladin and Antichrist. He predicted that Saladin would soon be defeated and the Holy Land restored to the faithful, assuring Richard that the Lord had reserved this magnificent victory for his army.[79] In his *Chronica*, however, Roger rendered a slightly different version, predicting that Saladin, presently persecuting the Church and holding the Sepulchre of the Lord, the Holy City of Jerusalem and the entire land where His feet had trodden, would soon lose it, seven years after the fall of Jerusalem, namely in 1194.[80] The extensive description of the fall of the Kingdom of Jerusalem and the *res gestae* of the Third Crusade (1189–1192) in the *Chronica* includes two versions of the prophecy of the Last Emperor, Adso's as well as one other ascribed to Pope Gregory I.[81]

In fact, Roger of Hoveden's versions of the prophecies of Joachim of Fiore are an excellent example of the insertion of Jerusalem into apocalyptic visions which did not originally refer to Earthly Jerusalem at all. Joachim, as his own works show, did not support the crusade for the recovery of Jerusalem, perceiving it as contrary to God's will. Following the failure of the Third Crusade, he argued that it had failed because its participants had not followed God and thus had caused Him injury. A future crusade for the recovery of Jerusalem was unnecessary, as the third age or *status*, that of the Spirit, would begin between 1200 and 1260. At that time a race of *viri spirituali* would emerge, convert the Jews and the Gentiles and gather them peacefully into Christ's fold. On the whole, Jerusalem appears in the works of Joachim of Fiore mainly as 'Heavenly Jerusalem' or the 'pilgrim city' (*civitas pellegrina*); it is identical with the Church, which is the 'other Jerusalem' (*altera Hierusalem*), as well the 'New Jerusalem' and the 'spiritual Jerusalem' (*Hierusalem spiritualis*). Thus the Church is the New Jerusalem; the papacy

78 *L'Estoire d'Eracles empereur et la conqueste de la terre d'Outremer*, *RHC.Hocc*, 2, pp. 57–9; *La Continuation de Guillaume de Tyre*, pp. 49–50; Edbury and Rowe, 'William of Tyre', pp. 6–7. See also the same prophecy in Sicardus Cremonensis, *Cronica*, *PL*, 213, col. 518.

79 Roger of Hoveden, *Gesta Regis Henrici Secundi, RS*, 49, vol. 2, pp. 151–3.

80 *Idem*, *Chronica*, *RS*, 51, 3, pp. 75–9.

81 *Ibid.*, pp. 80–86.

should care for 'its Jerusalem', namely, the Church.[82] Yet the idea of the recovery of Earthly Jerusalem was too attractive for Joachim's disciples to ignore. Thus, for example, Salimbene of Parma, writing before 1276, described a vision of the Angelic Pope (*pastor angelicus*) including a victory over the Saracens and the recovery of the Holy Sepulchre by Christendom.[83] Similarly, the general of the Franciscan Order, Bonaventura, although deeply influenced by Joachimite theories, placed the reconquest of Jerusalem, its rebuilding and the rebuilding of the Temple in the Joachimite seventh *tempus*. Only then, he argued, would 'the prophecy of Ezekiel 40–43 ... be fulfilled when the city comes down from heaven'.[84]

In sum: there is evidently a link between the religion of the rulers of Jerusalem and its place in visions of the End. In other words, as long as Christians were ruling in Jerusalem, its apocalyptic importance declined. The First Crusade departed under the influence of apocalyptic expectations; it moved in an apocalyptic atmosphere, and the conquest of Jerusalem was seen as a part of the Last Days. During the twelfth century Jerusalem was recognized as the scene of the Last Days. Nevertheless, apocalyptic expectations weakened. It is as if the Earthly City of Jerusalem as the gateway to the Heavenly one, or even the figure of the Heavenly Jerusalem,[85] took the place of the apocalyptic Jerusalem. This changed following the conquest of Jerusalem by Saladin, an event seen by many as part and parcel of the events of the End and as a realization of scriptural and other prophecies. The fall of Jerusalem produced new visions, which described both the disaster and its remedy, the recovery of Jerusalem, as part of the Last Days. After 1187 and for the rest of the Middle Ages, Earthly Jerusalem, ruled by the Moslems (except for the short period of 1229–1244), was to loom large in all types of late medieval apocalypticism,[86] introducing an optimistic tone into the terrifying events of the End: the *pastor angelicus*, together with the French king as the Second Charlemagne, would defeat the Antichrist, reform the church and regain Jerusalem, and 'His Sepulchre [would] be glorified by all' (Isaiah 11:10).[87]

82 M. Reeves, *Joachim of Fiore and the Prophetic Future*, London 1974, pp. 1, 8, 72; E.R. Daniel, pp. 126–54, esp. p. 153 n. 10; B.Z. Kedar, *Crusade and Mission. European approaches towards the Muslims*, pp. 112–16 and Appendix 5/6. For a different interpretation of the ideas of Joachim of Fiore see Siberry, *Criticism of Crusading*, pp. 201–4.

83 Salimbene de Adam, *Cronica, MGH.SS*, 32, pp. 492–3. For the introduction of the papacy into the apocalyptic scenario see B. McGinn, 'Angel Pope and Papal Antichrist', *Church History*, 47 (1978), pp. 155–73.

84 McGinn, *Visions of the End*, p. 200. For Bonaventura see also Reeves, *Joachim of Fiore*, pp. 36–3.

85 See above, ch. 7.

86 R.E. Lerner, *The Powers of Prophecy. The Cedar of Lebanon Vision from the Mongol Onslaught to the Dawn of the Enlightenment*, Berkeley 1983, pp. 9–134; Reeves, *Joachim of Fiore*, pp. 29–82; McGinn, *Visions of the End*, pp. 127–285.

87 McGinn, 'Angel Pope', pp. 171–3.

Chapter 9

'The Terrible News': The Reaction of Christendom to the Fall of Jerusalem (1187)

On 4 July 1187, the army of the Latin Kingdom suffered a catastrophic defeat at the battle of Hattin. Most of the kingdom's knights and nobles, including the king, were either killed or captured. During the summer and autumn of 1187 the now undefended cities and castles surrendered one after another to Saladin. Jerusalem was defended by Queen Sibylla, Patriarch Heraclius and Balian of Ibelin, the last having escaped from Hattin and managed to return to Jerusalem by way of Nablus. Since at the time there were only two knights in the city, Balian knighted those sons of knights who had reached the age of fifteen and in addition sons of the city's patriciate. During Saladin's siege of the city (20 September–20 October 1187) the most extraordinary rituals of penance took place in the city. Noble ladies undertook a rather peculiar ritual of penance: 'they brought basins and placed them before Mount Calvary and filled them with cold water. Then they put their children in up to their chins, cut off their hair and threw it away.'[1] The clergy led processions on the walls of the city, carrying the relic of the True Cross, which belonged to the Jacobite community, as the relic belonging to the crusaders had been lost at Hattin. When the great cross, which the crusaders had set up on the wall when they entered the city on 15 July 1099, fell down during one of Saladin's attacks, it was considered as bad an omen as the capture of the relic of the True Cross at Hattin. But all these efforts were in vain. The city, lacking military force, capitulated on 20 October 1187 after a fortnight's siege.[2] During the negotiations between the defenders and Saladin over the terms of surrender, some of those in

---

1 *L'Estoire*, I.23, ch. 58, pp. 87–8. *The Conquest of Jerusalem and the Third Crusade. Sources in Translation*, trans. P.W. Edbury, Aldershot 1996, p. 59.

2 For these events see Prawer, *Histoire du Royaume*, vol. 1, pp. 641–80; Mayer, *The Crusades* (1972 edn), pp. 128–33. There is no study of the reaction to the fall. For some aspects see A. Cartellieri, *Philipp II. August. König von Frankreich*, Leipzig 1921–1922, pp. 39–83, 268–73; Prawer, *Histoire du Royaume*, vol. 2, pp. 22–34; Siberry, *Criticism of Crusading*, pp. 52–62, 81–5; Cole, 'Christian Perceptions', pp. 9–39. Y. Katzir's article 'The Conquests of Jerusalem', pp. 103–13, does not deal, as the title suggests, with the Christian reaction to the fall of Jerusalem in 1187 but only with the Moslem reaction. For the True Cross of the Crusaders see Katzir, 'The Vicissitudes of the True Cross', pp. 243–53.

Jerusalem, such as the anonymous author of *Libellus de expugnatione Terrae Sanctae per Saladinum*, opposed the capitulation, arguing: 'Who has heard of such a thing, that a successor should pay to be expelled from his inheritance? ... Who can imagine such an act of wickedness as handing over to the Saracens the Sepulchre of the Resurrection of Christ, the most noble temple (*Templum Domini*), the most holy Mount Zion and other holy places of the Holy City? ... Did we ever hear of the Jews abandoning their Holy of Holies without bloodshed and heavy fighting, handing it over voluntarily? A curse upon those despicable merchants who of their free will are selling the Holy City and Christ!' The poor, he reported, 'preferred to die for Christ in the Holy City than to serve the impure and unclean Turks and Saracens in other parts of the Promised Land'.[3]

The news of the defeat at Hattin spread quickly all over Europe. It first reached Genoa and then, through that city, arrived at the papal court by October 1187.[4] The following month brought the news of the fall of Jerusalem.[5] From the papal court in Rome the tragic news was transmitted to all parts of Europe by papal encyclicals and letters of churchmen who were staying at the Curia; Peter of Blois, for example, brought the news to King Henry II of England.[6] In addition, letters from the Latin East reached the European courts. These were addressed in particular to the emperor and the king of England. From the royal courts these letters found their way into contemporary chronicles. Thus the letter of Terricus, the grand preceptor of the Order of the Temple, to Henry II of England, was copied into the chronicles of Roger of Hoveden, whereas the German *Historia de Expeditione Friderici I* contains a collection of letters sent to the popes of the period and to Duke Leopold V of Austria.[7] One can say that the loss of Jerusalem

---

3 *Libellus de expugnatione Terrae Sanctae per Saladinum*, *RS*, 66, p. 248. The *Libellus*, falsely ascribed to Ralph of Coggeshall, is based upon an eyewitness of the siege and fall of Jerusalem.

4 See the letter of a Genoese merchant to the pope in Roger of Hoveden, *Gesta*, vol. 2, pp. 11–13; *RRH*, *Adds.* no. 664a. The delay in the papal reaction to the news may be attributed to the death of Pope Urban III (20 October 1187); the newly elected Pope Gregory VIII, crowned on 25 October 1187, reacted four days later (29 October 1187) with *Audita tremendi*.

5 According to Cartellieri, *Philipp II.*, pp. 268–73, the news of the fall of Jerusalem had reached the papal court by the end of November. For this news see the letter of the Hospitaller Hermenger to Leopold V of Austria of October 1187 in *Historia de Expeditione Friderici Primi*, pp. 4–5; *RRH*, no. 678. See also *ibid.*, no. 677.

6 Roger of Hoveden, *Gesta*, vol. 2, p. 15.

7 See the letter of Terricus in Roger of Hoveden, *Gesta*, vol. 2, pp. 13–14; *idem*, *Chronica*, vol. 2, pp. 324–5; *RRH*, n. 660. Eng. trans., Barber, *The New Knighthood. A History of the Order of the Temple*, Cambridge 1994, pp. 115–16. Additional letters sent from the Latin East are the following: 1. A letter from the heads of the Latin Kingdom to Pope Urban III sent shortly after the battle of Hattin (4 July 1187). See *Hugonis Chronici Continuatio Weingartensis*, *MGH.SS*, 21, pp. 475–6; *RRH*, no. 658. 2. A letter written before the surrender of Ascalon (4 September) by Hospitallers to Archumbald, the

and the Holy Land resulted in a tremendous literary activity, similar to that engendered following its conquest by the First Crusade. Numerous letters from the Latin East, letters transmitting the news in Europe, detailed chronicle accounts, poems[8] and finally treatises[9] devoted to the event and its interpretation are all, if not a literary genre, a body of writings, treating the events from different aspects.

With the proclamation on 29 October 1187 of a new crusade whose major aim was the recovery of Jerusalem, the papal court and the courts of Henry II and Frederick I became not only centres of information regarding developments in the Latin East, but also centres of crusading propaganda and preaching. Pride of place belongs, obviously, to the encyclical *Audita tremendi*, which reverberated throughout Europe. The crusading propaganda of Peter of Blois is echoed in English accounts of the fall of Jerusalem, while Cardinal Henry of Albano's

---

Hospitaller master in Italy. See *Historia de Expeditione Friderici Primi*, pp. 2–4; *RRH*, no. 661; *The Conquest*, ed. Edbury, pp. 160–62. 3. A letter from Patriarch Heraclius to the pope written shortly before the siege of Jerusalem, that is between 5 and 20 September 1187. See B.Z. Kedar, 'Ein Hilfruf aus Jerusalem vom September 1187', *DA*, 38 (1982), pp. 120–22; Edbury, *The Conquest of Jerusalem*, pp. 162–3. 4. A letter of Bohemond III of Antioch to prelates in the West written after the surrender of Ascalon (4 September 1187) and after the beginning of the siege of Jerusalem (20 September 1187). See *RRH*, no. 663. 5. The letter of Aimery, patriarch of Antioch (1142–87) to King Henry II of England, written shortly after 20 September 1187. See *RRH*, no. 664. 6. The letter of the Hospitaller Knight Hermenger to Duke Leopold V of Austria (composed in 1188); *Historia de Expeditione Friderici Primi*, pp. 4–5. 7. Another letter sent by the Templar Terricus in 1188, this time to Henry II of England. See Roger of Hoveden, *Gesta*, vol. 2, pp. 40–41. Eng. trans., vol. 2, pp. 90–91.

8 The following poems refer to the fall: *Graves nobis ad modum dies effluxere* in Roger of Hoveden, *Chronica*, vol. 3, pp. 37–8; *Jerusalem, civitas inclita* in Hagenmeyer, 'Deux poésies', pp. 582–5; Berter d'Orléans, *Juxta trenos Jeremiae*, *Oxford Book of Medieval Verse*, no. 201, pp. 297–300 (also in Roger of Hoveden, *Chronica*, vol. 2, pp. 330–32); *Carmen Sangallense: Quid Dormis? Vigila!* in Hagenmeyer, 'Deux poésies', p. 585; *Indue cilicium, sedeas in pulvere, mater*, *AH*, 33, no. 266, pp. 316–19; *Christiani nominis corruit insigne*, *ibid.*, 15, no. 228, pp. 258–9; *Crucifigat omnes*, *Carmina Burana*, no. 47; Henri de Champagne, *Sede, Sion, in pulvere*, *AH*, 21, no. 234, p. 164; *Plange, Sion et Judaea*, *ibid.*, 33, no. 265, p. 315; *Jerusalem, luge, medio dolor orbis in orbe*, *PL*, 155, cols 1095–8; *Heu, voce flebili cogor enarrare*, *Carmina Burana*, no. 50; *Quod spiritu David precinuit*, *ibid.*, no. 48; *Miror, cur tepeat*, *AH*, 33, no. 267, pp. 319–20; *Venit Jesus in propria*, *ibid.*, no. 21, p. 164; *Lugent Sion et Judaea*, *ibid.*, no. 236, pp. 165–6; Conon of Béthanie, *Ahi Amors* in *Chansons de Croisades*, no. III, pp. 32–5; *Pour Son Peuple Resconforteir*, *ibid.*, no. VII, pp. 78–81; *Vos qui ameis de vraie amour*, *ibid.*, no. II, pp. 20–22; Albrecht von Johansdorf, *Die hinnen varn* in Müller, pp. 42–3; Heinrich von Rugge, *Ein tumber man I hat*, *ibid.*, pp. 47–51.

9 The most important treatises on the subject are: Henry of Albano's *De Peregrinante Civitate Dei* and Peter of Blois' *De Hierosolymitana peregrinatione acceleranda* and his *Passio Reginaldi*. Chronicles such as *Historia de Expeditione Friderici Primi*, Roger of Hoveden's *Gesta*, William of Newburgh's *Historia* as well as Ralph of Diceto, Magnus of Reichersberg and Arnold of Lübeck include extensive accounts of the event.

interpretation of the momentous event and preaching of the crusade influenced German and French chroniclers and poets. Links were thus forged between the reaction of the papacy, the rulers, apologists of the crusade, as well as crusade preachers and the European public.

Christendom, unprepared for the disaster, received the news with amazement, horror and grief. It was reported that Pope Urban III had died on 20 October of pain and grief upon hearing of the fall of Jerusalem;[10] when King Henry II heard in November that the city had been lost, he remained speechless for four days.[11] King William II of Sicily, upon hearing the news, donned sackcloth and sat in solitude for four days. The cardinals in Rome declared a truce of seven years among Christians, undertook to refrain from all riches and to preach the cross 'not solum verbo sed opere et exemplo', that is, by becoming itinerant preachers of the cross. They were the first to take the cross, swearing also not to mount a horse 'as long as the land in which His feet had trodden was under the domination of His enemies'.[12] Cardinal Henry of Albano refused the papal throne, declaring that he wished to devote himself to the preaching of the cross.[13] Abbot Samson of Bury St Edmunds, upon hearing of the fall of Jerusalem, donned a hairshirt and abstained from meat. A rare glimpse of how devastating was the fall of Jerusalem to Europeans at large – in this case to the inhabitants of Assisi – derives from a document from the archives of its cathedral. It reports that the people dressed in sackcloth; the name of Jerusalem was on everyone's lips, almost to the point of obsession; people were invited to weep openly and to do penance. The churches remained open night and day. The priests of Assisi continually repeated the mournful psalm of the people of Israel: 'O God, the heathen are come into their inheritance' (Psalms 78(79):1). Large paintings were made which depicted the Holy Sepulchre being trampled upon by horses and Jesus Christ oppressed by Mohammed.[14]

Like the crusader conquest of Jerusalem almost a hundred years earlier, the city's fall was associated with Christ's Passion and Crucifixion. According to an anonymous chronicler from the Austrian Cistercian abbey of Zwettl, the grief of

---

[10] William of Newburgh, pp. 267–8; Albericus de Trium Fontium, *Chronica*, *MGH.SS*, 23, p. 748; *Continuatio Gemblacensis*, pp. 424–5; Gervase of Canterbury, vol. 1, p. 388; Roger of Hoveden, *Chronica*, vol. 2, p. 322; Robertus Autissiodorensis, p. 252. As a matter of fact, Urban III died on 20 October 1187 before the news of the fall of Jerusalem could have reached him.

[11] Gervase of Canterbury, vol. 1, p. 389. On Henry II and the Holy Land see H.E. Mayer, 'Henry II of England and the Holy Land', pp. 721–39.

[12] Peter of Blois to Henry II in Roger of Hoveden, *Gesta*, vol. 2, p. 15.

[13] Albericus de Trium-Fontium, p. 748; Congar, 'Henri de Marcy', p. 43.

[14] Jocelinus de Brakelond, *De rebus gestis Samsonis abbatis monasterii sancti Edmundi*, ed. and Eng. trans. H.C. Butler, London 1951, p. 29; Lerner, *The Powers of Prophecy*, pp. 38–9. For Assisi see R. Pazzelli, *St Francis and the Third Order. The Franciscans and the Pre-Franciscan Movement*, Chicago 1982, p. 73.

all Christendom resulting from the loss of the 'City of the Lord' was as deep and great as if it had been caused by the death of Christ.[15] Walter Map also saw in the fall of Jerusalem an act of God which touched not only the faithful but Christ himself.[16] Henry of Albano, writing in 1187 to the German princes, claimed that: 'the earth shook when Christ lost the land which for our redemption was sanctified by His feet', making use of Matthew's description of the Crucifixion: '... and the earth shook and the rocks were split and many tombs were also opened ...' (Matthew 27:51–52). He thus described the disaster, and particularly the Moslems' capture of the Holy Cross, as a second Crucifixion of Christ.[17] This simile was also put forward by Gerald of Wales, the archdeacon of St David's in Wales, who accompanied Baldwin, a Cistercian and archbishop of Canterbury, on his campaign of crusade preaching in Wales in 1188. He presents an account of a dream he had, in which he saw the heavenly court attacked and the citizens of heaven killed and God himself dragged from his throne, his right side pierced with a spear. Gerald's explanation of the meaning of his dream runs as follows: 'He who formerly endured suffering in His own person ... is now showing us that He is suffering once more, but this time through the sufferings of His people ... in the holy land which He finally hallowed by shedding His own blood, after it had enjoyed so many wonderful manifestations of His bodily presence, His faithful followers had been suffering no longer on the cross, but in martial conflict. It has therefore been His wish that His own passion, which He is now enduring somehow through the suffering of His people, in that place where He reigns in majesty with His Father, should be similarly revealed, not on the Cross but amidst fighting and the tumult of war ...'[18]

However, the fall of Jerusalem was perceived, more than anything else, as both an injury to Christ and to Christendom, and as an act of pollution. As an injury it was not only a collective one inflicted by Islam upon Christendom but one inflicted individually upon every believer. The perception of Saladin's capture of Jerusalem as an act of pollution appears first in *Audita tremendi*, where Gregory VIII depicts Saladin's siege and capture of Jerusalem as a 'pagan invasion' directed against Christians; by 'barbarian' people desiring 'to profane the sacred places and obliterate the worship of God from the earth'. The entire atmosphere of the pope's message is dictated by Psalms 78(79):1: 'Deus venerunt gentes in haereditatem tuam; Polluerunt templum sanctum tuum; Posuerunt Jerusalem in pomorum custodiam.' And indeed Psalm 78(79):1 became the most used one within the reaction of the fall; it was used to describe the event, to explain it, as

15 *Continuatio Zwettlensis Altera*, *MGH.SS*, p. 543.

16 Walter Map, *De Nugis Curialium*, ed. M.R. James *et al.*, Oxford 1914 (new edn Oxford 1983), p. 42.

17 Henry d'Albano, *Epistolae*, no. 32, col. 250; *Historia de Expeditione Friderici Primi*, p. 11.

18 Gerald of Wales, *Expugnatio Hibernica*, pp. 215–16.

well as in a prophetic sense. Pope Gregory VIII used the verse to describe the event: 'On hearing with what severe and terrible judgement the land of Jerusalem has been smitten by the divine hand, we and our brothers have been confounded by such great horror and affected by such great sorrow that we could not easily decide what to do or say; over this situation the psalmist laments and says, *O God, the heathen are come into thy inheritance*.'[19]

The perception of the conquest as an act of pollution looms large in the accounts of the event. The Lombard academic, Prepositinus of Cremona, preaching in Paris shortly after the battle of Hattin, pictured the Moslems' celebration of their victory as a pagan bacchanal around Christ's tomb: 'they taunt Christians with the death of Christ: Where is the God of the Christians? The Saracens have not lost their God; the God of the Jews is asleep, and the God of the Christians is dead.'[20] King Henry II, writing in 1188 to the patriarchs of Jerusalem and Antioch and to Raimond of Antioch, argued that: '... Divine judgement permitted the land which was redeemed with His own blood to be polluted by the hands of unbelievers'.[21] King Richard, writing on 1 October 1191 from Acre to Garnier of Rochefort, the abbot of Clairvaux, referred to the fate of 'the places of His death that have been consecrated by His precious blood and which the enemies of the cross of Christ have hitherto shamefully profaned'.[22] The anonymous author of the *Itinerarium Peregrinorum et Gesta Regis Ricardi* lamented Jerusalem, saying: 'Glorious was Jerusalem, the city of God, where the Lord suffered and was buried, and where he displayed the glory of his resurrection; but she is now subject to contamination at the hands of her baseborn foe; nor is there any grief like the grief that those who persecuted him should possess the sepulchre in which he lies buried; and those who had despised the Crucified, have made themselves masters of his Cross ... When the city was taken, the proclaimer of the law of Mohammed proceeded to the summit of the rock of Calvary, and there published the false law, in the place where Christ had consummated the law of death upon the cross. The enemy then perpetrated another

---

[19] Gregorius VIII, *Audita tremendi*, pp. 8–9. There are three versions of this bull; two are dated 29 October and the third 3 November 1187. Though the bull was composed before the news of the fall of Jerusalem had reached the papal court (see above, n. 5), the event was already expected. See Peter of Blois in Roger of Hoveden, *Gesta*, vol. 2, p. 15, who, writing in *c*. September 1187 to Henry II, argued that he doubts that Jerusalem would be able 'to resist the dogs', that is, the Moslems. The wide circulation of *Audita tremendi* is demonstrated by its inclusion in a relatively large number of chronicles. See William of Newburgh, pp. 267–70; Roger of Hoveden, *Gesta*, vol. 2, pp. 15–19; *idem*, *Chronica*, vol. 2, pp. 326–9; *Historia de Expeditione Friderici Primi*, pp. 6–10. See also a summary in Arnoldus Lubecensis, pp. 169–70. Eng. trans. Riley-Smith, *Idea and Reality*, pp. 63–7.

[20] G. Lacombe, *La vie et les oeuvres de Prévostin*, Le Sauchoir, Kain 1927, pp. 199–200; Cole, '*O God*', pp. 106–7. See also *idem*, 'Christian Perceptions', p. 37, n. 61.

[21] Roger of Hoveden, *Gesta*, vol. 2, pp. 38–9.

[22] Roger of Hoveden, *Chronica*, vol. 3, pp. 129–30; Edbury, *The Conquest of Jerusalem*, p. 179. See also Albericus de Trium Fontium, p. 860.

diabolical act. They fastened ropes round a certain cross, which stood upon the pinnacle of the Church of the Hospitallers, and dragged it to the ground, where they spat upon it, and hacked it, and drew it, in derision of our faith, through all the filth of the city.'[23] The anonymous author, referring here probably to the cross of the *Templum Domini* (and not a Hospitaller church),[24] presented the Moslems' acts of purification of the holy places of Jerusalem as acts of desecration and pollution. A similar attitude, revealing a much better knowledge of the holy places of Jerusalem, was expressed by the anonymous author of the *Libellus de expugnatione Terrae Sanctae per Saladinum*, who seems to have been an eyewitness to the siege and surrender of Jerusalem. He referred to the destruction by Saladin's army of the Church of the Ascension on the Mount of Olives, the Church of St Mary in the Valley of Jehoshaphat and the church of Gethsemane; all those places including the Temple of the Lord were polluted. He presented the cleansing of the Temple of the Lord by the Moslems as a preposterous, barbaric act, accompanied by a 'filthy and horrible bellowing of Mohammed's law'; it was, he claims, 'not a cleansing but a pollution of the entire place'.[25] Actually, following his entry into the city, Saladin ordered the two major shrines on the Temple Mount to be restored to their original state. Churches were turned into mosques or Moslem religious or military institutions. The Church of St Anna was first of all made into a mosque and then into a law school; the Church of the Dormition on Mount Zion became the dwelling for Saladin's soldiers.

The same perception of the fall is revealed, although in less theologically sophisticated language, by poets lamenting the fall. Henry of Champagne in his 'Sit, Sion in sackcloth' (*Sede, Sion, in pulvere*) lamented 'the pollution of the Holy City, the desecration of the Holy Places'.[26] Another poet exhorting the Europeans to recover the 'House of God' (*domus Dei*) asks rhetorically: 'Cur gens non opprimitur / Vitae bestialis, / Qua sic Christi premitur / Sedes principalis? / Factum in prostibulum / Civitas regalis, / Versa est in stabulum / Domus specialis.'[27] The author of 'I wonder why lukewarm' (*Miror, cur tepeat*), a lament inspired by the Book of Lamentations, refers to the Holy Sepulchre and Jerusalem as held in the hand of the Turkish mob; the glory of the whole world is now polluted and desecrated.[28]

---

[23] *Itinerarium Peregrinorum et Gesta Regis Ricardi*, *RS*, 38, pp. 22–3. Eng. trans. Helen J. Nicholson, *Chronicle of the Third Crusade*, p. 39.

[24] See *Libellus de expugnatione Terrae Sanctae per Saladinum*, p. 249.

[25] *Ibid.*, pp. 239–41, 249–50; Cole, '*O, God*', pp. 107–8.

[26] Henri de Champagne, *Sede, Sion, in pulvere*, *AH*, 21, no. 234, p. 164. See also *Lugent Sion et Judae*, *ibid.*, no. 236, p. 165.

[27] *Christiani nominis, Curriut insigne*, AH, 15, no. 228, pp. 258–9.

[28] *Miror, cur tepeat*, *AH*, 33, no. 267, p. 319. For the pollution of the Holy Sepulchre see also *Quod spiritu David precinuit*, *Carmina Burana*, no. 47, p. 92; no. 48, p. 95; *Fides cum Ydolatria*, *ibid.*, no. 46, p. 91; 'Sepulchrum gloriosum / prophetis declaratum / impugnatur a canibus / quibus sanctum non dabimus, / nec porcis margarite / mittuntur deridende.' This

The fall itself is described, almost exclusively, in Old Testament phrases, in liturgical and semi-liturgical terms of a basically penitential character. The most frequently cited verse was Psalm 78(79):1: 'O God, the heathen have come into your inheritance …'. This was into a large extent the influence of *Audita tremendi*, which described the disaster by quoting this verse.[29] Contemporaries in general, crusade preachers, chroniclers and poets also evoked the Book of Lamentations and the words of other prophets of doom, such as Isaiah, to describe the fate of Jerusalem and their own grief. Peter of Blois, claiming that 'all Jeremiah [Lamentations] does not suffice to express the present misery', argued that the lamentable event was the fulfilment of Jeremiah's prophecy: 'The Lord hath accomplished his fury; he hath poured out his fierce anger ...' (Lamentations 4:11). The news of the fall was for him 'a voice of wailing [that] is heard out of Zion' (Jeremiah 10:19), whereas as a result: 'The crown is fallen from our head and our dance is turned into mourning' (Lamentations 5:15–16).[30] The author of *L'Estoire d'Eracles*, describing the entrance of Saladin to Jerusalem, wrote: 'who could recount for you the tears and the sorrow of so great a misfortune as befell the holy city of Jerusalem? She who was named the mistress of other cities had become a slave and handmaid. She who should have ruled as sovereign was now under tribute' (Lamentations 1:1).[31] Roger of Hoveden described Jerusalem as: 'Princess among the provinces ... became tributary. She weepeth sore in the night and her tears are on her cheeks; among all her lovers she hath none to comfort her; all her friends have dealt treacherously with her ...' (Lamentations 1:1–2).[32] Berter of Orléans, to take another example, composed a poem of lamentation inspired by Lamentations 1:4: 'The roads to Zion mourn, for none come to the appointed feasts; all her gates are desolate.' The poet wrote: 'In the strains of Jeremiah / the ways of Sion mourn indeed, / that no longer is there one upon the solemn day / to visit the Holy Sepulchre ...'[33] Writing in 1187,

---

is inspired by Isaiah, 11:10: '... and his dwellings shall be glorious' and Matthew 7:6: 'Do not give dogs what is holy; and do not throw your pearls before swine ...' as well as Bernard of Clairvaux. See his *Epistolae*, no. 363, *PL*, 182, col. 565B.

29 Gregorius VIII, *Audita tremendi*, pp. 6–7 and see on this subject A. Linder, '*Deus Venerunt Gentes*, Psalm 78(79) in the Liturgical Commemoration of the Destruction of Latin Jerusalem', *Medieval Studies in Honour of Avrom Saltman*, ed. Bat-Sheva Albert *et al.*, Ramat Gan 1995 (*Bar-Ilan Studies in History*, 4), pp. 145–71; Cole, '*O God*', pp. 84–111. See also above, n. 19.

30 Petrus Blesensis, *De Hierosolymitana Peregrinatione Acceleranda*, cols 1057–8; *idem*, *Dialogus*, p. 112. For the use of Lamentations see also *Historia peregrinorum*, p. 117; *Continuatio Zwetlensis Altera*, p. 543; Arnoldus Lubecensis, p. 169; *La Continuation de Guillaume de Tyre*, p. 73; *Libellus de expugnatione Terrae Sanctae*, p. 247; Albericus de Trium Fontium, p. 860; Berter d'Orléans, *Juxta threnos Jeremiae*, *The Oxford Book of Medieval Verse*, no. 201, p. 297.

31 *L'Estoire*, ch. 60, pp. 91–3. Eng. trans., Edbury, *The Conquest*, p. 65.

32 Roger of Hoveden, *Gesta*, vol. 2, p. 20.

33 Berter of Orléans, *Juxta threnos Jeremiae*, *The Oxford Book of Medieval Verse*, no. 201, pp. 297–300.

Walter Map, in his 'Courtiers' Trifles' (*De nugis curialum*) lamented Jerusalem as follows: 'For in the same year of calamity they tell us that Jerusalem, the holy city, has been taken and led captive by Saladin, the lord of the paynims, and is laid waste by a fiercer plague than that which Jeremiah bewailed in the Lamentations, where he says with tears, "Her priests sigh, her virgins are afflicted" [Lamentations 1:4]. Titus destroyed this same people down to a small remnant when (though he knew it not) he avenged the wrongs of the Lord; but Saladin has wholly annihilated them and absolutely blotted out every Christian there. The sepulchre and cross of the Lord are the prey of dogs ... Many as are the cries of woe, the plagues, disasters, and death which prophets have pronounced against this bitterly afflicted city, this time it seems as if the Lord had fulfilled their oracles to the uttermost ... Israel of old was chastized, not slain; in our day they are slain, not chastized.'[34]

Though the fall itself is described almost exclusively in Old Testament terms, within the reaction Jerusalem appears as both the City of the Old and New Testaments. Thus we find Pope Gregory VIII, in *Audita tremendi*, describing *terra Jerosolimitana* as the 'place where His feet had trodden' (Psalms 132:7; Zechariah 14:5) and where, through His Passion and Resurrection, 'He hath wrought salvation in the midst of the earth' (Psalms 74:12). It was the site of salvation not by chance but by God's will; it was the site chosen by Christ as the stage for the salvation of the human race.[35] Influenced by *Audita tremendi*, which carried the papal words all over Europe,[36] Gregory's contemporaries depicted Jerusalem, the site of salvation, as the city of Christ, His life, death, Resurrection and His sepulchre.[37] The place which He had chosen as His house (Psalms 122:1), which He had made, as His dwelling, into 'another heaven', where He performed His miracles, issued the sacraments, preached, was resurrected and ascended to heaven. It was *civitas Domini virtutum*, His inheritance, His *patrimonium*, His sanctuary and His *patria*: it was the earthly city which is 'our mother' (Galatians 4:26), as well as the Promised Land,[38] it was the 'glorious City of God', the 'holy of holies' of the Christians, the sanctuary of the Lord now profaned by the pagans.[39] It was also the marriage portion of the Virgin Mary, the site of the life

---

34 Walter Map, dist. I, c.15, pp. 40–47.

35 Gregorius VIII, *Audita tremendi*, pp. 7–8.

36 For the wide circulation of *Audita tremendi* see above, n. 19.

37 Henry, bishop of Strasbourg, in *Historia peregrinorum*, p. 124.

38 Petrus Blesensis, *Epistolae*, *PL*, 207, cols 533–4; *idem*, *Passio Reginaldi*, cols. 958, 963; *idem*, *De Hierosolymitana Peregrinatione Acceleranda*, cols 1057, 1063, 1070; Haymarus, p. 105.

39 Henry d'Albano, *De Peregrinante Civitate Dei*, cols. 350, 353; *idem*, *Epistolae*, *PL*, 204, cols 249–52. Cf. Bernardus Claraevallensis, *Opera*, VIII, no. 363, p. 312; no. 458, p. 435; Congar, 'Henri de Marcy', p. 85. For Jerusalem as 'terra illius in qua steterunt pedes Domini' (Psalms 132:7) see Petrus Blesensis, *Dialogus*, p. 106; *idem* to Henry of England (*c*. Sept. 1187) in Roger of Hoveden, *Gesta*, vol. 2, p. 15; *Historia de Expeditione Friderici Primi*, p. 5; Henry d'Albano, *Epistolae*, no. 32, col. 249; Ralph Niger, p. 96; Innocent III,

of the Mother of God and Martyr who had seen her son betrayed, captured, judged, flagellated, crucified and put to death.[40]

In comparison with European accounts, Frankish sources presented a different and much more complex picture. Conrad of Montferrat, writing in 1189 to Baldwin, the archbishop of Canterbury, referred to the fate of the Holy Sepulchre, Calvary and the Virgin's sepulchre in the Valley of Jehoshaphat.[41] The anonymous *Libellus de expugnatione Terrae Sanctae* mentioned the 'Sepulchre of the resurrection of Christ', the 'noble Temple of the Lord' called the 'House of God', as well as various holy traditions of the Temple Mount, the site of the Presentation of Christ in the Temple, the place where Christ had saved the adulterous woman and the *pinnaculum templi* from which Jacob the Just had been thrown down to his death. He further referred to the tradition of the Dormition of the Virgin on Mount

---

Register, I, no. 336, p. 499; no. 355, p. 531; no. 437, p. 661; *Annales Colonienses Maximi*, *MGH.SS*, 17, p. 793; *Historia peregrinorum*, p. 124; *Heu, voce flebili*, *Carmina Burana*, no. 50, p. 101; Ricardus Divisiensis, p. 78; Conrad of Montferrat in Radulfus de Diceto, p. 61; *Gesta Episcoporum Halberstadensium*, *MGH.SS*, 23, p. 110. For Jerusalem as *mater* see Petrus Blesensis, *De Hierosolymitana Peregrinatione Acceleranda*, cols 1059, 1063; *Historia peregrinorum*, p. 124 (*mater et nutrix fidei*). For Jerusalem as *patrimonium* see the letter of the Hospitaller Hermenger to Leopold V of Austria in *Historia de expeditione Friderici Primi*, p. 4; Gerald of Wales, *De principis Instructione*, *Opera*, *RS*, 21, p. 200; *Venit Jesus in propria*, *AH*, 21, p. 164; Petrus Blesensis, *Epistolae*, *RHGF*, 19, pp. 274, 280; Conrad of Montferrat in Ralph of Diceto, p. 61. For Jerusalem as *terra promissionis* see Petrus Blesensis, *Epistolae*, *PL*, 207, no. 232, col. 534; *idem*, *Passio Reginaldi*, cols 957, 958, 966, 967; Gerald of Wales, *De principis Instructione*, p. 235; *Historia de expeditione Friderici Primi*, p. 57; *Pour son Peuple Resconforteir*, *Chansons de Croisade*, no. VII, p. 80; Ricardus Divisiensis, p. 68; *Chronicon Universale Anonymi Laudunensis*, p. 46; *Grandes Chroniques*, p. 166; Rigord, p. 24. For Jerusalem as the place where the Lord 'hath wrought salvation in the midst of the earth' (Psalms 74:12) in accounts of the fall see *Audita tremendi*, p. 8; Rostangus, *De Translatione capitis S. Clementis papae et martyris ab Constantinopoli ad Cluniacum, PL*, 209, cols 905, 907; Petrus Blesensis, *Passio Reginaldi*, col. 958; *Continuatio Zwettlensis Altera*, p. 543. See also Innocent III, *Register*, vol. 1, no. 302, p. 432; no. 336, p. 499; no. 355, p. 531; no. 437, p. 661. For Jerusalem as *hereditas Christi* see Petrus Blesensis, *De Hierosolymitana Peregrinatione Acceleranda*, col. 1059. See also *ibid.*, cols 959, 1068; *Historia de Expeditione Friderici Primi*, p. 5; *Historia peregrinorum*, p. 124; Haymarus, p. 105; Alain de Lille, p. 281; Clement III in *La 'Collectio Seguntina'*, in W. Holtzmann, 'La "Collectio Seguntina" et les décrétales de Clément III et de Célestine III', *Revue d'histoire ecclésiastique*, 50 (1955), no. 38, p. 430; Innocent III, *Register*, vol. 1, no. 11, p. 90; no. 302, p. 430; no. 397, p. 596; II, no. 202, p. 395; no. 244, p. 468; no. 258, p. 492; *Libellus de expugnatione Terrae Sanctae*, p. 210; *Itinerarium peregrinorum*, pp. 247, 258; *Itinerarium Regis Ricardi*, p. 22; *Christiani nominis corruit insigne*, *AH*, 15, no. 228, pp. 258–9; Heraclius in *Oriens Pontificius*, III, no. 149, p. 323.

40 Ralph Niger, p. 96; Alain de Lille, p. 281; Conrad of Montferrat in Radulfus de Diceto, p. 61; *Heu voce flebili*, *Carmina Burana*, no. 50, p. 101.

41 See in Radulfus de Diceto, p. 61.

Zion, also locating there the tradition of Christ's salutation of *pax vobis* to the apostles after the Resurrection.[42]

Reacting to the fall of Jerusalem, contemporaries identified themselves with the biblical Israelites expelled from their Promised Land. As a result descriptions of the fall presented the Moslems as the enemies of the Israelites, and the Christians as the Children of Israel. Pope Gregory VIII exhorted the Christians in *Audita tremendi* to follow the example of the Maccabees, 'to perform without any fear what ... [they] did'.[43] Peter of Blois described Raynald of Châtillon as *alter Jonathan*.[44] Chroniclers and poets referred to the Saracens as Edomites, Egyptians, Canaanites, Midianites as well as Amalekites and Philistines. The Christians were the Children of Israel, punished by God by the loss of the Ark of the Covenant.[45] According to an anonymous poet, who wrote between the fall of Jerusalem and the Third Crusade, the Christians of the Holy Land were groaning under the same yoke that the Israelites had suffered in Egypt (Exodus 1:13–14).[46] Another poet expressed his hope that the Third Crusade would capture Jerusalem from the Jebusites.[47] However, though Jerusalem was not captured by the forces of the Third Crusade, in 1198 Pope Innocent III expressed his firm belief in the victory of Israel over the Philistines and of David over Goliath.[48]

Not only was the event described in biblical terms, but like other major events, including the crusaders' conquest of Jerusalem almost a century earlier, the fall of Jerusalem was presented as the fulfilment of scriptural prophecies and particularly those of the Old Testament. Roger of Hoveden wrote: 'Thus was fulfilled what Jeremiah prophesied concerning Jerusalem, saying: "Princess among the provinces ... became tributary. She weepeth sore in the night and her tears are on her cheeks; among all her lovers she hath none to comfort her; all her friends have dealt treacherously with her ..." (Lamentations 1:1–2). And another prophet: "O God, the heathen are come into thine inheritance, thy temple have they defiled"

---

42 *Libellus de expugnatione Terrae Sanctae*, pp. 248–50. It should be noted that the author referred to the 'Church of Resurrection' using the Frankish term rather than the European's 'Church of the Holy Sepulchre'. See above, ch. 5, pp. 70–3. See also Heraclius in *Oriens Pontificius*, III, p. 326.

43 Gregorius VIII, *Audita tremendi*, p. 9. For the crusaders exhorted to follow the example of the Maccabees see also Petrus Blesensis, *Epistolae*, *PL*, 207, no. 232, col. 533; *Christiani nominis curruit insigne*, *AH*, 15, no. 228, p. 258; *Miror, cur tepeat*, *ibid.*, 33, no. 267, p. 319.

44 Petrus Blesensis, *Passio Reginaldi*, col. 965–6.

45 Rostangus, col. 908; *Historia peregrinorum*, p. 119; *Libellus de expugnatione Terrae Sanctae*, p. 245; *Lugent Sion et Judaea*, *AH*, 21, no. 236, p. 165; *Miror, cur tepeat*, *ibid.*, 33, no. 267, p. 319; *Graves nobis admodum dies effluxere in* Roger of Hoveden, *Chronica*, vol. 3, pp. 37–8; Berter d'Orléans, *Juxta threnos Jeremiae*, *The Oxford Book of Medieval Verse*, p. 297.

46 *Crucifigat Omnes*, *Carmina Burana*, no. 47, p. 92; Sibery, *Criticism*, p. 57.

47 Henri de Champagne, *Sede, Sion, in pulvere*, *AH*, 21, no. 231, p. 164.

48 Innocentius III, *Register*, I, no. 12, p. 21. See also *ibid.*, no. 13, p. 22.

(Psalms 78(79):1).'[49] According to William of Newburgh, the fall of Jerusalem was predicted by Jeremiah 30:14: '... for I have wounded thee with the wound of an enemy, with the chastisement of a cruel one'.[50] Thus, lamenting the fall of Jerusalem, some of its contemporaries interpreted it as the fulfilment of biblical prophecies, whereas others believed the disaster to be a part of the events of the Last Days.[51]

The fall of Jerusalem presented a theological problem and an immediate solution was demanded worldwide. Namely, why God had allowed it to happen; why he had let His city, His inheritance be lost. Most people, pondering and dwelling upon the cause of the disaster, found the explanation in the affairs of this world, in the 'formula' of the *peccatis exigentibus hominum*. The basic idea underlying this explanation in the years immediately following the fall of Jerusalem was explicitly phrased by Henry of Albano: the victory of Islam over Christendom took place 'not because Mohammed could have caused it, but because Christ wished it'.[52]

The official, papal interpretation of the disaster reflects Bernard of Clairvaux's apologetics for the Second Crusade (rather than Bernard's concept of the crusade).[53] Gregory VIII's *Audita tremendi* blamed the whole of Christendom: the disaster had occurred because of 'not only the sins of [Jerusalem] inhabitants but also our own and those of the whole Christian people. ...'. God had punished his people severely because of their iniquity. 'We ought not to believe', wrote the pope, 'that these things have happened through the injustice of a violent judge, but rather through the iniquity of a delinquent people.'[54] It was a severe and terrible judgement, and yet a just one. Gregory's *Nunquam melius superni* of 29 October 1187 similarly presented the disaster as caused by the wrath of the

---

49 Roger of Hoveden, *Gesta*, vol. 2, p. 20. For Jerusalem as Christ's inheritance see above, ch. 3, pp. 44–5 and here above, n. 39. For Moslem rule of Jerusalem as polluting its holy places see here above, nn. 19–28.

50 William of Newburgh, p. 249. For the fall as the fulfilment of Jeremiah's prophecies see also Albericus de Trium-Fontium, p. 748 (Lamentations 2:4); *Libellus de expugnatione Terrae Sanctae*, p. 250 (Lamentations, 2:7); *Plange, Sion et Judaea, AH*, 33, no. 265, p. 315; *Indue cilicium Sedeas in pulvere, ibid.*, no. 266, p. 316; Henri de Champagne, *Sede, Sion, in pulvere, ibid.*, 21, no. 234, p. 164; *Miror, cur tepeat, ibid.*, 33, no. 267, pp. 315–20; Berter d'Orléans, *Juxta threnos Jeremiae, ibid.*, p. 297.

51 See above, ch. 8, pp. 155–7.

52 Henry d'Albano, *De peregrinate Civitate Dei*, col. 355. At the same time, he claimed that the disaster occurred because of *occulto judicio* of divine justice, namely its secret decision. See below, nn. 78–9. See also Albericus de Trium-Fontium, pp. 860–61.

53 See Bernardus Claraevallensis, *Opera*, VIII, no. 363, pp. 311–17; *idem*, *De Consideratione*, pp. 410–13. Bernard's influence is pointed out by Congar, 'Henri de Marcy', pp. 61–90. However, whereas this influence is recognizable in explanations of the disaster and the idea of crusade, there is no such influence as far as the idea of Jerusalem is concerned.

54 Gregorius VIII, *Audita tremendi*, pp. 7–9.

Supreme Judge at the sins of the inhabitants of the Land of Jerusalem and of all Christians.[55]

After the fall of Jerusalem this papal explanation of *peccatis exigentibus hominum*, which can be defined as the traditional ecclesiastical justification of all the defeats and disasters ever suffered by Christendom,[56] was taken up by some contemporaries. They located the fall of Jerusalem not in the framework in the history of the crusades,[57] but rather within the 'Holy History', the history of the 'New Israel', that, like the historical Israel, had now been punished for its sins by God with the loss of its holiest of holies – Jerusalem, the Holy Sepulchre and the Holy Cross. The divine punishment was considered even more terrible and severe than the retribution visited on the Israelites, and yet it was considered just. Thus, comparing the fall of Jerusalem to Saladin to its fall to the Romans in AD 70, Henry of Albano, echoing St Jerome, declared that God had punished Christendom even more severely than He had punished the Israelites in the past. The Jews held their holy places in much veneration and admiration and much lamented their violation: 'We, who in our time are seeing our holy places not only violated but held in captivity, what sort of consolation can we have?'[58]

---

55 Gregorius VIII, *Epistolae*, no. 3, col. 1539. For the same view see Clement III in *Oriens Pontificius*, III, no. 153, p. 330; Clement III in *La 'Collectio Seguntina'*, no. 2, p. 422; Celestinus III, *ibid.*, no. 79, p. 441; Innocentius III, *Register*, I, no. 302, p. 430; no. 487, p. 717; II, no. 202, pp. 395–6; no. 241, p. 460. For the loss as attributed to *ira Dei* see Henry d'Albano, *De Peregrinante Civitate Dei*, col. 356; Petrus Blesensis, *Passio Reginaldi*, col. 958; *idem, De Hierosolymitana Peregrinatione Acceleranda*, cols 1057–62; Genoese letter in Roger of Hoveden, *Gesta*, II, p. 12; letter of the Templar Terricus, *ibid.*, p. 13; letter of Patriarch Aimery of Antioch, *ibid.*, p. 37; *Historia peregrinorum*, p. 117; *Heu voce flebilie*, *Carmina Burana*, no. 50, p. 102; *Lugent Sion et Judaea*, *AH*, 21, no. 236, p. 166; *Jerusalem, Luge*, *PL*, 155, col. 1096.

56 For the concept and its origins see Siberry, *Criticism*, pp. 70–72.

57 For the Second Crusade see Constable, 'The Second Crusade', pp. 213–81; Siberry, *Criticism*, pp. 77–81.

58 For the explanation of *peccatis exigentibus hominum*, Petrus Blesensis, *Dialogus*, p. 112; *idem*, *Passio Reginaldi*, col. 968; Hermenger in *Historia de Expeditione Friderici Primi*, p. 4; *ibid.*, p. 6; Gerald of Wales, *De principis instructione*, p. 200; Walter Map, pp. 40–46; Genoese letter to Pope Urban in Roger of Hoveden, *Gesta*, vol. 2, p. 11; Henry II to the patriarch of Jerusalem, *ibid.*, p. 38; *Libellus de expugnatione Terrae Sanctae*, p. 246; *Chronicon Universale Anonymi Laudunensis*, p. 46; *Ottobonis Scribae Annales*, *MGH.SS*, 18, p. 101; *Historia peregrinorum*, p. 117; Arnoldus Lubecensis, pp. 162–4; *Itinerarium peregrinorum*, pp. 246–7, 258; *Itinerarium Peregrinorum et Gesta Regis Ricardi*, p. 5; *Parti de mal e a bien aturné in* Morris, 'Propaganda for War', p. 97; Albrecht von Johansdorf, *Ich unde ein wip, wir haben gestriten* in Müller, *Kreuzzugsdichtung*, pp. 45–6; *Heu voce flebili*, *Carmina Burana*, no. 50; Berter d'Orléans, p. 35; Ambroise, v.11–17, 29–31, 1355–78, 2558–65; *Miror, cur tepeat*, *AH*, 23, no. 267; *Indue cilicium sedeas in pulvere*, *ibid.*, no. 266, pp. 316–18; Clement III in *La 'Collectio Seguntina'*, no. 38, p. 430; *ibid.*, no. 79, p. 441.

Though most contemporary writers attributed the divine wrath to *peccatis exigentibus hominum*,[59] that is, the sins of Christians at large, some saw the inhabitants of the Holy Land, particularly of Jerusalem, as responsible for the loss. It is striking that it was first and foremost the Franks themselves who, already in their appeals for aid in the summer and autumn of 1187, assumed responsibility for the loss of their kingdom, of Jerusalem, the Holy Sepulchre and the Holy Cross, explaining it as a just divine judgement on the grounds of *peccatis nostris exigentibus*. Thus already the Templar Terricus writing to the West before 6 August 1187 claimed that: 'Quot quantisque calamitatibus ira Dei, nostris peccatis exigentibus, nos in praesenti flagellari permiserit, nec litteris nec flebili voce, proh dolor! explicare valemus.'[60] Conrad of Montferrat, writing in 1189 to Baldwin, the archbishop of Canterbury, refers explicitly to the sins of the inhabitants of Jerusalem.[61] The Hospitaller Knight Hermenger, writing in 1188 to Leopold V of Austria, referred to the sins of all Christians of the Outremer.[62] The author of the *Libellus de expugnatione Terre Sancte per Saladinum* accused with bitterness the besieged citizens of Jerusalem who peacefully surrendered to Saladin for being betrayers of Christ. They were Christian in name alone, hypocrites burdened with sin, praising Christ with their lips, creatures whose natures were so distorted with avarice, malevolence and wickedness that they possessed only the external appearance of men whose humanity had been extinguished. To treason the anonymous author added the sin of cowardice. Never, he claimed, in all accounts of the ancient Jewish wars do we read that they surrendered without a fight or that they willingly yielded. The Christian action was unprecedented. The shame was intolerable. Those Christians were like evil merchants who sold Christ and the Holy City and thus they resembled Judas; they, who should have refused to betray the inheritance of Christ. He concluded his denunciation of the inhabitants of Jerusalem with a curse (citing Psalms 108(109):9–10): 'May his children be fatherless / and his wife a widow! ... may they be driven out of the ruins they inhabit.' 'Who could ever think that such a wicked thing could be done by Christians' he asked, 'to surrender of their own free will the tomb of Christ's resurrection, the noble Temple, the most sacred Mount Zion and the other places of the sacred city into the hands of unbelievers? O, the pain of it! There is no pain like this pain of ours!'[63]

---

59 See also Henry d'Albano, *De Peregrinante Civitate Dei*, cols 350–60 esp. col. 353; Gerald of Wales, *Expugnatio Hibernica*, p. 207; *idem*, *De Principis instructione*, p. 234; Ralph Niger, pp. 193–4; Hermenger in *Historia de Expeditione Friderici Primi*, p. 4; Clement III in *La 'Collectio Seguntina'*, no. 2, p. 422; Arnoldus Lubicensis, pp. 163–4; Sicardus Cremonensis, p. 168. For comparison with Titus' conquest see also Arnoldus Lubicensis, p. 164.

60 See the Hospitaller's letter in *Historia de Expeditione Friderici Primi*, p. 3; Hermenger, *ibid.*, p. 4; Terricus in Roger of Hoveden, *Gesta*, vol. 2, p. 13.

61 Conrad of Montferrat in Radulfus de Diceto, p. 61.

62 See in *Historia de Expeditione Friderici Primi*, p. 4.

63 *Libellus*, p. 248; Cole, 'Christian Perceptions', pp. 21–2.

The idea that it was the sins of Outremer that caused the disaster appeared in Europe immediately following the battle of Hattin in the form of the Genoese letter to Pope Urban III (written between 9 July 1187 and 4 September 1187 and probably immediately following the fall of Acre on 9 July 1187). The Genoese referred to a report from Outremer about 'the judgement operated by God in those parts [that is, in Outremer] in these days, provoked by our sins'.[64] Europeans followed suit and blamed the iniquities of the Christians of Outremer for the fate of their kingdom. Pope Gregory VIII, though stressing in *Audita tremendi* the responsibility of all Christendom, observed rather fatalistically '"that the land devoured its inhabitants" (Numbers 13:33), nor could it remain in a quiet state for long, nor could it keep in check transgressors against divine law',[65] thus pointing out the sins of the inhabitants of the Latin Kingdom. The pope was more explicit in his *Nunquam melius superni* of November 1187, where he explained the fall as caused 'ex peccato maxime habitatorum terrae et totius populi Christiani'.[66] Writing in Europe probably a few years after the fall, the Continuator of William of Tyre summarized these feelings as follows: '... One should not wonder that the Land of Jerusalem was lost. Because there was so much sin in Jerusalem that our Lord was deeply angered.'[67] Moreover, he argued, the clergy and people of Jerusalem, lacking the example of a righteous patriarch (due to the election of Heraclius of Caesarea as patriarch in 1180) fell into sin; therefore the wrath of the Lord Jesus Christ was provoked so that he cleansed the city of its inhabitants, save for two old men whom Saladin had allowed to remain.[68]

William of Newburgh, in his reflections upon the fall of Jerusalem, claimed that, according to the Holy Scriptures, those who polluted the Promised Land were always swept away. Thus were the Israelites expelled, first by the Babylonians and then by the Romans. The Holy Land, restored to the Christians by Constantine, was wrested from them by the Saracens. At the time of the First Crusade, the Holy Land welcomed the Christians; now because they had sinned, it once more rejected them. The present predicament of the inhabitants of the Holy Land was the Lord's judgement; their fate was to be held up as an example and warning to the faithful.[69] The inhabitants were no longer devout and God-fearing men, as

---

64 Roger of Hoveden, *Gesta*, vol. 2, pp. 11–13.

65 Gregorius VIII, *Audita tremendi*, p. 8.

66 *Idem*, *Epistolae*, no. III, col. 1539.

67 *La Continuation*, p. 49. For the author see J.H. Pryor, 'The *Eracles* and William of Tyre: An Interim Report', pp. 284–8.

68 *La Continuation*, p. 67; Edbury and Rowe, p. 5. This was also the opinion of William of Tyre himself, who wrote in 1182, in the preface to the twenty-third book of his history, as follows: 'Truly, we can lament with the prophet that there has perished from our midst "law from the priest, counsel from the wise, and the world from the prophet" (Jeremiah 18:18); among us it has come to pass that "like people, like priest" (Hosea 4:9) ...'. *WT*, 23, p. 1061. Eng. trans., vol. 2, p. 506.

69 William of Newburgh, vol. 1, pp. 249–55.

were the participants of the First Crusade; they were 'criminals, followers of *luxuria*, drunkards and minstrels and actors'.[70] Moral decline occupied also the author of the *Itinerarium Peregrinorum et Gesta Regis Ricardi.* He argued that the spirit of corruption, which had diffused everywhere, 'especially took possession of Syria, so that other nations now drew an example of uncleanness from the same source which formerly had supplied them with the elements of religion. For this cause, therefore, Lord, seeing that the land of his birth and place of his passion had sunk into an abyss of turpitude, treated with neglect his inheritance, and suffered Saladin, the rod of his wrath, to put forth his fury to the destruction of that stiff-necked people ...'[71] Rostangus of Cluny held a similar view of the Franks. Writing after the Fourth Crusade, he described the Holy Land as a hybrid of people, languages and customs. The Franks were indulging in a luxurious way of life; they were even deceiving and robbing pilgrims who had come from all over the world to adore the place 'where His feet had trodden'. It was on account of the sins of its inhabitants that the Holy City of Jerusalem was captured.[72]

It was particularly the sinfulness of the Frankish clergy which preoccupied the author of the *Persecutio Saladini*, based upon a source contemporary with the fall and emanating probably from *Outremer*. Both the clergy and the people gave themselves over to the excesses of luxurious living, and the entire land was defiled by their dissolute crimes. Those who made a pretence or put on a show of religious living transgressed disgracefully the limits of moderation and the regular life: there were few either religious or secular who were not corrupted by the disease of either avarice or luxury.[73]

Even harsher were the accusations hurled against the inhabitants of Jerusalem by Ralph Niger. In winter 1187–1188 he claimed that the Saracens had captured the Promised Land, its king, princes and people, had taken into their polluted hands the most holy temple, the Holy of Holies, that is, the Holy Sepulchre, and the very wood of the Cross had been carried off. This was a result of divine judgement upon *peccatis enim Palestine exigentibus*. 'And no wonder', he wrote, 'as this land was more dissolute than others, there was no reverence for God in it and in luxury ... it surpassed all other lands.'[74] The Franks went so far as to make pacts with the infidels, and by their intrigues betrayed into the enemy's hand even crusaders who had come to relieve their plight. Therefore, the land had invited God's wrath upon itself. Not only should one not wonder at the punishment that God had inflicted upon the inhabitants of Jerusalem, but also one should shed no

70 *Ibid.*, p. 254.

71 *Itinerarium Peregrinorum et Gesta Regis Ricardi*, pp. 5–6. Eng. trans. Nicholson, *Chronicle*, p. 23.

72 Rostangus, cols 905, 906. This was already said about the Franks by Gerhoh of Reichersberg following the Second Crusade. See Gerhoh Reicherbergensis, *De investigatione Antichristi*, p. 377. See also for the thirteenth century, Jacques de Vitry, pp. 162–3.

73 Richard, 'An Account', pp. 175–6; Cole, 'Christian Perceptions', pp. 17–18.

74 Ralph Niger, pp. 186–7.

tears. God did this no doubt for the correction of the people's sins and therefore, ultimately, for their own good. He frequently permitted calamities to befall the Israelites, so that they might be brought to their senses and repent. In fact, Ralph Niger concluded, a crusade for the recovery of Jerusalem might actually interfere with God's inscrutable designs by shortening the time ordained for the punishment and correction of the inhabitants of the Holy Land. It is the Lord who controls the distribution of territories (Psalms 23:1) and only He decides when Moslem domination of the Holy Land should cease.[75]

The fact that the sins of the Franks became in Europe one of the main explanations of the disaster, if not the most common one, needs some qualification. First of all, it is possible that this opinion derived from the *Audita tremendi* of Pope Gregory VIII as well as the numerous letters dispatched from Outremer, where the Franks themselves presented the disaster as the result of their own sins. But whereas the Franks used this as a sort of apologetic, as well as to describe the dimensions of the disaster and the immediate need for aid from Europe, in the West their declaration was taken literally and they became scapegoats for the disaster. Second, as the inhabitants of the Holy Land, the Franks were identified with the Sons of Israel. The comparison between them and the sinning Sons of Israel, severely punished by the wrath of God, was too attractive for European writers not to use it in their accounts of the loss of Jerusalem and the Holy Land.

At the same time, however, there were other explanations based not upon the idea that the 'Lord was losing his land', but rather on the idea that the 'Lord was purifying his land'. This was the opinion of the author of the *Itinerarium Peregrinorum et Gesta Regis Ricardi*: 'the lord seeing that the land of His birth and the place of His passion had sunk into the abyss of turpitude, treated with neglect His inheritance, and suffered Saladin, the rod of his wrath, to put forth His fury to the destruction of that stiff necked people: for He would rather that the Holy Land should, for a short time, be subject to the profane rights of the heathen, than it should any longer be possessed by those men, whom no regard for what is right could deter from things unlawful'.[76] The Moslems, thus, are agents not of Satan but of God. They were sent to chastise God's believers who have given themselves to sin. Saladin is an agent of God sent to punish sinners. Rather than pollution, as presented by other sources, here the Moslem conquest was presented as God's act of cleansing His holy places.

Yet another explanation connected the sinfulness of the inhabitants of the Holy Land and Christians at large with God's will to grant his believers an occasion for

---

75 *Ibid.*, pp. 92, 194–7, 200, 219–20 and n. 1. For the criticism of the vices of the Franks see also *Historia peregrinorum*, p. 117; *Itinerarium Peregrinorum et Gesta Regis Ricardi*, p. 5; *Itinerarium Peregrinorum*, pp. 246–7; Arnoldus Lubicensis, pp. 164–8; *Grandes Chroniques*, p. 15; Flahiff, pp. 162–79; Siberry, *Criticism*, p. 84; Schein, 'The Image', pp. 704–17.

76 *Itinerarium Peregrinorum et Gesta Regis Ricardi*, pp. 22–3.

their salvation. This idea appears mainly within the preaching of the Third Crusade, where the loss of Jerusalem and the fate of its sinful inhabitants become part and parcel of God's plan attempted to engender in Christians spiritual strength and renewed relationship with God. Gerald of Wales summarized this idea as follows: 'by God's judgement which is inscrutable but never unjust, the pagans and Parthians were allowed to gain the victory over the Christians, either to punish the eastern church for being tepid in its devotion to Christ, or to test and try the fervent faith and devoted loyalty of the peoples of the West'.[77] This was also the opinion of others involved with the preaching of the crusade, namely Henry of Albano and Peter of Blois. Cardinal-bishop Henry of Albano was appointed by Gregory VIII in the autumn of 1187 as his legate *a latere* and charged with preaching the crusade in Germany and France. In his letter of 1187 to the German princes, Henry reflected doctrine and ideas expressed by Bernard of Clairvaux on the eve of the Second Crusade and by Gregory VIII in *Audita tremendi*. He insisted upon the importance of the Holy Land as the place of Christ's Incarnation, Crucifixion and Redemption. He saw the loss of the Outremer as both the product of God's judgement upon sinful Christians and on opportunity for salvation. The victory of Islam over Christendom took place 'not because Mohammed could have caused it, but because Christ wished it', wishing to give Christians an occasion 'to display zeal for the glory of God, to avenge the injury of the Father and of Christ's inheritance'.[78] In his *De Peregrinante Civitate Dei*, composed shortly after the *Curia Ihesu Christi* at Mainz on 27 March 1188, identifying the fall with the Crucifixion, Henry explained why God has permitted the triumph of Mohammed over Christ. Everything that occurred is a manifestation of God's will, his justice and his mercy. Through Hattin and the loss of Jerusalem God was testing men to discover who are his faithful believers. He offered Christians the opportunity to prove themselves worthy by 'showing their zeal for the glory of the Lord, by vindicating the injury which the Father has suffered, and by vindicating His inheritance'.[79]

Peter of Blois who, upon the fall of Jerusalem set about composing works intended to stimulate support for Pope Gregory VIII's crusade,[80] also presented the fall as ordained by God. It had been with God's help that the Christians conquered during the First Crusade and ruled the holy places. It had been also God's will, because of the sinfulness of the people whom God entrusted with the custody of his holy places that the disaster occurred. Peter compared the Franks to previous rulers of the Holy Land – Syrians, Persians, Egyptians, Medes,

---

77 Gerald of Wales, *Expugnatio Hibernica*, c. 28, pp. 206–7 and see also *ibid.*, c. 30, pp. 214–17. On the Welsh tour of Baldwin of Canterbury, see Cole, *The Preaching*, pp. 71–8.

78 See above, n. 52.

79 Henry d'Albano, *De Peregrinante Civitate Dei*, col. 355.

80 See Southern, 'Peter of Blois and the Third Crusade', pp. 207–18; Cole, *Preaching*, pp. 72–4.

Macedonians and Romans, and insisted that they had all been destroyed for their sinfulness. The disaster, in common with other previous ones, was engineered by God to test the spiritual strength of His people, each period of devastation being followed by recovery. It is through a crusade that the believers would be woken from the deep spiritual lethargy that engulfed them and that the Holy Land would be recovered and rebuilt.[81]

An explanation which is also concerned with human sinfulness, but mainly with that of the European clergy, is that proposed by the Benedictine abbot, Arnold of Lübeck. According to Arnold, who compared the fall of Jerusalem to its destruction by Titus, the disaster occurred not because of the enemy's zeal but because of the perversity of the prelates. The clergy 'destroyed the walls of [Celestial] Jerusalem and therefore were banished from the Temple of God'. They became as desolate as the people (Isaiah 24:2). He quoted Jeremiah: 'Your ways and your doing have brought this upon you ...' (Jeremiah 4:18) as well as: 'the prophets prophesy falsely, and the priests rule at their direction; my people love to have it so ...' (*ibid.*, 5:31). He concluded with: 'Let us lie down in our shame and let our dishonour cover us; for we have sinned against the Lord our God, we and our fathers ... and we have not obeyed the voice of the Lord our God' (*ibid.*, 3:25).[82]

The loss of Jerusalem and the Holy Land became for the rest of the twelfth century a living reality for the Europeans as the memory of the disaster was kept alive, first of all, by the constant efforts to recover the Holy Land and, above all, Jerusalem; and second, by the various measures of penitence imposed on all Christians by Pope Gregory VIII; as a result a new liturgy was created bearing directly on the liberation of Jerusalem. The highly charged crusading atmosphere intensified penitential fervour at all levels of European society. The inspiration for this came largely from Gregory VIII's bull, *Audita tremendi*, in which the pope exhorted the believers to amend their ways and to take the cross as proof of their repentance: 'It is incumbent upon all of us to consider and to choose to amend our sins by voluntary chastisement and to turn to the lord our God with penance and works of piety; and we should first amend in ourselves what we have done wrong and then turn our attention to the treachery and malice of the enemy.'[83] The pope also announced various measures for the reform of the clergy and lay society, including a prohibition of gambling, hunting, feasting and luxurious clothes. He ordered a fast every Friday for five years and abstinence from meat on Wednesdays and Saturdays, declaring that the entourage of the pope and the cardinals would fast on Mondays as well.[84]

---

81 Petrus Blesensis, *Passio Reginaldi*, cols 968–70, 972–3; Cole, 'Christian Perceptions', pp. 26–30.

82 Arnoldus Lubicensis, pp. 162–4.

83 Gregory VIII, *Audita tremendi*, pp. 8–9.

84 Gregorius VIII, *Epistolae*, no. 3, col. 1539; no. 22, col. 1561; no. 23, col. 1561.

This plea for repentance and call for a crusade was reiterated by crusading preachers and crusade apologists, as well as poets and troubadours. Henry of Albano, in 1187, exhorted the Germans to take the cross, calling them to turn from 'malitia' to 'militia', and thus echoing the pleas of Bernard of Clairvaux.[85] Preaching the crusade at the assembly of Strasbourg in December 1187, Henry, the bishop of the city, also cited Bernard of Clairvaux, urging the German laity to seize this opportunity to mend their ways by taking the cross.[86] Ralph Niger advised, in 1187–1188, those who planned to join the Third Crusade to ascertain that they underwent a spiritual pilgrimage to Heavenly Jerusalem as well as a physical pilgrimage to the earthly city; for all their labours would be in vain if they had not purified themselves. It was more important, he maintained, to follow the way of faith, grace and justice than to go on pilgrimage; anyone who failed to build Jerusalem – the vision of peace – within himself would derive no benefit from his journey to Earthly Jerusalem.[87]

As before the fall of Jerusalem, the journey to the Holy City was now considered a gateway to Heavenly Jerusalem.[88] This idea was clearly expressed by Pope Gregory in *Audita tremendi*. The disaster, he declared, 'should serve as a lesson and example to those who are making their way to the Heavenly Jerusalem and to those who may not reach it except by the exercise of good works and many temptations.' One of these 'good works' was 'the labour for the recovery of that land in which truth arose from the earth for our salvation ...'[89] Henry of Albano expressed much the same idea in his *De Peregrinante Civitate Dei*. Though influenced by Augustine of Hippo, the cardinal of Albano presented the relation between the two cities as complementary rather than contradictory. Both heavenly and earthly cities are cities of God; Earthly Jerusalem was a preparation for the other. God had founded both cities, the earthly one through the nativity of Christ, who was thus the 'founder of the city' (*conditor civitatis*). It was, in Henry's opinion, the desertion of Heavenly Jerusalem that caused the fall of the earthly city. In order to recover both, knighthood should fight in the service of the faith, while churchmen should renounce worldly vanities. Thus Henry of Albano, like Pope Gregory VIII, Bernard of Clairvaux and Peter the Venerable, and many others writing in the twelfth century, presented Jerusalem as a Christian 'Holy of Holies' (*Sancta Sanctorum*) and as such the source and stage of salvation for the laity. Their way to reach Celestial Jerusalem, the true 'holy of holies', leads through the earthly city, through participation in a crusade for its recovery.[90]

---

85 Henry d'Albano, *Epistolae*, no. 32, cols 249–52. Cf. Bernard Claraevallensis, *Opera*, VIII, no. 363, p. 315.

86 *Historia Peregrinorum*, pp. 123–4; *Annales Marbacenses*, p. 163.

87 Ralph Niger, pp. 92–3, 147, 179, 182–83, 218–19.

88 See above, ch. 7.

89 Gregorius VIII, *Audita tremendi*, p. 9.

90 Henry d'Albano, *De Peregrinatione*, cols 257, 261, 283, 285, 287, 289, 296, 299, 338, 329, 351–61, 551; Congar, 'Henri de Marcy', pp. 56–61; *idem*, 'L'église et Cité de Dieu', pp. 173–202; Prawer, *Histoire*, II, pp. 13–14.

This seemingly common *topos* of crusade propaganda reappeared in poems composed after the fall of Jerusalem. Poets, lamenting the disaster, described participation in the crusade and the expected recovery of Jerusalem as a doorway to the celestial city. The author of 'Jerusalem, the Glorious City' (*Jerusalem, civitas inclita*; 1187) promised that military aid to the earthly city will grant access to the celestial city.[91] Another poet, writing between 1187 and 1189, exhorted the faithful to abandon Babylon and to fight for the 'water of life, for the celestial region'.[92] Regarding the crusade as an opportunity for salvation, poets exhorted European nobility to take the cross as an act of penance. Berter of Orléans, writing in 1187, described the path to the Holy Sepulchre by quoting Lamentations, 1:4, a text used in the service of matins on Holy Saturday, and compared the crusade with the visit of the three Mary's to Christ's tomb.[93] The anonymous author of 'What in his spirit David predicted' (*Quod spiritu David precinuit*) referred to the word pronounced by the angel at the tomb (*surrexit*) on the morning of Easter Sunday (Matthew 28:6; Mark 16:6; Luke 24:6) by ending each stanza with the refrain: 'Exsurgat Deus'. The emphasis was thus on Christ's Easter victory, in order to assure the crusaders that Christ's victory was a guarantee that those fighting for Jerusalem in His name would ultimately triumph. Even though God in His omnipotence had no need for assistance, in the fall of Jerusalem He had created a situation in which the knights were offered the means of salvation.[94] Those who refused to fight for His Cross and Sepulchre could not hope for Heaven or for their salvation, claimed the German poet Albrecht von Johansdorf.[95]

Other poets argued that God had allowed Earthly Jerusalem to be lost so that the faithful might have an opportunity to gain salvation by fighting for its recovery. Their slogan was that those who assisted the Earthly Jerusalem would gain the Heavenly City; they would be saved from the tribulations of Hell and achieve the crown of martyrdom.[96] 'Whoever in His need fails Him now / Knows that, in Judgement, he in turn will fail. / So let the great and little all take note / that they should take the field in chivalry / Where paradise and honour can be won,' as written by Conon of Béthune on the eve of his departure on the Third Crusade.[97]

---

91 Hagenmeyer, 'Deux poésies', p. 584.

92 *Crucifigat omnes*, *Carmina Burana*, no. 47, p. 92.

93 Berter of Orléans, *Juxta threnos Jeremiae*, *The Oxford Book of Medieval Verse*, no. 201, pp. 297–300; Green, *The Millstätter Exodus*, pp. 409–11; Siberry, *Criticism*, p. 57.

94 *Carmina Burana*, no. 48, p. 96; Green, pp. 411–13. See also *Pour Son Peuple Resconforteir*, *Chansons de Croisade*, no. VII, pp. 78–81 esp. pp. 79–80; *Miror, cur tepent*, *AH*, 33, pp. 319–20.

95 Albrecht von Johansdorf, *Die hinnen varn* in Müller, *Kreuzzugsdichtung*, no. 27, pp. 42–3.

96 *Jerusalem, civitas inclita* in Hagenmeyer, 'Deux poésies', p. 584; *Indue cilicium, sedeas in pulvere*, *AH*, 33, no. 266, p. 317; Berter d'Orléans, *Juxta threnos Jeremiae*, *Oxford Book of Medieval Verse*, no. 201, pp. 299–300. For martyrdom see *Crucifigat omnes*, *Carmina Burana*, no. 47, p. 92; *Itinerarium Peregrinorum*, p. 55.

97 Ahi, Amors, *Chansons de croisade*, no. III, pp. 32–5.

Another poet (writing in *c.* 1189) sang: 'The day of peace has come / Which God, in his great tenderness, / Will give to those who of love of him / Take the cross and for their burden / Suffer pain both night and day. / Then he will see who truly love him.' He declared: 'Do you know what God has promised / To those who wish to take the cross? / God help me, a very fair wage: Paradise, by firm promise ...'.[98] The author of 'What David in his spirit predicted' (*Quod spiritu David precinuit*) claimed that God wished Jerusalem to be lost so that the believers, by undertaking its recovery, could redeem their sins.[99] The troubadour Ambroise argued that, if God did not punish His people severely for their transgressions, they would not amend their ways and only a few would come to Him after the Last Judgement.[100]

And, indeed, the appeal to launch a crusade had an almost immediate effect. The response was overwhelming. According to the chroniclers, a great multitude responded to the call, including many of the rulers of Christendom.[101] The future King Richard I of England, then count of Poitou, took the cross in November 1187, immediately after the news of the fall of Jerusalem had reached the West. King Henry II and Philip Augustus emulated his example at Gisors in January 1188,[102] an act ascribed by contemporaries to God's miraculous intervention.[103] The dukes of Burgundy and Flanders and the count of Champagne, as well as numerous knights from both kingdoms, followed both kings.[104] Emperor Frederick I, together with several German princes, bishops and knights,[105] took the cross on 27 March 1188 at the so-called *Curia Ihesu Christi*, the assembly of Mainz.[106]

---

98 *Vos qui ameis de vraie amour*, *Chansons de croisade*, no. II, pp. 20–22. Eng. trans. Riley- Smith, *Idea and Reality*, pp. 89–90.

99 *Carmina Burana*, no. 48, p. 95.

100 Ambroise, v. 2558–65.

101 Gerald of Wales, *De principis Instructione*, p. 241; *Annales Colonienses Maximi*, *MGH.SS*, 17, pp. 796–7; Arnoldus Lubicensis, p. 170; Rigord, p. 25; Gerard of Wales, *Expugnatio Hibernica*, pp. 207–9; *Chronicon Universale Anonymi Laudunenses*, pp. 46–7.

102 Radulfus de Diceto, vol. 2, pp. 50–51; Rigord, p. 25; Guillaume le Breton, *Gesta Philippi Augusti*, *RHGF*, 17, p. 146; *Grandes Chroniques*, pp. 167–8; Roger of Hoveden, *Gesta*, vol. 2, pp. 29–30. See also William of Newburgh, pp. 271–2; Gerard of Wales, *De principis Instructione*, pp. 239–40.

103 Rigord, p. 25; *Grandes Chroniques*, p. 168.

104 Gerard of Wales, *Expugnatio Hibernica*, p. 209; William of Newburgh, pp. 271–2; Roger of Hoveden, *Chronica*, vol. 2, p. 335.

105 *Historia de Expeditione Friderici Primi*, pp. 14–15; *Historia Peregrinorum*, pp. 125–6; *Annales Colonienses Maximi*, pp. 796–7; Arnoldus Lubicensis, p. 170; Magnus Reicherspergensis, p. 509. According to *Annales Colonienses Maximi*, pp. 796–7, 15,000 knights took the cross after the assembly at Mainz; the army of the emperor counted 30,000 men according to Rigord, p. 25; Guillaume le Breton, p. 146; Arnoldus Lubicensis, p. 170.

106 For this assembly as *Curia Jesu Christi*, or *Curia Dei* see *Historia Peregrinorum*, p. 123; *Historia de Expeditione Friderici Primi*, p. 14; Gerard of Wales, *De principis Instructione*, p. 241; *idem*, *Expugnatio Hibernica*, p. 209; *Annales Colonienses Maximi*, p. 794; *Itinerarium Peregrinorum et Gesta Regis Ricardi*, p. 43; *Annales Marbacenses*, p. 164; Otto de Sancto Blasio, p. 44.

The aim of the Third Crusade was seen by its contemporaries in relation to the recovery of what had been lost, namely, the Holy Land and Jerusalem but particularly the Holy Sepulchre and the most sacred of Christian relics – the Holy Cross.[107] Peter of Blois described the Holy Cross as 'our ark of New Testament', 'the standard of our salvation', the 'ladder to heaven' and 'gateway to paradise'.[108] For Henry of Albano, the Holy Cross was *inter alia* the 'glory of the Christian nation', 'the medicine of sinners' and 'the ark of the Covenant of the Lord'.[109] An anonymous poet lamenting the loss of Jerusalem referred to the Holy Cross as 'Ark of Zion',[110] while the Templar Terricus called it the 'Ark of David'.[111] Such descriptions of the relic of the Holy Cross were apparently directly inspired by the liturgy of the Feast of the Exaltation of the Holy Cross observed on 14 September.[112]

The Third Crusade itself, like the First, was mainly perceived by its contemporaries as the *via Sancti Sepulchri*.[113] It, too, aimed to avenge the injury done to God by His enemies and the insult to His Cross and Sepulchre. Saracen rule over Jerusalem was considered a pollution of its holy places and thus an insult to God.[114] Alan of Lille (in 1189) exhorted the faithful 'to avenge Christ's injuries and to set off to Jerusalem to liberate the land of our inheritance, Christ's

---

[107] On the Holy Cross see also Katzir, 'The Vicissitudes', pp. 251–3.

[108] Peter of Blois, *Passio Reginaldi*, cols 959, 968.

[109] Henry d'Albano, *De Peregrinante Civitate Dei*, col. 353; *idem*, *Ex quo vox*, p. 11.

[110] *Indue cilicium, sedeas in pulvere*, *AH*, 33, no. 26, p. 317.

[111] Roger of Hoveden, *Gesta*, vol. 2, p. 40.

[112] For the liturgy see the 'Sequence for the Exaltation of the Holy Cross' (c. 1100) in *The Oxford Book of Medieval Latin Verse*, no. 147, pp. 204–8. The Holy Cross is described as 'mundi salus' and 'medicina christiana' which is to 'salva sanos, aegros sana' (*ibid.*, p. 207). For the loss of the Holy Cross see also Sicardus Cremonensis, p. 168; *Carmen Sangallense* in Hagenmeyer, 'Deux Poésies', p. 585; *Crucifigat omnes*, *Carmina Burana*, no. 47, p. 92; Haymarus, p. 74; Gregorius VIII, *Audita tremendi*, pp. 7–8; Roger of Hoveden, *Gesta*, vol. 3, p. 361; Arnold Lubicensis, pp. 169, 170; *Annales Colonienses Maximi*, p. 793; *Historia peregrinorum*, p. 119; *Itinerarium peregrinorum*, p. 265; Berter d'Orléans, *Juxta threnos Jeremiae*, *The Oxford Book of Medieval Verse*, pp. 298–9; Rostangus, col. 907; *Lugent Sion et Judaea*, *AH*, 21, no. 236, pp. 165–6; *Miror, cur tepeat*, *ibid.*, 33, no. 267, pp. 319–20; *Heu, voce flebili*, *Carmina Burana*, no. 50, p. 100.

[113] *Annales Admontenses*, *MGH.SS*, 9, p. 586; *Continuatio Claustroneoburgensis secunda*, *MGH.SS*, 9, p. 618; Sicardus Cremonensis, p. 168; *Historia de Expeditione Friderici Primi*, p. 15; *Historia peregrinorum*, p. 122; Rigord, p. 25; *Annales Mellicenses*, *MGH.SS*, 9, p. 505; Magnus Reicherspergensis, p. 509; Haymarus, p. 74; Albrecht von Johansdorf, *Die Hinnen varn* in Müller, no. 27, pp. 42–3; *Quod spiritu David precinuit*, *Carmina Burana*, no. 48, p. 95. Ambroise states that King Henry II (who died on 6 July 1189) 'had made plans for visiting / The Holy Tomb at God's command / But was prevented by death's hand'. Ambroise, v. 168–70. On Henry's reaction to the fall of Jerusalem see also *Epistolae Cantuarienses*, *RS*, 38, no. 158, p. 134.

[114] For the Saracens' presence in Jerusalem as polluting its holy places see here above, nn. 19–28. See also Haymarus, p. 74; Otto de Sancto Blasio, p. 43.

inheritance, the Virgin's marriage portion'.[115] Of the holy places of Jerusalem, however, almost only the Holy Sepulchre figured in European reaction to the loss of Jerusalem. Except for a few references to the Virgin Mary,[116] Jerusalem was depicted as *Civitas Crucifixi*. As the dual goals of the Third Crusade were the liberation of the Holy Sepulchre and the Holy Cross, when King Philip Augustus returned to France, some Europeans complained that he had returned 'without achieving anything, without having, as his booty, either the Holy Sepulchre or the Holy Cross'.[117]

Problems concerning the French and English armies between 1188 and 1190 triggered off various reactions. For example, a poet known as 'Maistre Renaut', outraged by the delay of the departure to the East of Richard I and Philip Augustus, warned them both of the perils of the Last Judgement and urged them to avenge the loss of the Holy Sepulchre.[118] Others too, and among them Ralph Niger, reacted to the delays with criticism of the crusaders.[119] Some were even sceptical as to the very advisability of a crusade for the recovery of Jerusalem and the Holy Land. The Minnesinger, Albrecht of Johansdorf, reveals that some crusade preachers had encountered opposition from sceptics: 'those who leave this country tell us ... that Jerusalem, the pure city, and the country have never been in greater need of help. This appeal is greeted with derision by the simple ones. They all say, "If Our Lord was offended, he could avenge himself without all the soldiers having to cross the sea".'[120] The views of Ralph Niger and Albrecht of Johansdorf were, however, exceptional and in sharp contrast to the enthusiastic climate of 1187–1188. As a matter of fact, the accounts of the fall of Jerusalem contain no criticism and disapproval of the crusade other than of the delays in its departure. The criticism was thus levelled not against the idea of the crusade but against those who failed to promote it. Thus, for example, Henry of Albano claimed that those who refused to take the cross and fight for Jerusalem would be afflicted with the curse of Judas Iscariot.[121]

The memory of the fall of Jerusalem was kept alive also by penitential liturgical practices imposed on all Christians by Pope Gregory VIII. According to *Nunquam Melius Superni*, his letter of 29 October 1187, 'all persons shall, for the next five years, on every sixth of the week, at the very least, fast upon Lenten fare, and that,

---

115 Alain de Lille, p. 281. See also Henry II in Roger of Hoveden, *Gesta*, vol. 2, p. 39; Celestine III in *Oriens Pontificius* III, no. 230, p. 406; *Quod spiritu David precinuit*, *Carmina Burana*, no. 48, pp. 95–6.

116 *Heu voce Flebili*, *Carmina Burana*, no. 50, p. 101; Ralph Niger, p. 96; Alain de Lille, p. 281.

117 Ricardus Divisiensis, p. 58.

118 *Chansons de croisade*, pp. 59–63, 67–71, 77–81; Siberry, *Criticism*, pp. 59–60.

119 Flahiff, pp. 163–5; Siberry, *Criticism*, pp. 53–63.

120 Albrecht von Johansdorf, *Die hinnen varn* in Müller, no. 27, pp. 42–3. See also Heinrich von Rugge, *Ein tumber man iu hat*, *ibid.*, no. 32, pp. 47–51; Siberry, *Criticism*, pp. 192–3.

121 See here above, n. 17.

wherever mass is performed, it shall be chanted at ninth hour; and this we order to be observed from the Advent of our Lord until the Nativity of our Lord. Also, on the fourth day of the week, and on Saturdays, all persons without distinction, who are in good health, are to abstain from eating flesh ... This therefore we enact to be observed, and whosoever shall be guilty of transgressing the same, is to be considered as breaker of the fast in Lent.'[122]

In 1188 a new liturgy,[123] bearing directly on Psalms 78(79):1: 'Deus venerunt gentes', was introduced when the Curia issued general directives to hold fasts and celebrate public prayers. The implementation appears to have been immediate for, even in the same year, masses were recorded as being celebrated in London at Westminster or at St Paul's, 'pro pace et pro deliberatione terrae Jerusalem et Christianorum captivorum ...' Both the penitential measures and the new liturgy aimed to commemorate the loss of Jerusalem and the entire Holy Land, as well as to provide a backdrop to the propaganda campaign for the Third Crusade. Crusading and penance were always closely connected: in the new liturgy the two themes appeared together with that of the sanctity of Jerusalem. What is striking is that within this new liturgy the theme of Jerusalem's sanctity overshadows all the other themes. Thus, in common with the other genres of sources, the liturgy clearly reflected that the recovery of Jerusalem would become the main topic in preaching and propaganda for the Third Crusade. The main feature of the new liturgy was the introduction of a sequence of prayers into the canon of the mass, immediately before the *Agnus Dei*, each sequence consisting of a psalm accompanied by an antiphon, followed by eleven verses and a collect, and set out in a seven-day programme of daily masses.

All seven selected psalms referred directly to Jerusalem, implicitly linking it with the particular character of the liturgical occasion. The sequence began with Psalm 2, to be recited on Sunday, and describing Jerusalem as a scene of a struggle by the Gentiles against God and prophesying God's victory. The psalm for Monday, Psalm 53(54), combined the psalmist's cry for aid against his enemies with the story of David's flight from Saul. The psalm did not refer explicitly to Jerusalem but included the verses: 'with a freewill offering I will sacrifice to thee' (Psalms 53(54):6), verses traditionally linked with Jerusalem as the city of the Temple and the sacrifice of Christ. Tuesday's Psalm 59(60) referred to God's anger and the punishing of his people, thus alluding to the fall of Jerusalem and the Holy Land as God's punishment (Psalms 59(60):1–3). The verse 'thou has made the land to quake' (Psalms 59(60):2) provided inspiration to many of the contemporaries describing the event. Thus King Richard I, possibly also moved by Bernard of Clairvaux and Henry of Albano, wrote on 1 October

---

122 Gregory VIII, *Nunquam Melius Superni*, *PL*, 202, col. 1539. Eng. trans. Roger de Hoveden, *Chronicle*, vol. 2, pp. 74–5.

123 For the liturgy see Roger de Hoveden, *Chronica*, vol. 2, pp. 359–60. Eng. trans., vol. 2, pp. 103–4; Linder, '*Deus Venerunt Gentes*', pp. 145–52.

1191: 'After the tragic and universally lamented destructions of the holy city of Jerusalem, the city of the living God by which His name is invoked, the world was shaken and trembled because the King of Heaven has lost his land wherein His feet had stood.'[124] The text of this psalm also offered the promise of victory and retribution over the enemies of Israel such as Moab, Edom and Philistia, the traditional foes of the Sons of Israel, and to be identified, in the accounts of the fall, with the Moslems.

The Psalm for Wednesday was 73(74). This psalm was a perfect commemoration of the destruction of Jerusalem, including the desecration and the destruction of its Temple by God's enemy: 'the enemy has destroyed everything in the sanctuary! Thy foes have roared in the midst of thy holy place ... they set thy sanctuary on fire; to the ground they desecrated the dwelling place of thy name' (Psalms 73(74):3–7). The plea for aid against the enemy recalls the concept of Jerusalem as God's dwelling place and his inheritance: 'momento ... liberasti virgam hereditatis tua mons Sion' (Psalms 73(74):2). On Thursday Psalm 78(79) was recited. This psalm included the most commonly used verse in reaction to the fall, namely: 'O God, the heathen have come into thy inheritance; they defiled thy holy temple; they have laid Jerusalem in ruins.' The psalm narrated the conquest of Jerusalem by the Gentiles as well as the martyrdom of God's servants. It confessed the sins committed, and asked for both forgiveness and retribution against those who say: 'Where is their God?' (Psalms 78(79):10). On Friday Psalm 82(83) was recited. Like the preceding psalms it asked for help and harsh retribution against God's enemies described *inter alia* as Moab, Ammon, Philistia and Amalek. It insisted that those who said: 'hereditatem possideamus nobis sanctuarium Dei' (Psalms 82(83):12) would perish in disgrace. The series ended with Psalms 93(94), recited on Saturday; it was a cry for vengeance and justice against those who 'crush thy people ... and afflict they heritage' (Psalms 93(94):5); the psalm included a promise for the future: 'For the Lord will not forsake his people; he will not abandon his heritage' (Psalms 93(94):14).

The analysis of the liturgy reveals its close affinity to the sentiments and *topoi* expressed in other genres. Similarly to other sources, the fall is explained by God's punishment of human sinfulness. Jerusalem appears as the City of God, this dwelling place, the City of his Sanctuary or Temple, His inheritance. In its present state this City of God and its sanctuaries were being destroyed and polluted by God's enemies. It was, however, only a temporary state of affairs as God will not desert His city for long. He will avenge his enemies and Jerusalem will return to Christian hands.

While it is known that in 1188 the special liturgy was celebrated, perhaps in a shortened version, not only in England but also in other parts of Europe, for

---

[124] Roger of Hoveden, *Chronica*, vol. 3, pp. 130–33. Eng. trans. Edbury, *The Conquest*, pp. 179–81. For Bernard of Clairvaux see above, ch. 3, n. 44. For Henry d'Albano see above, n. 17.

instance, in Germany, the kingdom of France and Sicily,[125] by 1194, Psalm 78(79) had been inserted into the Mass for the Dead and other masses by the Cistercian Order which instituted a weekly 'oratio pro Terra Ierosolimitana'.[126] In April 1213, Innocent III outlined the programme of daily prayers in *Quia maior*, which proclaimed the Fifth Crusade. The pope commanded '[that] every day during the celebration of mass, when the moment has come after the Kiss of Peace, when the saving sacrifice is to be offered for the sins of the world or is about to be consumed, everyone, men and women alike, must humbly prostate themselves and the psalm "Oh God, the heathens are come into thy inheritance" [Psalms 78(79)] should be sung loudly by the clergy. When this has been ended reverently with this verse, "Let God arise, and let his enemies be scattered; and let those who hate him flee from before him" [Psalms, 68:2], the celebrant must chant this prayer over the altar: "God, who has disposed all things with marvellous providence, we humbly beseech Thee to snatch from the hands of the enemies the cross and the land which thine only begotten son consecrated with his own blood and to restore it to Christian worship ..."'[127]

To summarize: the vehement reaction to the fall of Jerusalem may be compared in its intensity with the response to its conquest by the First Crusade almost a century earlier, though that initial joy was now replaced with grief. This reaction was particularly striking in comparison with the rather mild response to Alexander III's appeals for aid to the kingdom of Jerusalem on the eve of its fall.[128] Like the crusading conquest of Jerusalem, the loss of the city was regarded as God's will and was associated with Christ's Passion, even being seen as a kind of Second Crucifixion. It was perceived as the realization of biblical prophecies, and as part and parcel of the events of the Last Days, in which Jerusalem's role was to be crucial, but here the resemblance ends. In European accounts of its fall, Jerusalem appears as a vastly different city from that of the 1090s. The City of the Holy Sepulchre[129] had become the place at which the Lord 'hath wrought salvation in the midst of the earth' (Psalms 74:12), the site as well as the source of salvation – the inheritance of Christ and his patrimony and above all the 'place where His feet had trodden' (Psalms 132:7).[130] Jerusalem was the city of the life of Christ and the apostles, and especially the stage He had chosen for His act of salvation. Though the aim of the Third Crusade was seen mainly in terms of the recovery of the Holy

---

[125] Arnoldus Lubicensis, p. 170; Guilelmus Andrensis, *Chronica Andrensis*, *MGH.SS*, 24, p. 719; Linder, '*Deus Venerunt Gentes*', pp. 150–51.

[126] *Statuta Capitulorum Generalium Ordinis Cisterciensis*, ad. an. 1194, statut. 10, p. 172; Linder, '*Deus Venerunt Gentes*', pp. 151–2. For the liturgy for the recovery of Jerusalem in the thirteenth century see Linder, *ibid.*, pp. 152–62; Cole, 'Purgatory', pp. 713–25.

[127] Innocent III, *Quia maior*, *PL*, 216, col. 821; Cole, 'Purgatory', p. 716.

[128] Alexander III, *Epistolae*, no. 360, cols 384–6; no. 404, cols 1294–6; Rowe, 'Alexander III and the Jerusalem Crusade', pp. 112–32.

[129] Above, n. 39.

[130] *Ibid.*

Sepulchre and the Holy Cross, the image of Jerusalem within the reaction to its fall was a far cry from what it was on the eve of its conquest by the First Crusade. Jerusalem could number among its many titles 'Holy City', 'mistress of other cities', 'princess among provinces', 'His house', 'His dwelling', 'His inheritance', 'His *patrimonium*', 'His Sanctuary', *Civitas Domini Virtutum*, 'the Glorious City of God', 'Holy of Holies of Christians', 'another heaven', the 'City of Living God'. Inspired largely by Old Testament phraseology, those twelfth-century writers on Crusader Jerusalem clearly demonstrate the transformation of the city's status – into capital, royal city, City of the Crucified and gateway to the Heavenly City.

As far as the explanation of the fall is concerned, the influence of church leaders was, it seems, quite limited. The fall was explicable in terms of God's just punishment for human sinfulness, *peccatis exigentibus*. Nevertheless, a controversy raged over the subject of 'whose sins'. The 'official' explanation, as propagated by the papacy and the official preachers of the crusade such as Henry of Albano and Peter of Blois, insisted that the sins were those of all Christians. On a more popular level, chroniclers and poets rejected this view, most of them tending to blame the sinfulness of the Franks.

Both explanations of the fall as well as the image of Jerusalem are universal in the sense that there are hardly any geographical or 'national' differences; the only differences were those existing between the inhabitants of the Latin Kingdom and those of the West. This showed how far the image of the Holy City had become a universal and homogeneous one throughout Western Christendom. If the image of Jerusalem itself was a homogeneous one, there are differences of opinion in the descriptions of what was lost as well as in regard to the aims of the Third Crusade. The contemporaries at large saw the loss as well as the aim of the Third Crusade almost exclusively in terms of the Holy Sepulchre and Jerusalem. The papacy, however, as well as the official preachers of the Third Crusade, referred to the Holy Land or the 'Land of Jerusalem' rather than to Jerusalem or the Holy Sepulchre. It should be pointed out that neither Gregory VIII nor his successor on the papal throne, Clement III, referred to the Holy Sepulchre in their crusade propaganda. Nor did the leading figures among the preachers of the Third Crusade[131] make the Holy Sepulchre the main theme of their exhortations for the recovery of Jerusalem. The exception is Celestine III who, in *Cum ad propulsandam* of 11 January 1193, described the aims which the Third Crusade failed to achieve as: '... of repelling

---

[131] The statement of Penny J. Cole (*The Preaching*, p. 69) that 'the Holy Sepulchre figures largely in Henry's exhortation ...' is rather vague as Henry d'Albano hardly mentioned the Holy Sepulchre at all; it was absent from his circular letter to the prelates of the Church, the *Publicani et peccatores* of 1187 (see *PL*, 204, cols 247–74) as well as his circular letter to the German prelates and magnates summoning them to the Court of Mainz, the *Ex quo vox* (*PL*, 204, cols 249–52). In his *De Peregrinante Civitate Dei*, he mentioned the Holy Sepulchre once, together with the Holy Cross (*PL*, 204, col. 353). Peter of Blois referred to the Holy Sepulchre only in his *Dialogus*, *PL*, 207, col. 987.

the injuries done to the people of Christendom, and wiping away that stain to our common faith which the pollutions of the pagans, in the capture of the Holy Land, which has been named the inheritance of the Lord, had perniciously and violently imprinted there upon, and for cleansing the Holy City and the Sepulchre of our Lord from the abominations of the Saracens ...'. This pope urged the faithful to 'liberate the land and the Sepulchre of the Lord'.[132]

The idea of the crusade and its recovery of Jerusalem was emphasized by churchmen such as Henry of Albano and became a huge success. The preaching of the Third Crusade presented Jerusalem as the source and stage of salvation for the laity, as the gateway to Celestial Jerusalem. The reaction to the fall was characterized by a strengthening of the bond between the earthly city and its celestial counterpart. It seems that never before – save perhaps at the time of the First Crusade – was Earthly Jerusalem considered by so many and so widely to be the gateway to the Heavenly Jerusalem as it was immediately after its fall. The massive response by those taking the cross even by those who had been strongly warned against it, namely, by members of the monastic orders and the clergy,[133] demonstrates the great attraction and power of the idea at all levels of society. Typical of this state of mind are the verses of the poets of the Third Crusade.[134] Lamenting those who had died on this crusade, 'those [who] returned not', Ambroise, himself a participant in this crusade, wrote: '[They] shall be with God at his right hand / In high Jerusalem's heavenly land / These, through the good deeds done by them / Won that fairer Jerusalem'.[135]

---

132 Celestine III, *Epistolae*, no. 102, cols 970–73; Roger of Hoveden, *Chronica*, vol. 3, pp. 200–203. Eng. trans. vol. 2, pp. 282, 285.

133 For their participation see *Historia de Expeditione Friderici Primi*, pp. 14–15. For a list of 'prominent people' who took the cross see also Henry d'Albano, *De Peregrinante Civitate Dei*, cols 357–8.

134 See here above, nn. 8, 28, 39, 45, 46–50, 55, 58.

135 Ambroise, v. 12252–5.

Conclusion

During the twelfth century the attitude of Christendom to Jerusalem underwent a meaningful transformation which was partly the result of changes in Western spirituality and partly the outcome of the establishment of Christian rule in Jerusalem. What is striking, first of all, is the quantitative increase in the very references to Jerusalem. Jerusalem looms large in all the literary genres of the period, including sermons, theological treatises, chronicles and annals, letters, crusading *exitatoria*, literature and especially poetry.

The biblical atmosphere of the First Crusade led to the revival of the biblical past of Jerusalem, a past which had been intentionally bypassed during the previous period of Christian rule of the city in the Byzantine period. Already during the first decade of the twelfth century, Jerusalem became for Christendom not only the City of the New Testament but also that of the Old Testament. Though the basis of its holiness remained its special link with Christ, what was stressed now, more than ever before, was Jerusalem's biblical past, particularly as the 'City of David'.

As the City of the New Testament, Jerusalem became now the City of the Humanity of Christ, the place where He had lived as a child, as a boy and then as a man. It was also the place where He had suffered as a man. Contemporaries, whenever they referred in hierarchical terms to the holiness of Jerusalem, linked it first directly to Christ, then to the Virgin Mary and finally to the apostles including St James. At the same time, one of the main features characterizing the early medieval attitude to Jerusalem had vanished before the First Crusade, namely, the almost exclusive identification of the city with the Holy Sepulchre, the site of Christ's burial. Jerusalem now became the scene of the life of Christ. During the crusader period Jerusalem's significance as the symbol of the supreme event in the history of the Redemption, the place of the Resurrection of Christ, became predominant. Both the perception of the Holy Sepulchre as the site of the Resurrection of Christ and the emphasis on Jerusalem's biblical past were peculiar, especially to the 'Franks', the inhabitants of the crusader kingdom. The kings of Jerusalem identified themselves with the kings of the House of David and with the Maccabees, whereas their subjects were identified with the Children of Israel. Under the rule of the crusaders the 'Church of the Holy Sepulchre' became the 'Church of the Resurrection of Christ' and the patriarch of Jerusalem changed his title from *Dei gratia patriarcha Jherosolimitanus* to *patriarcha sancti Resurrectionis Christi ecclesie*.

The character of the holiness of Jerusalem had not only undergone a transformation but, one may say, it also increased quantitatively during the twelfth century. A massive topographical identification of holy traditions, together with the renovation or rebuilding of the sites where they were located, led to an extraordinary enrichment of the 'Holy Geography of the Holy City'. As a result, never before had the holy geography of Jerusalem been as rich and heterogeneous as it was during the twelfth century. This was the Jerusalem of King David and Christ, of the Virgin, of both the biblical prophets and the apostles. Centred on its two major shrines, the newly built Church of the Holy Sepulchre and the restored *Templum Domini*, Crusader Jerusalem embodied the entire history of the Christians as depicted in both the Old and the New Testaments. Yet what was cardinal was Jerusalem's position as the 'city of the King of Kings'. This was defined by a theological theme, through which the contemporaries succinctly expressed their image of the Holy City: 'the place where His feet had trodden'. Originating in the papal chancery, this so-called formula became the most popular image as well as an expression of that which linked the believers to the Holy City.

Held as a miracle and a manifestation of divine will, the conquest of Jerusalem also brought about the establishment and consolidation of the political claim of Christendom to Jerusalem. Justification of the Christian right to possess Jerusalem reflects the city's new status within Christendom. The right to dominate Jerusalem was formulated in different terms from those applied to other conquered territories, including the rest of the Holy Land. Thus, even the canonists who normally used the Augustinian concept of 'Just War' to justify Christian conquests, argued now that it was Jerusalem's sanctity that justified its rule by Christians. Considered as Christ's inheritance or his fief, the Christian rule in Jerusalem was, in the twelfth century, considered as axiomatic, as a self-evident truth. Any other rule was regarded as an insult to God, an offence to Christianity, and the pollution of its most holy places.

No less remarkable was the change that occurred in the wake of the crusader conquest in the relations between the Earthly and the Heavenly Jerusalem. As a goal of both pilgrimages and crusades, Earthly Jerusalem emerged in the twelfth century, with almost no opposing voices, save for the members of the monastic orders, and established itself as a gateway to the Heavenly City, a gateway to salvation, and the way by which the faithful Christian might achieve the salvation of his soul. It is quite clear from letters, popular literature and surviving testaments that Heavenly Jerusalem was foremost in the minds of both pilgrims and crusaders, when undertaking their journey. The crusade was perceived mainly in terms of pilgrimage to Jerusalem, an entrance to the Heavenly City. Opposition to Earthly Jerusalem, common enough in the early Middle Ages, appeared in the twelfth century only rarely and, when it did so, it was mainly within the framework of the general opposition of monastic leaders to pilgrimage by members of their communities. This resulted from a greater insistence upon the ideals of stability and obedience, as well as a shift in monastic moral teaching from external behaviour to internal attitudes. The sharp increase in monastic circles of the twelfth century in the opposition to pilgrimages as such, including those to Jerusalem, might also have been a reaction to the actual increase in pilgrimages by regulars and their participation in the crusades to Jerusalem. This opposition, however, was limited to monastic circles and should not be regarded as the negation of the value of the Earthly Jerusalem.

What is characteristic of this state of affairs is the position taken by the two luminaries of the monastic world, Bernard of Clairvaux and Peter of Cluny. Both obviously preferred the Celestial Jerusalem. Both, while preaching the merits of the crusade to the European knighthood, described fighting for Jerusalem as the gateway to the Heavenly City. Both at the same time vigorously prohibited their monks from participating in a crusade or a pilgrimage to Jerusalem. In fact, it seems as if the rejection of the Earthly City in favour of the Heavenly became, in the twelfth century, one of the self-made characteristics of the monastic orders as a group in the face of the laity. On the one hand, there was the increasing belief, sometimes verging on obsession, in the salutary value of Earthly Jerusalem as the goal of both pilgrimages and crusades of laity. On the other, the total negation of this concept by monastic orders points to a change that occurred during the twelfth century in European society. Now the name 'Jerusalem' no longer held the same meaning for all. It meant one thing to the nobility, another to monks and, to some extent, something else to the clergy. Whereas the link, established as early as the eleventh century between the European nobility and the Earthly Jerusalem, had meaningfully strengthened in the twelfth century and incorporated the belief that the road to salvation led through Earthly Jerusalem, the monastic orders had rejected this road altogether. Thus, only the final aim remained the same for all – the Heavenly Jerusalem. This decline in the opposition to the Earthly Jerusalem, which dominated early Christianity and the early Middle Ages, seems to have been an outcome of practical political considerations. The political claim to Jerusalem was constantly reiterated on account of the continuous need for military aid to the crusader kingdom. Clearly then, one could not preach a crusade to the aid of Jerusalem and at the same time deny it a meritorious position in the framework of salvation, as opposed to its celestial counterpart, the Heavenly Jerusalem. On the whole, the twelfth century should be noted for its promotion of the great popularity of the idea of the special value of Jerusalem as a goal of either pilgrimage or crusade among the laity and especially among the nobility who comprised the European ruling class. Never before and never afterwards did the Earthly Jerusalem approach so closely to its celestial counterpart, being regarded by so many as its image and the gateway to the Heavenly City. Moreover, it was now often described in terms used in the previous century for the Heavenly City only. This special link between the Earthly Jerusalem and the European nobility, already forged in the eleventh century, grew considerably stronger during the twelfth. One of the manifestations of this phenomenon is the 'feudalization' of the links between Earthly Jerusalem and European nobility, namely the presentation of the obligation of the crusaders towards Christ in feudal terms. Jerusalem is thus presented as Christ's patrimony, His heritage, His fief, which the crusaders, as Christ's vassals, should guard and liberate for Him.

Jerusalem, the 'City of the Humanity of Christ', 'Inheritance of Christ' and the 'Promised Land' of all Christians as well as the 'gateway to the Heavenly City', was also considered as the geographical centre of the world and the site at which the events of the Last Judgement would take place. Both ideas, expressions of the centrality of Jerusalem in Christian spirituality, were already current in the early Middle Ages. In the twelfth century, however, the theological concept of Jerusalem as the centre of the world became almost universal and was translated into real geographical terms, as witnessed by the world-maps of the period. Similarly, the presentation of Jerusalem as the scene of the End became a regular and constant feature of twelfth-century eschatology. Its conquest and rule by Christians was regarded as a necessary introduction of the Last Days, as well as the realization of the prophecies of both the Old and the New Testaments. On the whole, however, it seems that during the twelfth century the importance of apocalyptic Jerusalem decreased and was overtaken by Earthly Jerusalem as the gateway to the Heavenly one or even the image of the Heavenly Jerusalem. What reflects more than anything else the closeness between the two is the description of the Earthly Jerusalem in terms which traditionally described the Heavenly one. This change came about following the fall of the city in 1187 when, once again, the apocalyptic character of Jerusalem increased. The fall of Jerusalem in 1187 was explained in eschatological terms and the event was interwoven into old and new prophecies that had foretold not only the disaster but its remedy, the future recovery of the Holy City.

The reaction of Christendom to the fall of Jerusalem may be compared in its intensity to that of the conquest by the armies of the First Crusade almost a century earlier. Both were seen by their contemporaries as part and parcel of sacred history and thus as God's will. Following the fall of Jerusalem, the shock and grief at the severity of the punishment were so great that the disaster could only have been explained, and in a sense justified, through examples of God's divine judgement of His people and His chosen capital in the past. Thus, parallels were drawn between the previous conquest of the city by the Babylonians, Romans, Persians and Moslems and the last's recent conquest of Jerusalem in 1187. It was God's will, a divine judgement upon human sinfulness, the *peccatis exigentibus* of all Christendom but most particularly the sins of the inhabitants of the Holy Land. Christendom, confronting the disaster, compared itself to biblical Israel facing the loss of its Jerusalem. Consequently, contemporaries of the fall of Jerusalem described the events and their feelings in biblical terms, mainly by making use of Psalm 78(79):1: 'O God, the heathen are come into thine inheritance' and Lamentations 1:1: 'How doth the city sit solitary ...'.

In the period following the fall of Jerusalem to Saladin, Christendom's attitude to Jerusalem gradually reverted to what it had been before its crusader conquest.

Under the influence of the Franciscans of the *Custodia Terrae Sanctae*, Jerusalem emerged in the course of the fourteenth century as the city of the Passion of Christ whose centre was the 'Way of the Cross', the *Via Dolorosa*, leading from the Prison of Christ near St Stephen's Gate to Calvary. Consequently, Jerusalem became now not only the City of the New Testament but more specifically the City of the Passion of Christ where Christians could, more effectively than in any other place on earth, participate in this Passion, in tune with the Christocentric *Devotio moderna* of the late Middle Ages. Gradually, during the fourteenth century, Christendom abandoned, if not its claim, then actually its attempts to re-establish its rule in Jerusalem. By the end of the fifteenth century, Felix Faber of Ulm could claim that 'at this present day the Christian care little about the Saracens' bearing their rule in Jerusalem, provided only that we (the pilgrims from the West) were allowed freedom to pass in and out of our temple of the Lord's Sepulchre without fear, and without vexations and extortionate payments.'[1]

Writing at the beginning of the thirteenth century, Jacques of Vitry described Jerusalem in the following terms: 'The city of cities, the holy of holies, great among nations, princess among the provinces, by special prerogative called the city of the great king. She stands as if in the midst of the earth, in the centre of the world so all the nations shall flow into her. She is the possession of the patriarchs, the nursing mother of the prophets, the teacher of the apostles, the cradle of our salvation; the native country of the Lord, the mother of the faith, whereas Rome is the mother of the faithful; she had been chosen and sanctified by God, trodden by His feet, honoured by angels and frequented by every nation under heaven.'[2] Jacques of Vitry's statement was a paean of praise to Crusader Jerusalem, lost in 1187 together with its crusader rulers, the Franks, who had attempted to rebuild the earthly city in the likeness of that above. It was this Jerusalem that – for less than a century in the history of Christendom – foreshadowed its Celestial counterpart.

1 Felix Faber, *The Book*, *PPTS*, 9–10, p. 262; Ricoldo da Monte Croce, pp. 105–12; Delaruelle, 'La spiritualité aux XIVe et XVe siècles', pp. 401–12; *idem*, 'Deux guides de Terre Sainte aux XIVe et XVe siècles', pp. 547–53; Pixner, 'Via Dolorosa', pp. 7–10, 51–2; B. Dansette, 'Les Pèlerinages occidentaux en Terre Sainte: une pratique de la dévotion moderne à la fin du Moyen Age', *Archivum Franciscanum Historicum*, 72 (1979), pp. 106–33.

2 Jacques de Vitry, c. 55, pp. 92–3; Y. Friedman, 'Francescinus of Pontremoli: A Pilgrim's Path to Pardon', *Franciscan Studies*, 43 (1983)', pp. 279–311; *idem*, 'The City of the King of Kings', pp. 212–16.

Bibliography

Sources

Acardus de Arroasia, *Tractatus super Templo Salomonis*, ed. P. Lehmann, *Corona Quernea. Festgabe Karl Strecker*, Leipzig 1941 (*Schriften der Monumenta Germaniae Historica*, 6).

Adam de Perseigne, 'Correspondance', ed. J. Bouvet, *Archives Historiques du Maine*, 13 (1951).

Adam de S. Victore, *De tripartito tabernaculo*, *PL*, 198.

Adelgot von Magdeburg, ed. W. Wattenbach, 'Handschriftliches', *Neues Archiv*, 7 (1882).

Ademarus Cabannensis, *De Apostolatu Sancti Martialis*, *PL*, 141.

—*Chronicon*, ed. J. Chavanon, Paris 1867 (*Collection de textes pour servir à l'étude et à l'enseignement de l'histoire*, 20).

Adso Dervensis, *De Ortu et Tempore Antichristi*, *CCCM*, 45.

Alain de Lille, 'Sermo de cruce domini', ed. M.T. d'Alverny, *Alain de Lille: Textes inédits*, Paris 1965 (*Etudes de philosophie médiévale*, 52).

Albericus de Trium-Fontium, *Chronica*, *MGH.SS*, 23.

Albertus Aquensis, *Historia Hierosolymitana*, *RHC.HOcc*, 4.

Alcuini Epistolae, *MGH. Epistolae Aevi Karolini*, 4.

Alcuinus, *Expositio in Johannis Apocalypsin*, *PL*, 117.

Alexander III, *Epistolae*, *PL*, 200.

Ambroise, *The Crusade of Richard Lion-Heart*, Eng. trans. M.J. Hubert and J.L. La Monte, New York 1941.

Analecta Hymnica medii aevi, XXI, XXXII, Leipzig 1895–1899.

Annales Admontenses, *MGH.SS*, 9.

Annales Altahenses Majores, *MGH.SS*, 20.

Annales Augustani, *MGH.SS*, 3.

Annales Colonienses Maximi, *MGH.SS*, 17.

Annales Corbeienses, *MGH.SS*, 3.

Annales Herbipolenses, *MGH.SS*, 16.

Annales S. Iacobi Leodiensis, *MGH.SS*, 16.

Annales Marbacenses, *MGH.SS*, 17.

Annales Mellicenses, *MGH.SS*, 9.

Annales Pegavienses, *MGH.SS*, 16.

Annales Reicherspergenses, *MGH.SS*, 17.

Annales Rosenveldenses seu Annales Harsfeldenses, *MGH.SS*, 16.
Annales Sancti Albani Moguntini (dicti Wirziburgenses), *MGH.SS*, 2.
Annales Sancti Albani Moguntini, *MGH.SS*, 3.
Annales Sancti Dionysii Rememses, *MGH.SS*, 13.
Annales Sancti Disibodi, *MGH.SS*, 17.
Annales Sancti Pauli Virdunensis, *MGH.SS*, 16.
Annales Seligenstadenses, *MGH.SS*, 17.
Annales Stadenses, *MGH.SS*, 16.
Annalista Saxo, *MGH.SS*, 6.
Annalium Hildesheimensium Continuatio A. 1041–1108, *MGH.SS*, 3.
Anonymous II, PPTS, 4.
Anselmus Cantuariensis, *Opera Omnia*, ed. F.S. Schmitt, Edinburgh 1946–1961.
De Antichristo. Quomodo Nasci Debeat auctore Albuini, *CCCM*, 45.
Antoninus Placentinus, *Itinerarium*, *CCSL*, 175.
Arculfus, *De Locis Sanctis*, *CCSL*, 175.
Arnoldus Lubecensis, *Chronica Slavorum*, *MGH.SS*, 21.
Augustinus, *De Civitate Dei*, *CSEL*, 40/1–2.
— *Contra Faustum*, *PL*, 42.
— *Enarrationes in Psalmos*, *PL*, 37.
— *Enchiridion*, ed. J. Krabinger, Tübingen 1961.
Baldricus Dolensis, *Historia Jerosolimitana*, *RHC.Hocc*, 4.
Bartolfus de Nangis, *Gesta Francorum Jherusalem Expugnantium*, *RHC.Hocc*, 3.
Bede, *De Locis Sanctis*, *CCSL*, 175.
— *De Tabernaculo*, ed. D. Hurst, *Bedae Venerabilis Opera*, *CCSL*, 119A.
— *De Templo*, *Bedae Venerabilis Opera*, *CCSL*, 119A.
— *Sibilinorum verborum interpretatio*, *PL*, 90.
Benedict of Peterborough, *Gesta Regis Henrici Secundi*, *see* Roger of Hoveden, *Gesta Regis Henrici Secundi.*
Benzo Albensis, *Ad Heinricum IV Imperatorem Libri IV*, *MGH.SS*, 11.
Bernardus Claraevallensis, *Apologia ad Guillelmum Abbatem*, *Opera*, III, ed. J. Leclercq *et al.*, Rome 1963.
— *De Consideratione*, *Opera*, III, ed. J. Leclercq *et al.*, Rome 1963.
— *De Laude Novae Militiae*, *Opera*, III, ed. J. Leclercq *et al.*, Rome 1963. Eng. trans. M. Conrad Greenia, *Bernard of Clairvaux: Treatises III*, Kalamazoo (*Cistercian Fathers Series*, 12).
— *Letters*, Eng. trans. B.S. James, Chicago 1953.
— *Opera*, ed. J. Leclercq *et al.*, Rome 1957–1977.
Bernardus Clunensis (=Morlanensis), *De Contemptu Mundi*, *RS*, 59/2.
Bernardus Monachus, *Itinerarium*, ed. T. Tobler, *Descriptiones Terrae Sanctae*, Leipzig 1874.
Bernoldus de Sancti Blasio, *Chronicon*, *MGH.SS*, 5.
Bonizo de Sutri, *Liber ad Amicum*, *MGH.Ldt*, 1.
— *Liber de Vita Christiana*, ed. E. Perels, Berlin 1930.

Breviarius de Hierosolyma, *CCSL*, 175.

Burchardus Wormaciensis, *Decretorum Liber*, *PL*,140.

Caffaro de Caschifelone, *De Liberatione Civitatum Orientis Liber*, *RHC.Hocc*, 5.

Carmina Burana, ed. A. Hilka *et al.*, Heidelberg 1930.

Cartulaire de l'abbaye de Saint-Aubin d'Angers, ed. A. Bertrand de Brouissilon, 3 vols, Paris 1903.

Cartulaire de l'abbaye de Saint-Père de Chartres, ed. B.E.C. Gérard, *Collection des Cartulaires de France*, 2, Paris 1840.

Cartulaire de l'abbaye de Saint-Vincent du Mans, ed. R. Charles and S.M. d'Elbenne, vol. 1, Le Mans 1886.

Cartulaire du Chapitre du Saint Sépulchre de Jérusalem, ed. D.G. Bresc-Bautier, Paris 1984 (*Documents relatifs à l'histoire des croisades*, 15).

Cartulaire général de l'ordre des Hospitaliers de St Jean de Jérusalem 1100–1310, ed. J. Delaville le Roulx, Paris 1894–1906.

Cartulaire Noir de la Cathédrale d'Angers, ed. C. Urseau, Paris–Angers 1908.

The Cartulary of the Knights of St John of Jerusalem in England. Secunda Camera. Essex, ed. M. Gervers, Oxford 1982 (*Records of Social and Economic History. New Series*, 6).

Celestinus III, *Epistolae et Privilegia*, *PL*, 206.

La Chanson d'Antioche, ed. S. Duparc-Quioc, 2 vols, Paris 1917–1918 (*Documents relatifs à l'histoire des croisades*, 11–12).

Chansons de croisade, ed. J. Bédier *et al.*, Paris 1909 (repr. Geneva 1974).

Chartes de Terre Sainte provenant de l'Abbaye de N.-D. de Josaphat, ed. H. François Delaborde, Paris 1880.

Chronica Casinensis auctore Petro diacono, *MGH.SS*, 7.

Chronica Monasterii Casinensis, *MGH.SS*, 34.

Chronica S.Petri Erfordensis, *MGH.SS. rer. Germ.*, 42.

Chronica regia Coloniensis et continuationes, *MGH.Scr.rer.Germ.*, 11.

Chronicle of the Third Crusade. A Translation of the Itinerarium Peregrinorum et Gesta Regis Ricardi, trans. Helen J. Nicholson, Aldershot 1997.

Chronicles of the Crusades, trans. H.G.B. Giles, London 1848, (repr. New York 1969).

Chronicon Malleacense, *Chroniques des églises d'Anjou*, ed. P. Marchegay *et al.*, Paris 1869.

Chronicon S. Andreae Castri Cameracensii, *MGH.SS*, 7.

Chronicon S. Huberti Andaginensis, *MGH.SS*, 8.

Chronicon S. Maxentii Pictavensis, *Chroniques des églises d'Anjou*, ed. P. Marchegay *et al.*, Paris 1869.

Chronicon S. Petri Amiciensis, *RHGF*, 12.

Chronicon Universale Anonymi Laudunensis, ed. A. Cartellieri *et al.*, Leipzig–Paris 1909.

La Chronique d'Ernoul et de Bernard le Trésorier, ed. L. de Mas Latrie, Paris 1871.

La Chronique de Saint-Hubert dite Cantatorium, ed. K. Hanquet, Bruxelles 1906.
Chronique de Zimmern, ed. H. Hagenmeyer, *AOL*, 2 (1884).
The Cistercian World: Monastic Writings of the Twelfth Century, trans. and ed. Pauline Matarasso, London 1993.
Citez de Iherusalem, ed. T. Tobler, *Descriptiones Terrae Sanctae*, Leipzig 1874.
Codice diplomatico della Republica di Genova, ed. C. Imperiale di Sant' Angelo, 3 vols, Rome 1936–1942.
La 'Collectio Seguntina', in W. Holtzmann, 'La "Collectio Seguntina" et les décrétales de Clément III et de Célestine III', *Revue d'histoire ecclésiastique*, 50 (1955).
Commemoratorium de Casis Dei, ed. Tobler *et al.*, *Itinera.*
Commentarius Cantabrigiensis in epistolas Pauli e schola Petri Abaelardi. In epistolam ad Hebraeos, ed. A. Landgraf, Notre Dame 1945.
Conciliorum Oecumenicorum Decreta, ed. G. Alberigo *et al.*, Bologna 1972.
The Conquest of Jerusalem and the Third Crusade. Sources in Translation, trans. Peter W. Edbury, Aldershot 1996.
Continuatio Claustroneoburgensis secunda, *MGH.SS*, 9.
Continuatio Gemblacensis, *MGH.SS*, 6.
Continuatio Zwettlensis Altera, *MGH.SS*, 9.
La Continuation de Guillaume de Tyre, dite du Manuscript de Rothelin, ed. M.R. Morgan, Paris 1982 *(Documents relatifs à l'histoire des croisades*, 14).
Corpus Inscriptionum Crucesignatorum Terrae Sanctae, ed. S. Sandoli, Jerusalem 1974.
Corpus Iuris Civilis. III. Novellae, ed. R. Schoell and G. Kroll, Berlin 1895.
Daniel, *Pilgrimage*, *PPTS*, 4.
Decretales Pseudo-Isidorianae, ed. P. Hinschius, Leipzig 1863 (repr. Aalen 1963).
Descriptio Cuiusdam Sapientis de Antichristi nativitate, *CCCM*, 45.
Early Latin Hymns, ed. A.S. Walpole, Cambridge 1922 (repr. Hildesheim 1966).
Egeria, *Itinerarium*, *CCSL*, 175.
Ekkehardus Uraugiensis, *Chronicon Universale*, *MGH.SS*, 6.
— *Hierosolymita*, *RHC.Hocc*, 5.
Elizabeth von Schönau, *Die Visionen der Elisabeth und die Aebte Ekbert und Emecho von Schönau*, ed. F.W.E Roth, Brünn 1884.
Enchiridion Locorum Sanctorum, ed. D. Baldi, Jerusalem 1945.
Epistola A. Dapiferi Militiae Templi ad Ebrandum magistrum ejusdem militiae, *RHGF*, 15.
Epistolae Cantuarienses, *RS*, 38.
Epistolae Lamberti Astrebatensis Episcopi, *RHGF*, 15.
L'Estat de la cité de Jherusalem, *Itinéraires.*
L'Estoire d'Eracles empereur et la conqueste de la terre d'Outremer, *RHC.Hocc*, 2.
Eucherius, *De Situ Hierosolimae*, *CCSL*, 175.
Eugenius III, *Epistolae*, *PL*, 180.

Eugenius III, *Quantum Praedecessores*, ed. P. Rassow, 'Kreuzzugsbulle', *Neues Archiv*, 45 (1924).

Eusebius, *Demonstratio Evangelica*, ed. J.A. Heikel, *Die griechischen christlichen Schriftsteller der ersten drei Jahrhunderte*, Leipzig–Berlin 1913.

— *Historia Ecclesiastica*, *PG*, 20. Eng. trans. *Eusebius Pamphili*, *Ecclesiastical History*, trans. Roy J. Defferrari, Washington, DC 1955.

— *Vita Constantini*, *PG*, 20.

Evangelia Apocrypha, ed. C. von Tischendorf, Leipzig 1876.

De Expugnatione Terrae Sanctae Libellus, ed. J. Stevenson, *Radulphi de Coggeshall Chronicon Anglicanum*, *RS*, 66.

Felix Faber, *Evagatorium in Terrae Sanctae*, *Arabiae et Egypti Perigrinationem*, ed. C.J. Hassler, Stuttgart 1843. Eng. trans. A. Stewart, *PPTS*, 7–10.

Fragmentum historiae Andegavensis, *Chroniques des comtes d'Anjou et des seigneurs d'Amboise*, ed. L. Halphen *et al.* Paris 1913.

Fulcher Carnotensis, *Historia Hierosolymitana*, ed. H. Hagenmeyer, Heidelberg 1913.

Fulcher of Chartres, *A History of the Expedition to Jerusalem 1095–1127*, trans. Frances Rita Ryan, Knoxville 1969.

Fulco et Gilo, *Historia Gestarum viae nostri temporis Hierosolymitanae*, *RHC.Hocc*, 5.

Gaufredus Grossus, *Vita B. Bernardi de Tironio*, *RHGF*, 9.

Gaufridi Dictamen de Primordiis Ecclesiae Castaliensis, *RHC.Hocc*, 5.

Geoffroy de Vendôme, *Epistulae* in *Oeuvres*, ed. and trans. Geneviève Giordanengo, Turnhout 1996.

Gerald of Wales [Giraldus Cambrensis], *Expugnatio Hibernica*, ed. A.B. Scott and F.X. Martin, Dublin 1978.

— *De principis Instructione*, *Opera*, *RS*, 21.

Gerberti acta concilii Remensis, *MGH.SS*, 3.

Gerhoh Reichersbergensis, *De aedificio Dei*, *PL*, 194.

— *Commentarius in Psalmos*, *MGH.Ldt*, 3.

— *Libellus de investigatione Antichristi*, *MGH.Ldt*, 3.

— *Libellus de Ordine donorum Sancti Spiritus*, *MGH.Ldt*, 3.

— *Liber de Simoniacis*, *MGH.Ldt*, 3.

— *Opusculum ad Cardinales*, *MGH.Ldt*, 3.

Gervase of Canterbury, *Opera historica*, *RS*, 73.

Gesta Adhemari episcopi Podiensis, *RHC.Hocc*, 5.

Gesta Episcoporum Halberstadensium, *MGH.SS*, 23.

Gesta Episcoporum Tullensium, *MGH.SS*, 8.

Gesta Francorum et aliorum Hierosolimitanorum, ed. R. Hill, London 1961.

Gilbertus Lumnicensis, *De Statu Ecclesiae*, *PL*, 159.

Gilo, *Historia de via Hierosolymitana*, *RHC.Hocc*, 5. Eng. trans. Gilo, *The Historia vie hierosolymitane of Gilo of Paris and a Second, Anonymous Author*, ed. and trans. C.W. Grocock and J.E. Siberry, Oxford 1997.

Giraldus Cambrensis, *see* Gerald of Wales.
Goffridus Vindocinensis, *Epistolae*, *PL*, 157.
— *Oeuvres*, ed. and trans. Geneviève Giordanengo, Brepols 1996.
Gotifredus Viterbiensis, *Pantheon*, *MGH.SS*, 22.
Grandes Chroniques de France, ed. J. Viard, Paris 1920–1953.
Gratianus, *Decretum*, *Corpus Juris Canonici*, ed. A. Friedberg, 2 vols, Leipzig 1879–1881.
Gregorius Nyssenus, *Epistolae*, *PG*, 46.
— *Lettres*, ed. and trans. Pierre Maraval, Paris 1990 (*Sources Chrétiennes*, *3*63).
Gregorius I, *In Hiezechielem*, *CCSL*, 142.
— *Homiliae, PL*, 76.
— *Liber Antiphonarius*, *PL*, 78.
Gregorius VII, *Register*, ed. E. Caspar, *MGH.Epistolae Selectae in usum scholarum*, 2/1.
Gregorius VIII, *Audita tremendi*, *Historia de Expeditione Friderici I*, *MGH.Scr. rer.Germ*, 5.
— *Epistolae*, *PL*, 202.
Guibertus Novigentis, *Gesta Dei per Francos*, *RHC.Hocc*, 4. Eng. trans. *The Deeds of God through the Franks. Gesta Dei per Francos. Guibert de Nogent*, trans. Robert Levine, Woodbridge 1997.
— *Liber quo Ordine Sermo Fieri Debeat*, *PL*, 156.
Guido de la Chartreuse, *Meditationes*, ed. André Wilmart, Paris 1936 (*Etudes de philosophie médiévale*, 22).
Guillaume le Breton, *Gesta Philippi Augusti*, *RHGF*, 17.
Guillelmus Andrensis, *Chronica Andrensis*, *MGH.SS*, 24.
Gunther de Pairis, *Historia Constantinopolitana*, *Exuviae Sacrae Constantinopolitanae*, ed. P. Riant, Geneva 1877.
Haymarus Monachus, *De Expugnata Accone liber*, ed. P. Riant, Paris 1865.
Haymo Halberstatensis, *Exposito in Epistolas S. Pauli*, *PL*, 117.
Heinrici Chronicon Lyvoniae, *MGH.SS*, 23.
Henry IV, *Briefe*, ed. F.J. Schmale, Darmstadt 1963 (*Ausgewählte Quellen zur Deutschen Geschichte des Mittelalters*, 12).
Henry d'Albano, *Epistolae*, *PL*, 204.
— *De Peregrinante Civitate Dei*, *PL*, 204.
— *Ex quo vox*, *in Historia de Expeditione Friderici I*, *MGH.SS rer.Germ*, 5.
Henry of Huntingdon, *De captione Antiochiae*, *RHC.Hocc*, 5.
Hieronymus, *De Antichristo in Danielem [IV]*, *CCSL*, 75A.
— *Commentariorum in Danielem Liber*, *CCSL*, 75B.
— *Commentariorum in Hieremiam*, *CCSL*, 59.
— *Commentariorum in Hiezechielem*, *CCSL*, 75.
— *Commentariorum in Isaiam*, *CCSL*, 73A.
— *Epistulae*, *CSEL*, 54.
— *Praefatio in librum Paralipomenon*, *PL*, 29.

Hilarius Pictaviensis, *Tractatus Super Psalmos*, *CSEL*, 22.
Hildegard von Bingen, *Epistolae*, *PL*, 197.
— *Scivias*, Eng. trans. Columba Hart and Jane Bishop, New York 1990.
Hincmar Rhemensis, *Opuscula et Epistolae*, *PL*, 126.
Historia Comitum Ghisnensium auctore Lamberto Ardensi, *MGH.SS*, 24.
Historia de Expeditione Friderici Primi, *MGH.SS. rer.Germ.*, 5.
Historia Compostellana, *CCCM*, 70.
Historia de translatione sanctorum magni Nicolai..., *RHC.Hocc*, 5.
Historia Peregrinorum, *MGH.SS.rer.Germ.*, 5.
Historia Peregrinorum, *RHC.Hocc*, *3*.
The History of the Holy Rood-Tree, ed. A.S. Napier, London 1894 (*EETS*, 103).
Honorius Augustodunensis, *De Animae Exsilio et Patria*, *PL*, 172.
— *Elucidarium*, *PL*, 172.
— *Speculum Ecclesiae*, *PL*, 172.
Hugo Flaviniacensis, *Chronicon*, *MGH.SS*, 8.
Hugo Floriacensis (de S. Maria), *Modernorum Regum Francorum Actus*, *MGH.SS*, 9.
Hugo de Folieto, *De Claustro animae*, *PL*, 176.
Hugo de S. Maria, *Itineris Hierosolimitani Compendium*, *RHC.Hocc*, 5.
Hugo de S. Victore, *Miscellanea*, *PL*, 177.
— *De Sacramentis*, *PL*, 176.
— *Speculum Ecclesiae*, *PL*, 177.
— *De Vanitate Mundi*, *PL*, 176.
Hugonis Chronici Continuatio Weingartensis, *MGH.SS*, 21.
Innocentius III, *Acta*, ed. T. Haluščynskyj, Vatican City 1949 (*Pontificia Commissio ad Redigendum Codicem Juris Canonici Orientalis Fontes*).
— *Opera Omnia*, *PL*, 214–17.
— *Quia maior*, ed. G. Tangl, *Studien zum Register Innocenz III.*, Weimar 1929.
— *Die Register Innocenz III., 2. Band, 2. Pontifikatsjahr 1199/1200 : Texte*, ed. O. Hageneder *et al.*, Rome–Vienna 1979.
Innominatus I, ed. T. Tobler, *Descriptiones Terrae Sanctae*, Leipzig 1874.
Innominatus IV, ed. M. de Vogüé, *Les Eglises de la Terre Sainte*, Paris 1860 (repr. Jerusalem 1973).
Innominatus VI, ed. T. Tobler, *Descriptiones Terrae Sanctae*, Leipzig 1874.
Innominatus VIII, ed. T. Tobler, *Descriptiones Terrae Sanctae*, Leipzig 1874.
Isidorus Hispalensis, *Etymologiarum sive Originum Libri XX*, ed. W.M. Lindsay, Oxford 1911 (repr. 1971).
Italia Pontificia, ed. P.F. Kehr, Berlin 1908.
Itinera Hierosolymitana et Descriptiones Terrae Sanctae, ed. T. Tobler and A. Molinier, Geneva 1879 (repr. Osnabrück 1966).
Itinera Hierosolymitana Crucesignatorum, ed. S. de Sandoli, 4 vols, Jerusalem 1980–1984.
Itinéraires à Jérusalem et descriptions de la Terre Sainte rédigés en français aux XIe, XIIe et XIIIe siècles, ed. Henry V. Michelant and Gaston Raynaud, Geneva 1882.

Itinerarium Burdigalense, *CCSL*, 175.
Itinerarium Peregrinorum, ed. H.E. Mayer, Stuttgart 1962 (*Schriften der Monumenta Germaniae Historica*, 18).
Itinerarium Peregrinorum et Gesta Regis Ricardi, *RS*, 38. Eng. trans. Helen J. Nicholson, *Chronicle of the Third Crusade.*
Ivo Carnotensis, *Decretum*, *PL*, 161.
— *Epistolae*, *PL*, 162.
— *Panormia*, *PL*, 161.
Jacques de Vitry, *Historia Orientalis*, ed. F. Moschus, Douai 1597.
Jean d'Ibelin, *Livre*, *RHC.Lois*, *1.*
Jocelin of Brakelond, *De rebus gestis Samsonis abbatis monasterii sancti Edmundi*, ed. and Eng. trans. H.C. Butler, London 1951.
Johannes Beleth, *Summa de Ecclesiasticis Officiis*, *CCCM*, 41A.
Johannes Cassianus, *Conlationes*, *CSEL*, 13.
— *Expositio in Psalmos*, *CSEL*, 13.
Johannes Chrysostomus, *Homiliae*, *PG*, 49.
Johannes Wirziburgensis, *Descriptio Terrae Sanctae*, ed. T. Tobler, *Descriptiones Terrae Sanctae*, Leipzig 1874.
— *Descriptio Terrae Sanctae*, *Peregrinationes Tres.*
Jotsaldus Sylviniacensis, *Vita S. Odilonis*, *PL*, 142.
Justinus Martyr, *Dialogus cum Tryphone Judaeo*, *PG*, 6.
Kreuzzugsbriefe aus den Jahren 1088–1100, ed. H. Hagenmeyer, Innsbruck 1901.
Lambertus Atrebatensis, *Epistolae*, *PL*, 162.
Landulfus, *Historia Mediolanensis*, *MGH.SS*, 8.
Lettres de Gerbert (983–997) (=Silvester II), ed. J. Havet, Paris 1889 (*Collection de textes pour servir à l'étude et à l'enseignement de l'histoire*, 6).
Libellus de expugnatione Terrae Sanctae per Saladinum, *RS*, 66.
Liber Anselmi de Antichristo, *CCCM*, 45.
Liber de poenitentia et tentationibus religiosorum, *PL*, 213.
Lupus Protospatharius, *Annales Barenses*, *MGH.SS*, 5.
Magnus Reicherspergensis, *Chronicon*, *MGH.SS*, 17.
Manegoldus, *Ad Gebehardum Liber*, *MGH.Ldt*, 1.
Minnesangs Frühling, ed. K. Lachmann *et al.*, Stuttgart 1965.
Monachus Anonymus Scaphusensis, *De Reliquiis Sanctissime Crucis et Domini Sepulchri*, *RHC.Hocc*. 5.
Monumenta Vizeliacensia, *CCCM*, 42.
Monumentorum Boicorum Collectio Nova, 28.
Notitia Lemovicensis II, *RHC.Hocc* 5.
Orderic Vitalis, *Historia Ecclesiastica*, ed. and Eng. trans. M. Chibnall, 6 vols, Oxford 1969–1981.
Ottobonis Scribae Annales, *MGH.SS*, 18.
Otto de Sancto Blasio, *Chronica*, *MGH.Scr.rer.Germ.*, 47.
Otto Frisingensis, *Chronica sive Historia de Duabus Civitatibus*, *MGH.SS*, 20.

—*Gesta Friderici Imperatoris*, *Ottonis et Rahewini Gesta Friderici I Imp. MGH. Scr.rer.Germ.*, 46.

The Oxford Book of Medieval Verse, ed. F.J.E. Raby, Oxford 1959.

Papsturkunden für Templer und Johanniter. Archivberichte und Texte, ed. Rudolf Hiestand, *Vorarbeiten zum Oriens Pontificius* I, Göttingen 1972.

Papsturkunden für Templer und Johanniter. Neue Folge, ed. Rudolf Hiestand, *Vorarbeiten zum Oriens Pontificius* II, Göttingen 1984.

Papsturkunden für Kirchen im Heiligen Lande, ed. Rudolf Hiestand, *Vorarbeiten zum Oriens Pontificius* III, Göttingen 1985.

Paschal II, *Epistolae*, *PL*, 163.

Paschasius Rudbertus, *Expositio in Mattheum*, *PL*, 120.

Passio S. Cholomanni, *MGH.SS*, 4.

Paulinus de Nola, *Epistulae*, *CSEL*, 29.

Petrus Abelardus, *Theologia Christiana*, *PL*, 178.

Petrus Alphonsus, *Dialogi*, *PL*, 157.

Petrus Blesensis, *Dialogus inter Regem Henricum Secundum et Abbatem Bonevallis*, ed. R.B.C. Huygens, *Revue Bénédictine*, 68 (1958).

— *Epistolae*, *PL*, 207.

— *Epistolae*, *RHGF*, 19.

— *De Hierosolymitana Peregrinatione Acceleranda*, *Opuscula*, *PL*, 207.

— *Passio Reginaldi*, *Opuscula*, *PL*, 207.

Petrus Cellensis, *Sermones*, *PL*, 202.

— *Pierre de Celle. L'Ecole du Cloître, Introduction, Texte Critique, Traduction et Notes*, by Gérard de Martel, Paris 1977.

Petrus Comestor, *Historia Scholastica*, *PL*, 198.

Petrus Diaconus, *Chronica Casinenses*, *MGH.SS*, 7.

— *De Locis Sanctis*, ed. S. de Sandoli, *Itinera*, 2.

Petrus Tudebode, *Historia de Hierosolymitano itinere*, eds J.H. and L.L. Hill, Paris 1977 (*Documents relatifs à l'histoire des croisades*, 12).

Petrus Venerabilis, *Adversus Iudeorum inveteratam duritiem*, *CCCM*, 58.

— *De Miraculis Libri Duo*, *CCCM*, 83.

— *Letters*, ed. G. Constable, Cambridge, Mass. 1967.

— *Sermo in Laudem Sepulchri Domini*, ed. G. Constable, 'Petri Venerabilis Sermones Tres', *Revue Bénédictine*, 64 (1954).

— *Statuta*, *Corpus consuetudinum monasticarum*, vol. 6, Siegburg 1970.

Philippus de Harveng, *Epistolae*, PL, 203.

Poésies populaires latines, ed. M.E. du Méril, Paris 1843.

Prosper Aquitanus, *Expositio Psalmorum*, *PL*, 51.

Qualiter Sita est Civitas Jerusalem, ed. Tobler and Molinier, *Itinera*.

Rabanus Maurus, *Homiliae*, *PL*, 110.

— *De Universo*, *PL*, 111.

Radulfus Cadomensis, *Gesta Tancredi in Expeditione Hierosolymitana*, *RHC.Hocc*, 3.

Radulfus Glaber, *Historiarum sui temporis libri V*, ed. M. Prou, Paris 1885.
Radulfus de Diceto, *Opera Historica*, *RS*, 68/2.
Radulfus Niger, *Chronica universalis*, *MGH.SS*, 27.
—*De re militari et triplici via peregrinationis Ierosolimitane*, ed. L. Schmugge, Berlin 1977.
Rahewini Gesta Friderici I Imperatoris, *MGH.Scr. rer.Germ.*, 46.
Raimond d'Aguilers, *Le 'Liber'*, eds. J.H. and L.L. Hill, Paris 1969 (*Documents relatifs à l'histoire des croisades*, 9).
Rangerius, *Vita Anselmi Lucensis*, *MGH.SS*, 30/2.
Recueil de Chartes de l'Abbaye de Cluny, ed. A. Bernard et A. Bruel, Paris 1904.
Reginaldus Dunelmensis, *Vita S. Godrici Eremitae Finchalae*, *AASS*, May, vol. V.
Reimbald of Liège, *Itineraria*, *CCCM*, 4.
Reineri Annales, *MGH.SS*, 16.
Ricardus de S Victore, *Benjamin major*, *PL*, 196.
— *Mystical Ark*, Eng. trans. G. Zinn, New York 1979.
— *Sermones*, *PL*, 177.
— *De tabernaculo*, *PL*, 196.
Ricardus Divisiensis, *Chronicon de tempore Regis Ricardi Primi*, ed. J.T. Appleby, London 1963.
— *Chronicon*, *RHGF*, 12.
Ricardus Pictaviensis, *Chronicon*, *RHGF*, 12.
Ricoldo da Monte Croce, *Liber Peregrinationis*, ed. J.C.M. Laurent, *Peregrinatores Medii Aevi Quatuor*, Leipzig 1864.
Rigord, *Gesta Philippi Augusti*, *RHGF*, 17.
Robertus Autissiodorensis, *Chronicon*, *MGH.SS*, 26.
Robertus Monachus, *Historia Jherosolimitana*, *RHC.Hocc*, 3.
Roger de Hoveden, *Gesta Regis Henrici Secundi*, *RS*, 49.
— *Chronica*, *RS*, 51.
— *The Annals*, Eng. trans. Henry T. Riley, 2 vols, London 1853.
Rolandus Bandinelli, *Summa*, ed. F. Thaner, Innsbruck 1874.
Rorgo Fretellus de Nazareth, *Description de la Terre Sainte*, ed. P.C. Boeren, Amsterdam 1980.
Rostangus, *De Translatione capitis S. Clementis papae et martyris ab Constantinopoli ad Cluniacum*, *PL*, 209.
Rufinus, *Summa*, ed. J.F. von Schulte, Giessen 1892.
Rupertus Tuitiensis (von Deutz), *De Victoria Verbi Dei*, ed. R. Haacke, *MGH. Quellen zur Geistesgeschichte der Mittelalters*, 5, Weimar 1970.
Saewulf, *Peregrinatio de Hierosolyma et Terra sancta, Peregrinationes Tres*. Eng trans. A. Rogers, *PPTS*, 4.
Salimbene de Adam, *Cronicae*, *MGH.SS*, 32.
Sedulius Scottus, *In Epistolam ad Galatas*, *PL*, 103.
Sicardus Cremonensis, *Cronica*, *PL*, 213.
— *Mitrale seu De Officiis Ecclesiasticis Summa*, *PL*, 213.

Sigebertus Gemblacensis, *Chronographia*, *MGH.SS*, 6.
— *Epistola Leodicensium adversus Paschalem Papam*, *MGH.Ldt*, 2.
— *De Situ Urbis Jerusalem et de locis sanctis*, ed. M. de Vogüé, *Les églises de la Terre Sainte*, Paris 1860.
Statuta Capitulorum Generalium Ordinis Cisterciensis, ed. J. Canivez, Louvain 1933–1935.
Stephanus Grandimontensis, *Sententiarum Liber*, *PL*, 204.
Stephen de Tournai, *Lettres*, ed. J. Desilve, Paris 1893.
— *Summa*, ed. J.F. von Schulte, Giessen 1891.
Suger, *Vie de Louis le Gros*, ed. A. Molinier, Paris 1887.
Sulpicius Severus, *Vita Sancti Martini*, *CCEL*, 1.
Tanhuma Kedoshim, ed. Buber, Vilna 1885 (Hebrew).
The Summa Parisiensis on the Decretum Gratiani, ed. T.G. McLaughlin, Toronto 1952.
Tertullianus, *Adversus Marcionem*, *CSEL*, 47.
'The Text of the Account of the Capture of Jerusalem in the Ripoll Manuscript, Bibliothèque Nationale (Latin) 5132', ed. J. France, *EHR*, 103 (1988).
Theobaldi vita S. Guilelmi eremitae, *AASS*, Feb. 2.
Theodericus, *Libellus de Locis Sanctis*, ed. M.L. and W. Bulst, Heidelberg 1976.
— *Libellus de Locis Sanctis*, *Peregrinationes Tres*.
Theodosius, *De Situ Terrae Sanctae*, *CCSL*, 175.
Theotonius, *Duo Itinera ad Terram Sanctam*, ed. de Sandoli, *Itinera*.
Thibaut d'Etampes, *Improperium cuiusdam in monachos*, eds R. Foreville and J. Leclercq, *Studia Anselmiana*, 41 (1957).
Tractatus Eboracenses II, *MGH.Ldt*, 3.
Tractatus de locis et statu sancte terre ierosolimitane, ed. G.M. Thomas, *Sitzungsberichte der philosophisch–historischen Klasse der K. Bayrischen Akademie der Wissenschaften*, 3 (1865).
Triumphus Sancti Lamberti de Castro Bullonio, *MGH.SS*, 20.
Venjance Nostre Seigneur, ed. Loyal A.T. Gryting, *The Oldest Version of the Twelfth-Century 'Venjance Nostre Seigneur'*, Ann Arbor, Michigan 1949 (repr. 1997).
Victorinus Petavionensis, *Commentarii in Apocalypsim*, *CSEL*, 49.
Vie de Louis le Gros par Suger suivie de l'Histoire du Roi Louis VII, ed. Auguste Molinier, Paris 1887.
Vita Altmanni Episcopi Pataviensis, *MGH.SS*, 12.
Vita Anselmi Lucensis Episcopi auctore Rangerio, *MGH.SS*, 30/2.
Vita Meinwerci, *MGH.SS*, 11.
Walafridus Strabo, *De Subversione Jerusalem*, *PL*, 114.
Walter Map, *De Nugis Curialium*, ed. M.R. James *et al.*, Oxford 1914 (new edn Oxford 1983).
Walteri Vita Karoli Comitis Flandriae, *MGH.SS*, 12.
Widrici Vita S. Gerardi, *MGH.SS*, 4.

Willelmi Tyrensis Archiepiscopi Chronicon, ed. R.B.C. Huygens *et al.*, *CCCM*, 63–63A.

Eng. trans. William, archbishop of Tyre, *A History of Deeds beyond the Sea*, trans. Emily Atwater Babcock and A.C. Krey, 2 vols, New York, 1943 (repr. 1976).

William of Malmesbury, *De Gestis Regum Anglorum Libri Quinque*, *RS*, 90.

William of Newburgh, *Historia rerum Anglicarum*, *RS*, 82/1.

Willibaldus, *Vita seu potius hodoeporicon*, ed. T. Tobler, *Descriptiones Terrae Sanctae*, Leipzig 1874.

Secondary Works

Alexander, P.J., 'Byzantium and the Migration of Literary Works and Motifs. The Legend of the Last Roman Emperor', *Medievalia et Humanistica*, 2 (1961).

———, *The Oracle of Baalbek: The Tiburtine Sibil in Greek Dress*, Washington, DC 1967.

———, 'The Diffusion of Byzantine Apocalypses in Medieval West and the Beginnings of Joachimism', *Prophecy and Millennarism. Essays in Honour of Marjorie Reeves*, ed. H. Williams, London 1980.

Alphandéry, P., 'Les citations bibliques chez les historiens de la première croisade', *Revue de l'Histoire des Religions*, 99 (1929).

———, Dupront, A., *La Chrétienté et l'idée de croisade*, 2 vols, Paris 1954–1959.

Arce, A., 'El sepulcro de David en un texto de Benjamin de Tudela (1169)', *Sefarad*, 23 (1963).

Ashworth, H., '*Urbs beata Jerusalem*, Scriptural and Patristic Sources', *Ephemerides Liturgicae*, 70 (1956).

Autour de la Première Croisade. Actes du Colloque de la 'Society for the Study of the Crusades and the Latin East' (Clermont-Ferrand, 22–25 juin 1995), ed. Michel Balard, Paris 1996.

Avi-Yonah, M., *The Madaba Map*, Jerusalem 1954.

Bachrach, Bemard S., 'The Pilgrimages of Fulk Nerra, Count of the Angevins, 987–1040', *Religion, Culture and Society in the Early Middle Ages. Studies in Honor of Richard S. Sullivan*, ed. T.F.X. Noble and J.J. Contreni, Kalamazoo 1987.

———, *Fulk Nerra, the Neo-Roman Consul, 987–1040. A Political Biography of an Angevin Count*, Berkeley 1993.

Barber, Malcolm, *The New Knighthood. A History of the Order of the Temple*, Cambridge 1994.

Beazley, C.R., *The Dawn of Modern Geography*, 2 vols, 2, London 1901.

Beech, George T., 'Urban II, the Abbey of Saint-Florent of Saumur and the First Crusade', *Autour de la Première Croisade*.

Beer, R., 'Die Handschriften des Klosters Santa Maria de Ripoll', *Sitzungsberichte der Kais. Akademie der Wissenschaften in Wien. Philosophisch-Historische Klasse*, 155 (1907); 158 (1908).

Benson, R.L., *The Bishop-Elect. A Study in Medieval Ecclesiastical Office*, Princeton 1968.

———, 'Provincia–Regnum', *Prédication et Propagande au Moyen Age. Islam, Byzance, Occident*, Paris 1983 (*Penn–Paris–Dumbarton Oaks Colloquia*, 3).

Benzinger, J., *Invectiva in Romam. Romkritik im Mittelalter vom 9. bis zum 12. Jahrhundert*, Lübeck–Hamburg 1968.

Berlière, U., 'Les pèlerinages judiciaires au moyen âge', *Revue Bénédictine*, 7 (1880).

Berry, V.G., 'Peter the Venerable and the Crusades', *Studia Anselmiana*, 40 (1956).

Bettex, A., *The Discovery of the World*, New York 1960.

Birch, Debra, J., 'Selling the Saints: Competition among Pilgrimage Centres in the Twelfth Century', *Medieval History*, 2 (1992).

Blake, E.O., 'The Formation of the Crusade Idea', *Journal of Ecclesiastical History*, 21 (1970).

——— and Morris, Colin, 'A Hermit Goes to War: Peter and the Origins of the First Crusade', *Studies in Church History*, 22 (1985).

The Blessings of Pilgrimage, ed. Robert Ousterhout, Urbana and Chicago 1990.

Bloomfield, M.W., *The Seven Deadly Sins*, East Lansing, Michigan 1952 (new edn 1967).

Bolton, Brenda, M., 'The Cistercians and the Aftermath of the Second Crusade', *The Second Crusade and the Cistercians*, ed. Michael Gervers, New York 1992.

Borg, A., 'Observations on the Historical Lintels of the Holy Sepulchre, Jerusalem', *Journal of the Warburg and Courtauld Institutes*, 32 (1969).

Bousset, W., *Der Antichrist in der Überlieferung des Judentums, des Neuen Testaments und der Alten Kirche*, Göttingen 1895.

Bredero, A.H., 'Jérusalem dans l'Occident médiéval', *Mélanges offerts à René Crozet*, ed. P. Gallais *et al.*, vol. 1, Poitiers 1966.

———, 'Jerusalem in the West', *Christendom and Christianity in the Middle Ages. The Relations between Religion, Church and Society*, trans. Reinder Bruinsma, Grand Rapids, Michigan 1994.

Bresc-Bautier, Geneviève, 'Les imitations du Saint-Sépulchre de Jérusalem (IXe–XVe siècles). Archéologie d'une dévotion', *Revue d'histoire de la spiritualité*, 50 (1974).

Brinken, van den, A.D., 'Mappa mundi und Chronographia. Studien zur *imago mundi* des abendländischen Mittelalters', *DA*, 24 (1968).

Brown, E.A.R., 'La notion de la légitimité et la prophétie à la cour de Philippe Auguste', *La France de Philippe Auguste – Le Temps des Mutations*, ed. R.-H. Bautier, Paris 1982.

——— and Cothren, M.W., 'The Twelfth-Century Crusading Window of the Abbey of Saint-Denis: Praeteritorum enim Recordatio Futurorum est Exhibitio', *Journal of the Warburg and Courtauld Institutes*, 49 (1986).

Brown, Peter, *The Cult of the Saints: Its Rise and Function in Latin Christianity*, Chicago 1981.

Brundage, J.A., *The Crusades: A Documentary Survey*, Milwaukee 1962.
———, 'Crucesignari: The Rite for Taking the Cross in England', *Traditio*, 22 (1966).
———, *Medieval Canon Law and the Crusader*, Madison 1969.
———, 'A Transformed Angel (X 3.31.18): The Problem of the Crusading Monk', *Studies in Medieval Cistercian History Presented to J.F. O'Sullivan*, Shannon 1971 (*Cistercian Studies Series*, 13).
Bull, Marcus, *Knightly Piety and the Lay Response to the First Crusade. The Limousin and Gascony c. 970–c. 1130*, Oxford 1993.
———, 'The diplomatic of the First Crusade', *The First Crusade*.
Busse, H., 'Vom Felsendom zum Templum Domini', *Das heilige Land in Mittelalter: Begegnungsraum zwischen Orient und Okzident*, ed. W. Fischer *et al.*, Neustadt 1982.
Bynum, Caroline Walker, *Jesus as Mother. Studies in the Spirituality of the High Middle Ages*, Berkeley 1982.
Canard, M., 'La destruction de l'église de la résurrection par le Calife Hakim et l'histoire de la descente du feu sacré', *Byzantion*, 35 (1965).
Cantor, N.F., *Church, Kingship and Lay Investiture in England 1089–1135*, Princeton 1958.
Cardini, F., '"Crusade" and "Presence of Jerusalem" in Medieval Florence', *Outremer*.
Cardman, Francine, 'The Rhetoric of Holy Places. Palestine in the Fourth Century', *Studia Patristica*, 17 (1982).
Cartellieri, A., *Philipp II. August König von Frankreich*, 2 vols, Leipzig 1921–1922.
Catalano, G., *Impero, Regni e Sacerdozio nel Pensiero di Uguccio da Pisa*, Milano 1956.
Chenu, M.D., *Nature, Man and Society in the Twelfth Century*, Eng. trans. J. Taylor and L.K. Little, Chicago 1968.
Clark, Ann L., *Elizabeth of Schönau. A Twelfth-Century Visionary*, Philadelphia 1992.
Cohn, N., *The Pursuit of the Millennium*, London 1970.
Cole, Penny J., *The Preaching of the Crusade to the Holy Land, 1095–1270*, Cambridge, Mass. 1991.
———, 'Christian Perceptions of the Battle of Hattin (583/1187)', *Al-Masaq*, 6 (1993).
———, '"O God, the Heathen have come into your inheritance" (Ps. 78:1). The Theme of Religious Pollution in Crusade Documents, 1095–1188', *Crusaders and Muslims in Twelfth-Century Syria*, ed. Maya Shatzmiller, Leiden 1993.
———, 'Purgatory and Crusade in St Gregory's Trental', *The International History Review*, 17 (1995).
Congar, Y.-M., 'Henri de Marcy, Abbé de Clairvaux, Cardinal-Evêque d'Albano et légat pontifical', *Studia Anselmiana*, 43 (1958).
———, 'L'église et Cité de Dieu chez quelques auteurs cisterciens à l'époque des croisades', *Mélanges offerts à Etienne Gilson*, Paris 1959.

Constable, G., 'The Second Crusade as Seen by Contemporaries', *Traditio*, 9 (1953).

———, 'The Vision of Gunthelm and other Visions attributed to Peter the Venerable', *Revue Bénédictine*, 66 (1956).

———, 'Opposition to Pilgrimage in the Middle Ages', *Studia Gratiana*, 19 (1976).

———, 'Monachisme et pèlerinage au moyen âge', *Revue historique*, 258 (1977).

———, 'Renewal and Reform in Religious Life: Concepts and Realities', *Renaissance and Renewal in the Twelfth Century*, ed. R.L. Benson *et al.*, Oxford 1977.

———, 'The Financing of the Crusades in the Twelfth Century', *Outremer*.

———, 'Papal, Imperial and Monastic Propaganda in the Eleventh and Twelfth Centuries', *Prédication et Propagande au Moyen Age. Islam, Byzance, Occident*, Paris 1983 (*Penn–Paris–Dumbarton Oaks Colloquia*, 3).

———, 'Medieval Charters as a Source for the History of the Crusades', *Crusade and Settlement*.

———, *Three Studies in Medieval Religion and Social Thought*, Cambridge 1995.

———, 'Attitudes toward self-inflicted suffering in the Middle Ages', in Giles Constable, *Culture and Spirituality in Medieval Europe*, Aldershot 1996, no. 9.

———, *The Reformation of the Twelfth Century*, Cambridge 1996.

Cowdrey, H.E.J., 'Pope Urban II's Preaching of the First Crusade', *History*, 55 (1970).

———, 'Cluny and the First Crusade', *Revue Bénédictine*, 83 (1973).

———, 'Abbot Pontius of Cluny', *Studi Gregoriani*, 11 (1978).

———, 'Pope Gregory VII's "Crusading" Plans of 1074', *Outremer*.

———, 'Martyrdom and the First Crusade', *Crusade and Settlement*.

Crone, G.R., *Maps and their Makers. An Introduction to the History of Cartography*, London 1964.

Crusade and Settlement. Papers Read at the First Conference of the Society for the Study of the Crusades and the Latin East and presented to R.C. Smail, ed. P.W. Edbury, Cardiff 1985.

The Crusades and Their Sources: Essays Presented to Bernard Hamilton, ed. John France and William G. Zajac, Aldershot 1998.

Daniel, E.R., 'Apocalyptic Conversion: The Joachimite Alternative to the Crusades', *Traditio*, 25 (1969).

Dansette, B., 'Les pèlerinages occidentaux en Terre Sainte: une pratique de la dévotion moderne à la fin du Moyen Age', *Archivum Franciscanum Historicum*, 72 (1979).

Delaruelle, E., 'The Crusading Idea in Cluniac Literature of the Eleventh Century', *Cluniac Monasticism in the Central Middle Ages*, ed. N. Hunt, London 1971.

———, 'La spiritualité aux XIVe et XVe siècles', *La Piété Populaire au Moyen Age*, Turin 1975.

———, 'Deux guides de Terre Sainte aux XIVe et XVe siècles', *La Piété Populaire au Moyen Age*, Turin 1975.

Delius, W., *Geschichte der Marienverehrung*, Munich 1963.

Dichter, B., *Sigurd Jorsalafare, King of Norway in Acre*, Acre 1970.

Diers, Michaela, *Bernhard von Clairvaux. Elitäre Frömmigkeit und begnadetes Wirken*, Münster 1991 (*Beiträge zur Geschichte der Philosophie und Theologie der Mittelalters*, 34).
Dronke, P. , *Women Writers of the Middle Ages*, Cambridge 1984.
Duby, Georges, 'The Diffusion of Cultural Patterns in Feudal Society', *Past & Present*, 39 (1968).
Dunbabin, J., 'The Maccabees as Exemplars in the Tenth and Eleventh Centuries', *The Bible in the Medieval World. Essays in Memory of Beryl Smalley*, ed. K. Walsh *et al.*, Oxford 1985 (*Studies in Church History. Subsidia*, 4).
Dupront, A., 'La spiritualité des croisés et des pèlerins d'après les sources de la première croisade', *Convegni del Centro di Studi sulla Spiritualità Medievale*, 4 (1963).
Edbury, P.W. and Rowe, J.G., 'William of Tyre and the Patriarchal Election of 1180', *EHR*, 93 (1978).
Edgington, Susan, 'The First Crusade: Reviewing the Evidence', *The First Crusade.*
Erdmann, C., 'Endkaiserglaube und Kreuzzugsgedanke im 11. Jahrhundert', *Zeitschrift für Kirchengeschichte*, 51 (1932).
———, *The Origin of the Idea of Crusade*, Eng. trans. M.W. Baldwin and W. Goffart, Princeton, NJ 1977.
The First Crusade. Origins and Impact, ed. Jonathan Phillips, Manchester 1997.
Flahiff, G.B., 'Deus non vult: A Critic of the Third Crusade', *Medieval Studies*, 9 (1947).
Flori, Jean, 'Guerre sainte et rétribution spirituelle dans la 2e moitié du XIe siècle', *Revue d'Histoire Ecclésiastique*, 85 (1990).
———, 'Mort et martyre des guerriers vers 1109. L'exemple de la première croisade', *Cahiers de Civilisation Médiévale*, 34 (1991).
———, 'Une ou plusieurs "première croisade"? Le message d'Urbain II et les plus anciens pogroms d'Occident', *Revue historique*, 285 (1991).
Focillon, H., *L'An Mil*, Paris 1952.
Folda, Jaroslav, *The Art of the Crusaders in the Holy Land 1098–1187*, Cambridge 1995.
Folz, R., *L'idée d'Empire en Occident du Ve au XVe siècle*, Paris 1953.
France, John, 'An Unknown Account of the Capture of Jerusalem,' *EHR*, 87 (1972).
———, 'The Destruction of Jerusalem and the First Crusade', *JEH*, 47 (1996).
———, 'Les origines de la Première Croisade. Un nouvel examen', *Autour de la Première Croisade*.
French, Dorothea R., 'Journeys to the Center of the Earth: Medieval and Renaissance Pilgrimages to Mount Calvary', *Journeys Toward God. Pilgrimage and Crusade*, ed. Barbara N. Sargent-Baur, Kalamazoo 1992.
Friedman, Y., 'Francescinus of Pontremoli: A Pilgrim's Path to Pardon', *Franciscan Studies*, 43 (1983).
———, 'The Status of the Church of the Holy Sepulchre in the Crusader Period', *Sefer Z. Vilanai*, vol. 2, Jerusalem 1987 (Hebrew).

———, 'The City of the King of Kings: Jerusalem in the Crusader Period', *The Centrality of Jerusalem. Historical Perspectives*, ed. M. Poorthuis and C. Safrai, Kampen 1996.

Frolow, H., *La relique de la Vraie Croix*, Paris 1961.

Fuhrmann, H., 'Studien zur Geschichte mittelalterlicher Patriarchate', *Zeitschrift der Savigny Stiftung für Rechtsgeschichte, kanonistische Abteilung*, 39 (1953); 40 (1954); 41 (1955).

Gieysztor, A., 'The Genesis of the Crusades: The encyclical of Sergius IV (1009–1012)', *Medievalia et Humanistica*, 5 (1948); 6 (1950).

Grabois, A., 'Anselme, l'Ancien Testament et l'idée de Croisade', *Les Mutations Socio-Culturelles au Tournant des XIe–XIIe siècles. Études Anselmiennes (IVe Session)*, Paris 1984.

———, 'The Crusade of King Louis VII: A Reconsideration', *Crusade and Settlement.*

———, 'Medieval pilgrims, the Holy Land and its image in European Civilization', *Pillars of Smoke and Fire. The Holy Land in History and Thought*, ed. M. Sharon, Johannesburg 1986.

———, 'Louis VII pèlerin,' *Revue d'Histoire de l'Eglise de France*, 74 (1988).

Green, D.H., *The Millstätter Exodus*, Cambridge 1966.

Grenia, C., 'Bernard of Clairvaux, "In Praise of the New Knighthood". A Note and English translation', *Bernard of Clairvaux: Treatises III* in *Cistercian Fathers Series*, 19 (1977).

Grill, L., 'Die Kreuzzugs-Epistel St Bernards: *Ad peregrinantes Jerusalem*', *Studien u. Mitteil. zur Geschichte des Benediktineer Ordens*, 67 (1956).

Hagenmeyer, H., 'Deux poésies latines relatives à la IIe croisade', *AOL*, 1.

Hahn, Cynthia, 'Loca Sancta Souvenirs: Sealing the Pilgrim's Experience', *The Blessings.*

Halphen, L., *Le Comté d'Anjou au XIe siècle*, Paris 1906.

Hamilton, B., 'Rebuilding Zion: The Holy Places of Jerusalem in the Twelfth Century', *Studies in Church History*, 14 (1977).

———, 'Reynald of Châtillon: The Elephant of Christ', *Studies in Church History*, 15 (1978).

———, *The Latin Church in the Crusader States. The Secular Church*, London 1980.

———, 'The Impact of Crusader Jerusalem on Western Christendom', *The Catholic Historical Review*, 80 (1994).

———, 'Ideals of Holiness: Crusaders, Contemplatives, and Mendicants', *The International History Review*, 17 (1995).

——— and McNulty, P.A., 'Orientale Lumen et Magistra Latinitas: Greek Influences on Western Monasticism (900–1100)', *Le Millénaire du Mont Athos, 963–1963. Études et Mélanges*, vol. 1, Chevetogne 1963; repr. Bernard Hamilton, *Monastic Reform, Catharism and the Crusades (900–1300)*, London 1979, no. 5.

Heers, J., 'Bourgs et faubourgs en Occident: Les pèlerinages et dévotions au Saint-Sépulchre', *Jérusalem, Rome, Constantinople. L'image et le mythe de la ville au Moyen Age*, ed. D. Poirion *(Cultures et Civilisations Médiévales, 5)*.

Heifetz, M., 'The Political Significance of a Ceremonial Occasion *Laetare Jerusalem*', *Criticism and Interpretation. Journal for Literature, Linguistics, History and Aesthetics*, 21 (1986) (Hebrew).

Hen, Y., 'Labels of the Holy Land Eulogiae in Frankish Gaul', *Cathedra*, 82 (1996) (Hebrew).

Heyman, Avital., 'The Representation of the Holy Sepulchre in Aubergnat Romanesque Sculpture: A Reflection of Crusader Patrons?', *Autour de la Première Croisade.*

Hiestand, R., 'Un centre intellectuel en Syrie du Nord? Notes sur la personnalité d'Aimery d'Antioche, Albert de Tarse et Rorgo Fretellus', *Le Moyen Age*, 100 (1994).

———, Mayer, H.E., 'Die Nachfolge des Patriarchen Monachus von Jerusalem', *Basler Zeitschrift für Geschichte und Altertumskunde*, 74 (1974).

Hill, Joyce, 'From Rome to Jerusalem: An Icelandic Itinerary of the Mid-Twelfth Century', *Harvard Theological Review*, 76 (1983).

Hoffman, H., *Gottesfriede und Treuga Dei*, Stuttgart 1964.

Holum, Kenneth, G., 'Hadrian and St. Helena: Imperial Travels and the Origins of Christian Holy Land Pilgrimage', *The Blessings*.

Hormayer, J. von, *Die Bayern im Morgenlande*, Munich 1832.

The Horns of Hattin. Proceedings of the Second Conference of the Society of the Crusades and the Latin East, ed. B.Z. Kedar, London and Jerusalem 1992.

Housley, Norman, *The Italian Crusades*, Oxford 1982.

———, 'Crusades Against Christians: Their origins and early development, c.1000–1216', *Crusade and Settlement*.

———, 'Jerusalem and the Development of the Crusade Idea, 1099–1128', *HH*.

Hunt, E.D., *Holy Land Pilgrimage in the Later Roman Empire, AD 312–460*, Oxford 1982.

Jacqueline, B., *Papauté et Episcopat selon Saint Bernard de Clairvaux*, Paris 1963.

Jeffery, G., *The Holy Sepulchre*, Cambridge 1919.

Jerusalem Pilgrimage, see J. Wilkinson and W. F. Ryan.

Jestice, Phyllis, G., *Wayward Monks and the Religious Revolution of the Eleventh Century*, Leiden 1997.

Joranson, E., 'The Great German Pilgrimage of 1064–1065', *The Crusades and Other Historical Essays Presented to Dana Munro*, ed. L.J. Paetow, New York 1928 (repr. New York 1968).

———, 'The Palestine Pilgrimage of Henry the Lion', *Medieval and Historiographical Essays in Honor of James Westfall Thompson*, ed. J.E. Cate *et al.*, Port Washington, N.Y. 1938 (repr. 1966).

Jotischky, A., *The Perfection of Solitude. Hermits and Monks in the Crusader States*, Philadelphia 1995.

———, 'St Gerard of Csanád and the Carmelites: Apocryphal Sidelights on the First Crusade', *Autour de la Première Croisade*.

Journeys Toward God. Pilgrimage and Crusade, ed. Barbara N. Sargent-Baur, Kalamazoo 1992.

Katzir, Y., *Papal Attitude to Jerusalem and the Crusades in the 12th Century*, unpublished Ph.D. thesis, University of Jerusalem 1982 (Hebrew).

———, 'The Patriarch of Jerusalem, Primate of the Latin Kingdom', *Crusade and Settlement*.

———, 'The Conquests of Jerusalem, 1099 and 1187. Historical Memory and Religious Topology'. *The Meeting of Two Worlds. Cultural Exchange between East and West during the Period of the Crusades*, ed. V.P. Goss *et al.*, Kalamazoo 1986.

———, 'The Vicissitudes of the True Cross of the Crusaders', *The Crusaders in their Kingdom, 1099–1291*, ed. B.Z. Kedar, Jerusalem 1987 (Hebrew).

Kedar, B.Z., 'Ein Hilferuf aus Jerusalem vom September 1187', *DA*, 38 (1982).

———, 'Gerard of Nazareth. A Neglected Twelfth-Century Writer in the Latin East', *Dumbarton Oaks Papers*, 37 (1983).

———, *Crusade and Mission. European approaches toward the Muslims*, Princeton 1984.

———, 'The Patriarch Eraclius', *Outremer*.

———, 'The Battle of Hattin revisited', *HH*.

———, 'Intellectual Activities in a Holy City: Jerusalem in the Twelfth Century', *Sacred Space. Shrine, City, Land*, ed. Benjamin Kedar and Zvi Werblowsky, New York 1998.

———, 'A Twelfth-Century Description of the Jerusalem Hospital', *The Military Orders, vol. 2: Welfare and Warfare*, ed. Helen Nicholson, Aldershot 1998.

———, 'The *Tractatus de locis et statu sancte terre ierosolimitane*', *The Crusades and their Sources*.

——— and Westergard-Nielsen, C., 'Icelanders in the Crusader Kingdom of Jerusalem: A Twelfth-Century Account', *Mediaeval Scandinavia*, 11 (1978–1979).

Keenan, J.M., 'The Cistercian Pilgrimage to Jerusalem in Guillaume de Deguiville's *Pèlerinage de la Vie Humaine*', in *Studies in Medieval Cistercian History*, vol. 2, ed. J.R. Sommerfeldt, Kalamazoo 1976 (*Cistercian Studies Series*, 24).

Kenaan-Kedar, N., 'Local Christian Art in Twelfth-Century Jerusalem', *Israel Exploration Journal*, 23 (1973).

———, 'Symbolic Meaning in Crusader Architecture. The Twelfth-Century Dome of the Holy Sepulcher Church in Jerusalem', *Cahiers Archéologiques*, 34 (1986).

Khitrowo, B. de, 'Pèlerinage en Palestine de l'Abbesse Euphrosine Princesse de Polotsk', *ROL*, 3 (1895).

Kingsford, C., 'Some Political Poems of the Twelfth Century', *EHR*, 18 (1890).

Kohler, C., 'Un rituel et un bréviaire du Saint Sépulchre de Jérusalem (XIIe–XIIIe siècles)', *ROL*, 8 (1900–1901).

———, 'Un sermon commémoratif de la prise de Jérusalem par les croisés, attribué à Foucher de Chartres', *ROL*, 8 (1900–1901).

Konrad, R., 'Das himmlische und das irdische Jerusalem im mittelalterlichen Denken', *Speculum historiale. Festschrift J. Spörl*, ed. C. Bauer *et al.*, Munich 1965.

Kötting, B., *Peregrinatio religiosa. Die Wallfahrten in der Antike und das Pilgerwesen in der alten Kirche*, Münster 1950.

Krautheimer, R., *Studies in Early Christian, Medieval and Renaissance Art*, New York 1969.

Krinsky, C.H., 'Representations of the Temple of Jerusalem before 1500', *Journal of the Warburg and Courtauld Institutes*, 33 (1970).

Kühnel, Bianca, *From the Earthly to the Heavenly Jerusalem. Representations of the Holy City in Christian Art of the First Millennium*, Rome–Freiburg–Vienna 1987 (*Römische Quartalschrift für Christliche Altertumskunde und Kirchengeschichte*, 42. *Supplementheft*).

———, *Crusader Art of the Twelfth Century. A Geographical, an Historical or an Art Historian Notion*, Berlin 1994.

———, 'Geography and Geometry of Jerusalem', *City of the Great King. Jerusalem from David to the Present*, ed. Nitza Rosovsky, Cambridge, Mass. 1996.

Labande, E.-R., 'Recherches sur les pèlerins dans l'Europe des XIe et XIIe siècles', *Cahiers de civilisation médiévale*, 1 (1958). repr. in *idem*, *Spiritualité et vie littéraire de l'Occident, Xe–XIVe s.*, London 1974.

———, 'Eléments d'une enquête sur les conditions de déplacement du pèlerin aux Xe–XIe siècle', *Pellegrinaggi e culto dei Santi in Europa fino alla la Crociata. Convegni del Centro di Studi sulla Spiritualità Medievale, 8–11 Ottobre 1961*, Todi 1963. repr. in *idem*, *Spiritualité et vie littéraire de l'Occident, Xe–XIVe s.*, London 1974.

———, '"Ad limina": Le pèlerin médiéval au terme de sa démarche', *Mélanges offerts à René Crozet*, 2 vols, ed. P. Gallais *et al.*, vol. 1, Poitiers 1966. repr. in *idem*, *Spiritualité et vie littéraire de l'Occident, Xe–XIVe s.*, London 1974.

Lacombe, G., *La vie et les oeuvres de Prévostin*, Le Sauchoir 1927.

Ladner, G, '*Homo Viator*: Medieval Ideas on Alienation and Order', *Speculum*, 42 (1967).

Lalanne, L., 'Des pèlerinages en Terre Sainte avant les croisades', *BEC*, 7 (1845/6).

Lamirande, E., 'Jérusalem céleste', *Dictionnaire de Spiritualité*, vol. 7.

Langosch, K., *Geistliche Spiele*, Basel–Stuttgart 1957.

Le Bras, G., 'Institutions ecclésiastiques de la chrétienté médiévale', *Histoire de l'Eglise*, ed. A. Fliche *et al.*, vol. 12, Paris 1959.

Leclercq, J., 'Une élévation sur les gloires de Jérusalem', *Mélanges J. Lebreton*, vol. 2, Paris 1952 (*Recherches de science religieuse*, 40).

———, 'La lettre de Gilbert Crispin sur la vie monastique', *Studia Anselmiana*, 31 (1953).

———, 'Gratien, Pierre de Troyes et la Seconde Croisade', *Studia Gratiana*, 2 (1954).

———, 'De Saint Grégoire à Saint Bernard', *Histoire de la Spiritualité chrétienne*, ed. L. Boyer *et al.*, vol. 2, Paris 1961.

———, Monachisme et pérégrinations du IXe et XIIe siècles, *Studia Monastica*, 3 (1961).

———, *Aux Sources de la Spiritualité Occidentale. Étapes et Constantes*, Paris 1964.

———, 'L'encyclique de Saint Bernard en faveur de la croisade', *Revue Bénédictine*, 81 (1971).

Le Goff, J., 'Babylone ou Jérusalem: La ville dans l'imaginaire collectif au Moyen Âge', *Critique*, 34 (1978).

Lerner, R.E., *The Powers of Prophecy. The Cedar of Lebanon Vision from the Mongol Onslaught to the Dawn of the Enlightenment*, Berkeley 1983.

Levy-Rubin, Milka, 'Medieval Maps of Jerusalem', *The History of Jerusalem. Crusaders and Ayyubids (1099–1250)*, ed. Joshua Prawer and Haggi Ben-Shammai, Jerusalem 1991 (Hebrew).

——— and Rubin, Rehav, 'The Image of the Holy City in Maps and Mapping', *City of the Great King. Jerusalem from David to the Present*, ed. Nitza Rosovsky, Cambridge, Mass. 1996.

Ligato, Guiseppe, 'The Political Meanings of the Relic of the Holy Cross among the Crusaders and in the Latin Kingdom of Jerusalem: An example of 1185', *Autour de la Première Croisade*.

Limor, O., 'King David's Tomb on Mount Zion: The Origins of a Tradition', *Jews, Samaritans and Christians in Byzantine Palestine*, ed. D. Jacoby and Y. Tsafrir, Jerusalem 1988 (Hebrew).

Linder, A., 'Ecclesia and Synagoga in the Medieval Myth of Constantine the Great', *BBPH*, 54 (1976).

———, 'Jerusalem as a Focal Point in the Conflict between Judaism and Christianity', *Jerusalem in the Middle Ages. Selected Papers*, ed. B.Z. Kedar *et al.*, Jerusalem 1979 (Hebrew).

———, 'Jerusalem between Judaism and Christianity in the Byzantine Period', *Cathedra*, 11 (1979) (Hebrew).

———, 'The Liturgy of the Liberation of Jerusalem', *Medieval Studies*, 52 (1990).

———, '*Deus Venerunt Gentes*: Psalm 78 (79) in the Liturgical Commemoration of the Destruction of Latin Jerusalem', *Medieval Studies in Honour of Avrom Saltman*, ed. Bat-Sheva Albert *et al.*, Ramat-Gan 1995 (*Bar-Ilan Studies in History*, 4).

Lindner, Molly, 'Topography and Iconography in Twelfth-Century Jerusalem', *HH*.

Lourie, E., 'The Will of Alfonso I, "El Batallador", King of Aragon and Navarre: A Reassessment', *Speculum*, 50 (1975).

Lubac, H. de, *Exégèse Médiévale. Les quatre sens de l'Ecriture*, Paris 1959.

Luscombe, D., 'Peter Comestor', *The Bible in the Medieval World. Essays in Memory of Beryl Smalley*, ed. K. Walsh *et al.*, Oxford 1985 (*Studies in Church History. Subsidia*, 4).

MacCormack, Sabine, '*Loca Sancta*: The Organization of Sacred Topography in Late Antiquity', *The Blessings*.

MacGregor, A.J., *Fire and Light in the Western Church*, Collegeville 1992.

Mähl, S., 'Jerusalem im mittelalterlicher Sicht', *Die Welt als Geschichte*, 22 (1962).
Mayer, H.E., 'Das Pontificale von Tyrus und die Krönung der Lateinischen Könige von Jerusalem', *Dumbarton Oaks Papers*, 21 (1967).
———, *The Crusades*, Eng. trans. J. Gillingham, London 1972 (2nd edn London 1987).
———, *Bistümer, Klöster und Stifte im Königreich Jerusalem*, Stuttgart 1977 (*Schriften der Monumenta Germaniae Historica*, 26).
———, 'Die Stiftung Herzog Heinrichs des Löwen für das Heilige Grab', *Heinrich der Löwe*, ed. W.D. Mohrmann, Göttingen 1980.
———, 'Henry II of England and the Holy Land', *EHR*, 97 (1982).
———, *Mélanges sur l'histoire du royaume latin de Jérusalem*, Paris 1984 (*Mémoires de l'Académie des Inscriptions et Belles Lettres*, 5).
McGinn, B., 'Saint Bernard and Eschatology', *Bernard of Clairvaux. Studies Presented to Dom Jean Leclercq*, Washington D.C. 1973 (*Cistercian Studies Series*, 23).
———, '*Iter Sancti Sepulchri*: The Piety of the First Crusaders', *The Walter Prescott Webb Lectures: Essays on Medieval Civilization*, ed. R.E. Sullivan *et al.*, Arlington, Texas 1978.
———, 'Angel Pope and Papal Antichrist', *Church History*, 47 (1978).
———, *Visions of the End*, New York 1979.
Micheau, F., 'Les itinéraires maritimes et continentaux des pèlerinages vers Jérusalem', *Occident et Orient au Xe siècle. Actes du IXe Congrès de la Société des Historiens Médiévistes de l'Enseignement Supérieur Public (Dijon 2–4 juin 1978)*, Dijon 1979.
Mierow, C.C., Evans, A.P., Knapp, C., *Otto of Freising, The Two Cities*, New York 1966.
Miles, M.R., *Image as Insight. Visual Understanding in Western Christianity and Secular Culture*, Boston Mass. 1985.
Möhring, H., *Der Weltkaiser der Endzeit*, Stuttgart 2000.
Morris, Colin, *The Discovery of the Individual 1050–1200*, London 1972.
———, 'Medieval Media. Mass Communication in the Making of Europe,' *An Inaugural Lecture. University of Southampton*, Southampton 1972.
———, 'Propaganda for War: The Dissemination of the Crusading Ideal in the Twelfth Century', *Studies in Church History*, 20 (1983).
———, *The Papal Monarchy. The Western Church from 1050 to 1250*, Oxford 1989.
———, 'Bringing the Holy Sepulchre to the West: S. Stefano, Bologna, from the fifth to the Twentieth Century', *Studies in Church History*, 33 (1997).
———, 'Peter the Hermit and the Chronicles', *The First Crusade*.
———, 'Picturing the Crusades: The Uses of Visual Propaganda, c. 1095–1250', *The Crusades and their Sources*.
——— and Blake, O.E., 'A Hermit Goes to War: Peter and the Origins of the First Crusade', *Studies in Church History*, 22 (1985).
Morse, Ruth, *Truth and Convention in the Middle Ages. Rhetoric, Representation and Reality*, Cambridge 1991.

Müller, U., *Kreuzzugsdichtung*, Tübingen 1979.

Munz, P. , *Frederick Barbarossa*, London 1969.

Murray, Alan, V., '*Mighty Against the Enemies of Christ*: The Relic of the True Cross in the Armies of the Kingdom of Jerusalem', *The Crusades and their Sources.*

Nibley, H., 'Christian Envy of the Temple', *Jewish Quarterly Review*, 50 (1959/1960).

Nineham, R., 'Anonymous of York, the So-Called', *Journal of Ecclesiastical History*, 14 (1963).

Ousterhout, Robert, 'The Church of Santo Stefano: A "Jerusalem" in Bologna', *Gesta*, 20 (1981).

———, 'Rebuilding the Temple: Constantine Monomachus and the Holy Sepulchre', *Journal of Architectural Historians*, 48 (1989).

———, 'Loca Sancta and the Architectural Response to Pilgrimage', *The Blessings.*

Outremer. Studies in the History of the Crusading Kingdom of Jerusalem. Presented to Joshua Prawer, ed. B.Z. Kedar *et al.*, Jerusalem 1982.

Patrides, C.A. and Wittreich, J. (eds), *The Apocalypse in English Renaissance Thought and Literature. Patterns, Antecedents and Repercussions*, Ithaca N.Y. 1984.

Pazzelli, Raffaele, *St Francis and the Third Order. The Franciscans and the Pre-Franciscan Movement*, Chicago 1982.

Pegg, Marc Gregory, 'Le Corps et l'Autorité: La Lèpre de Baudouin IV', *Annales ESC*, 45 (1990).

Pennington, K., 'The Rite for Taking the Cross in the Twelfth Century', *Traditio*, 30 (1974).

Peters, F.E., *Jerusalem. The Holy City in the Eyes of Chroniclers, Visitors and Prophets from the Days of Abraham to the Beginning of Modern Times*, Princeton 1985.

Pettorelli, J.P., 'Le thème de Sion, expression de la théologie de la rédemption dans l'oeuvre de Saint Hilaire de Poitiers', *Hilaire et son temps. Actes du Colloque de Poitiers. 29 Septembre–30 Octobre 1968*, Paris 1969.

Phillips, Jonathan, *Defenders of the Holy Land. Relations between the Latin East and the West, 1119–1187*, Oxford 1996.

Pixner, B., 'Where was the original via Dolorosa?', *Christian News from Israel*, 27 (1979).

Prawer, J., *Histoire du Royaume Latin de Jérusalem*, 2 vols, Paris 1969–1970.

———, *The Latin Kingdom of Jerusalem: European Colonialism in the Middle Ages*, London 1972.

———, 'Jerusalem in the Christian and Jewish Perspectives of the Early Middle Ages', *Settimane di studio del centro Italiano di studi sull'alto medioevo. XXVI. Gli Ebrei nell'Alto Medioevo*, Spoleto 1980.

———, 'The Roots of Medieval Colonialism', *The Meeting of Two Worlds. Cultural Exchange between East and West during the Period of the Crusades*, ed. V.P. Goss *et al.*, Kalamazoo 1986.

Prinz, O., 'Eine frühe abendländische Aktualisierung der lateinischen Übersetzung des Pseudo-Methodios', *DA*, 41 (1985).
Pryor, John, H., *Geography, Technology and War. Studies in the Maritime History of the Mediterranean, 649–1571*, Cambridge 1988.
———, 'The *Eracles* and William of Tyre: An Interim Report', *HH*.
Raedts, Peter, 'Jerusalem: Purpose of History or Gateway to Heaven? Apocalypticism in the First Crusade', *Church, Change and Revolution*, ed. J. van den Berg *et al.*, Leiden 1991.
———, 'St Bernard of Clairvaux and Jerusalem', *Prophecy and Eschatology*, ed. M. Wilks, Oxford 1994 (*Studies in Church History. Subsidia*, 10).
Rassow, P., 'Der Text der Kreuzzugsbulle Eugens III.', *Neues Archiv*, 45 (1924).
Reeves, M., *Joachim of Fiore and the Prophetic Future*, London 1974.
Renna, Thomas, 'The Idea of Jerusalem: Monastic to Scholastic', *From Cloister to Classroom. Monastic and Scholastic Approaches to Truth*, ed. E. Rozanne Elder, Kalamazoo 1989 (*Cistercian Studies Series*, 90).
———, 'Bernard of Clairvaux and the Temple of Solomon', *Law, Custom and the Social Fabric in Medieval Europe. Essays in Honor of Bryce Lyon*, eds Bernard S. Bachrach and David Nicholas, Kalamazoo 1990 (*Studies in Medieval Culture*, 28).
Richard, Jean, 'An Account of the Battle of Hattin referring to the Frankish Mercenaries in Oriental Moslem States', *Speculum*, 27 (1957).
Riley-Smith, J., *What were the Crusades?*, London 1977.
———, 'Peace Never Established: The Case of the Kingdom of Jerusalem', *Transactions of the Royal Historical Society*, 28 (1978).
———, 'The Title of Godfrey of Bouillon', *Bulletin of the Institute of Historical Research*, 52 (1979).
———, 'Crusading as an Act of Love', *History*, 65 (1980).
———, 'An Approach to Crusading Ethics', *Reading Medieval Studies*, 6 (1980).
———, 'Death on the First Crusade', *The End of Strife*, ed. D. Loades, Edinburgh 1984.
———, 'The First Crusade and St Peter', *Outremer*.
———, *The First Crusade and the Idea of Crusading*, London 1986.
———, *The First Crusaders 1095–1131*, Cambridge 1997.
——— and Riley-Smith, L., *The Crusades: Idea and Reality, 1095–1274*, London 1981.
Robinson, I.S., *Authority and Resistance in the Investiture Contest. The Polemical Literature of the Late Eleventh Century*, Manchester 1978.
Röhricht, R., *Beiträge zur Geschichte der Kreuzzüge*, Berlin 1874.
———, 'Karten und Pläne zur Palästinakunde aus dem 7. bis 16. Jahrhundert', *ZDPV*, 15 (1892).
———, *Die Deutschen im Heiligen Lande 650–1291*, Innsbruck 1894 (repr. Aalen 1968).
Rousset, P., *Les origines et les caractéristiques de la première croisade*, Neuchâtel 1945.

Routledge, Michael, 'Songs', *The Oxford Illustrated History of the Crusades*, ed. J. Riley-Smith, Oxford 1995.

Rowe, J.G., 'Paschal II and the relations between the spiritual and temporal powers in the Kingdom of Jerusalem', *Speculum*, 32 (1957).

———, 'The Papacy and the Ecclesiastical Province of Tyre (1100–1187)', *Bulletin of the John Rylands Library*, 43 (1960/61).

———, 'The Origins of the Second Crusade: Pope Eugenius III, Bernard of Clairvaux and Louis VII of France', *The Second Crusade and the Cistercians*, ed. Michael Gervers, New York 1992.

———, 'Alexander III and the Jerusalem Crusade. An overview of problems and failures', *Crusaders and Muslims in Twelfth-Century Syria*, ed. Maya Shatzmiller, Leiden 1992.

Runciman, S., *A History of the Crusades*, 3 vols, Cambridge 1951–1954.

———, 'The Pilgrimages to Palestine before 1095', *A History of the Crusades*, ed. Kenneth M. Setton *et al.*, vol. 1, Madison 1955.

Russell, F.H., *The Just War in the Middle Ages*, Cambridge 1975.

Sackur, E., *Sibyllinische Texte und Forschungen*, Halle 1898.

Sacred Space. Shrine, City, Land, ed. Benjamin Z. Kedar and Zvi Werblowsky, New York 1998.

Schaller, Hans Martin, 'Zur Kreuzzugsenzyklika Papst Sergius' IV, *Papsttum, Kirche und Recht im Mittelalter. Festschrift für Horst Fuhrmann zum 65. Geburstag*, ed. Hubert Mordek, Tübingen 1991.

Schein, Sylvia, 'La Custodia Terrae Sanctae Franciscaine et les Juifs de Jérusalem à la fin du Moyen Age', *Revue des études juives*, 141 (1982).

———, 'Between Mount Moriah and the Holy Sepulchre. The Changing Traditions of the Temple Mount in the Central Middle Ages', *Traditio*, 60 (1985).

———, 'The Image of the Crusader Kingdom of Jerusalem in the Thirteenth Century', *Revue Belge de Philologie et d'Histoire*, 64 (1986).

———, 'Die Kreuzzüge als volkstümlich-messianische Bewegungen', *DA*, 47 (1991).

———, 'The Miracula of the Hospital of St John and the Carmelite Elianic tradition – Two Medieval Myths of Foundation?', *Cross-Cultural Convergences in the Crusader Period. Essays Presented to Aryeh Grabois on his Sixty-Fifth Birthday*, eds Michael Goodich, Sophia Menache and Sylvia Schein, New York 1995.

———, 'Jerusalem: Objectif originel de la première croisade?', *Autour de la Première Croisade*.

Schmid, K., 'Aachen und Jerusalem: Ein Beitrag zur historischen Personenforschung der Karolingerzeit', *Das Einhardkreuz. Vorträge und Studien der Münsteraner Diskussion zum arcus Einhardi*, ed. K. Hauck, Göttingen 1974.

Schmidt, K.L., 'Jerusalem als Urbild und Abbild', *Eranos*, 18 (1950).

Schöber, S., *Die altfranzösiche Kreuzzugslyrik des 12. Jahrhunderts*, Vienna 1976.

Seidel, L., 'Images of the Crusades in Western Art: Models as Metaphors', *The Meeting of Two Worlds. Cultural Exchange between East and West during the Period of the Crusades*, ed. V.P. Goss *et al.*, Kalamazoo 1986.
Shepard, Jonathan, 'Cross purposes: Alexius Comnenus and the First Crusade', *The First Crusade*.
Siberry, E., *Criticism of Crusading 1095–1274*, Oxford 1985.
Sivan, Hagith, S., 'Pilgrimage, Monasticism and the Emergence of Christian Palestine in the fourth century', *The Blessings*.
Simpson, W., 'The Middle of the World in the Holy Sepulchre', *Palestine Exploration Fund Quarterly*, 20 (1888).
Smalley, B., *The Study of the Bible in the Middle Ages*, Oxford 1952.
Smith, Jonathan, Z., *To Take Place. Toward Theory in Ritual*, Chicago and London 1987.
———, 'Constructing a Small Place', *Sacred Space. Shrine, City, Land*, ed. Benjamin Z. Kedar and Zvi Werblowsky, New York 1998.
Snoek, G.J.C., *Medieval Piety From Relics to the Eucharist. A Process of Mutual Interaction*, Leiden 1995.
Somerville, R., *The Councils of Urban II*, vol. I. *Decreta Claromontensia*, Amsterdam 1972 (*Annuarium Historiae Conciliorum*, Suppl. 1).
———, 'The Council of Clermont (1095) and the Latin Christian Society,' *Archivum Historiae Pontificiae*, 12 (1974).
———, 'The Council of Clermont and the First Crusade', *Mélanges G. Fransen*, vol. 2, Rome 1976 (= *Studia Gratiana*, 20).
Southern, R.W., *The Making of the Middle Ages*, London 1953.
———, 'Peter of Blois and the Third Crusade', *Studies in Medieval History Presented to R.H.C. Davis*, eds H. Mayr-Harting and R.I. Moore, London 1985.
Spreckelmeyer, G., *Das Kreuzzugslied des Lateinischen Mittelalters*, Munich 1974 (*Münstersche Mittelalter-Schriften*, 21).
Stookey, L.H., 'The Gothic Cathedral as the Heavenly Jerusalem: Liturgical and Theological Sources', *Gesta*, 8 (1969).
Suchier, W., 'Über der alt französische Gedicht von der Zerstörung Jerusalems', *Zeitschrift für romanische Philologie*, 24 (1900); 25 (1901).
Sumberg, L.A.M., 'The "Tafurs" and the First Crusade', *Medieval Studies*, 21 (1959).
Sumption, J., *Pilgrimage. An Image of Medieval Religion*, London 1975.
Taylor, J.E., *Christians and the Holy Places. The Myth of Jewish–Christian Origins*, Oxford 1993.
Torrell, J.P. and Bouthillier, D., *Pierre le Vénérable et sa vision du monde*, Louvain 1986.
Tritton, A.S., 'The Easter Fire at Jerusalem', *Journal of the Royal Asiatic Society*, (1963).
Ullmann, W., *The Growth of Papal Government in the Middle Ages*, London 1955.
The Use and Abuse of Eschatology in the Middle Ages, ed. Werner Verbeke *et al.*, Leuven 1988 (*Mediaevalia Lovaniensia*, Series 1, *Studia* 15).

Vancourt, R., 'Patriarchates', *DTC*, 11 (1932).
Vandenbroucke, F., 'Nouveaux milieux, Nouveaux problèmes: du XIe au XVIe siècles', *Histoire de la Spiritualité Chrétienne*, ed. L. Boyer *et al.*, vol. 2, Paris 1961.
Van der Meer, F., *Atlas de l'ordre cistercien*, Paris 1965.
Verhelst, Daniel, 'Les textes eschatologiques dans *Le Liber Floridus*', *The Use and Abuse of Eschatology in the Middle Ages*, ed. Werner Verbeke *et al.*, Leuven 1988.
Vessey, D.W.T.C., 'William of Tyre: Apology and Apocalypse', *Hommages à A. Boutemy*, ed. G. Camlier, Brussels 1976 (*Collection Latomus*, 145).
Viola, Coloman, 'Jugements de Dieu et Jugement Dernier. Saint Augustin et la scholastique naissante (fin XIe–milieu XIII siècles)', *The Use and Abuse of Eschatology in the Middle Ages*, ed. Werner Verbeke *et al.*, Leuven 1988.
Vogel, C., 'Le pèlerinage pénitentiel', *Pellegrinaggi e Culto dei Santi in Europa fino alla prima Crociata. Convegni del Centro di studi sulla spiritualità medievale*, vol. 4, Todi 1963.
———, 'Les pélerinages pénitentiels', *Revue des sciences religieuses*, 38 (1964).
Vogüé, M. de, *Les Églises de la Terre Sainte*, Paris 1860 (repr. Jerusalem 1973).
———, 'Achard d'Arrouise, Poème sur le Templum Domini', *AOL*, 1 (1881).
Waddell, H., *Medieval Latin Lyrics*, London 1929 (1952 edn).
Walker, P.W.L., *Holy City, Holy Places? Christian Attitudes to Jerusalem and the Holy Land in the Fourth Century*, Oxford 1990.
Ward, B., *Miracles and the Medieval Mind*, London 1982.
Ward, John, 'Some Principles of Rhetorical Historiography in the Twelfth Century', *Classical Rhetoric and Medieval Historiography*, ed. E. Breisach, Kalamazoo 1986.
———, 'The First Crusade as Disaster: Apocalypticism and the Genesis of the Crusading Movement', *Medieval Studies in Honour of Avrom Saltman*, Ramat Gan 1995 (*Bar Ilan Studies in History*, 4).
Warner, Marina, *Alone of All Her Sex. The Myth and the Cult of the Virgin Mary*, New York 1983.
Werblowsky, R.J.Z., 'Jerusalem: Holy City of Three Religions', *Ex Oriente Lux*, 23 (1973–1974).
Wiederhold, W., 'Papsturkunden in Florenz', *Nachrichten von der Gesellschaft der Wissenschaften zu Göttingen. Phil. Hist. Kl.*, Göttingen 1901.
Wilken, Robert, L., *The Land Called Holy. Palestine in Christian History and Thought*, New Haven and London 1992.
Wilkinson, J., 'The Tomb of Christ. An Outline of its Structural History', *Levant*, 4 (1972).
———, *Jerusalem Pilgrims before the Crusades*, Jerusalem 1977.
———, *Egeria's Travels to the Holy Land*, Jerusalem 1981.
———, 'Jewish Holy Places and the Origins of Christian Pilgrimage', *The Blessings*.
——— and Ryan, W.F., *Jerusalem Pilgrimage, 1099–1185*, London 1988.

Williams, E., 'Cîteaux et la seconde croisade', *Revue d'histoire ecclésiastique*, 49 (1954).

Williams, G.H., *The Norman Anonymous of 1100 A.D.*, Cambridge, Mass. 1951 (*Harvard Theological Studies*, 18).

Wolfram, G., 'Kreuzpredigt und Kreuzlied', *Zeitschrift für Deutsches Alterthum*, 30 (1886).

Young, K., *The Drama of the Medieval Church*, 2 vols, Oxford 1933.

Index

Aachen 63 n.3
Abbild 5, 60, 61
Abraham 3, 40, 43, 47, 125, 142
Achard of Arrouaise 100, 106
Acre 156, 164, 173
Acroceraunia 111
Adam 142
Adam of Perseigne 137
Adam of St Victor 105
Adam the monk 128
Adhemar of Chabannes
Adhémar of Le Puy 87, 119
Adso of Montier-en-Der 18, 35, 114, 147, 148, 151, 154, 156
  De Ortu et Tempore Antichristi 35
  Prophecy of the Last Emperor 35
  Tractatus de Antichristo ('On the Antichrist') 18, 147
Agareni 26
Alain of Lille 48, 181
Alaric 15
Alberic 68
Albert of Aix (Aachen) 16, 19, 27, 37, 39, 119
Albrecht von Johansdorf 179, 182
Albuin of Gonze 148
Alcuin 95, 114
Alexander III 56, 57, 59, 60, 185
  Inter Omnia 57
  Quantum praedecessores nostri 57
Alexander of Lincoln 128, 129
Alexandria 49, 50, 52, 53, 84
Alexius Comnenus 149
allegorical interpretation of scripture 145, 153
allodial property 44
Amalric of Nesle, patriarch of Jerusalem 46
Amalric, King of Jerusalem 72, 84
Amanieu of Loubens 11
Ambroise 180, 187
Amorites 47
Amos 104
anagogy
  horizontal 109
  vertical 109
Anastasius, Emperor 74 n.49
Andrew, apostle 52, 85
Andrew of Montbard 80
Andrew, Master of the Temple 130
angel(s) 129–30, 154
Angers 10
Annalista Saxo 22
Anonymous of Laon 76
Anselm of Canterbury 96, 127–8, 136
Anselm of Ribemont 120
Antichrist 17, 18, 35, 43, 107, 146, 147, 148, 149, 150, 151, 152, 153, 154, 156, 157
  Play of (*Ludus de Antichristo*) 154
  plays 154
Antioch 12, 36, 49, 50, 52, 53, 54, 65, 119
Antonia, palace 86
Anzy-le-Duc 125
Apocalypse, beast(s) of 60, 156
Apocalypses, illustrated 143
apocalyptic expectations 147, 150, 157
apocalyptic literature 107, 146–8, 152, 153
apocalypticism 149, 157
apostle(s) 50, 52, 53, 54, 55, 80, 81, 85, 88, 104, 105, 108, 132, 169, 185, 189, 190
Apulia 148
Aquila 51
Aquileia 53
Arles 51
Arnold of Lübeck 177
Arnold of Morimond 128
Arnold of Seschaves 122

Arnoul of Flanders 26
Ascalon 98
Asia Minor 12
Assisi 162
Athens 1 n.2, 111
Augustine of Hippo 4, 59, 94, 111, 112, 145, 178
Augsburg 82
Austria 124
avarice 60, 138, 172, 174

Babylon 58–61, 107, 124, 127, 148, 154, 179
Babylonians 47, 114, 192
Baldric of Dol 14, 16, 18, 22, 23, 38, 41, 70 n.37, 100, 119, 120, 132, 152
Historia Jerosolimitana 100
Baldwin of Canterbury 69, 168, 172
Baldwin I of Jerusalem 65, 84, 96, 97, 118
Baldwin II of Jerusalem 132
Balian of Ibelin 159
Baltic 116
baptism 45, 112
baptisteries/y 82, 135
Bari, Pontifical of 66
Bartholomew, apostle 85
Bartold of Nangis 41
Baruch 4
Bavaria 76, 84
Bavarians 44
Beatus Apocalypse 43
Beaulieu 64
Bec 127
Bede 95, 105
Belli Locus 64
bellum iustum 36
Benedict of Nursia 133
Benedictine(s) 14, 22, 27, 38, 88, 126, 134, 136, 177
Benzo of Alba 35, 148
Berachiah 99
Bernard of Clairvaux xvi, 39, 44, 60, 61, 67, 73, 75, 77, 84, 105–6, 121–3, 128, 132, 133, 136, 152 n.62, 153, 170, 176, 178, 183, 191
Ad peregrinates Iherusalem 77, 121, 122
De laude Novae Militiae (*In Praise of the New Knighthood*) 73, 106, 121, 153
Bernard of Cluny 136
De Contemptu mundi 136
Bernard the Monk 142
Berter of Orléans 166, 179
Bertolf of Sperberseck (Sparwarisegge) 85
Bethany 81, 102, 135
Bethel 91, 99
Bethlehem 71, 97, 135
biblical interpretation, revolution in 25–7, 32
Birgitta of Sweden viii
bishop(s) 50, 51, 147, 180
Bobbio 63
Bohemia 39, 44, 75, 77
Bologna 10, 11, 82–3
Bonaventura 157
Bonizo of Sutri 26
Bordeaux 125
Bordeaux Pilgrim 5, 91, 93
Borgo San Sepolcro 63
Bourges 51
Brevarius de Hierosolyma 142
Britain 111
Brown, Peter 12
Bruno of Cologne 128
Burgundy 180
Busdorf 64
Byzantine Empire 1
Byzantium 127

Caesarea 80, 155
Caffaro of Caschifellone 42
De Liberatione Civitatum Orientis 42
Calabria 148, 156
Calixtus II 128
Calvary 81, 82, 85, 87, 89, 93, 104, 107, 113, 141, 142, 148, 154, 164, 168, 193
Cambrai 64, 116
Cambridge 82
canonists 46, 47, 53, 137, 190
Canterbury 51
Capernaum 71
Cappadocia 111
cardinal(s) 53, 162, 177
Cariat sepher 61
Carthage 51
Carthusian(s) 133
Cassian 114

Catalonia 30
Catholic Church 49
causa iuxta 39
Celestial City (Jerusalem) 58, 111, 114, 116, 117, 122, 133, 135, 139, 177, 179, 187, 191
Celestine III 56, 89, 186
  Cum ad propulsandam 186
  Quot et quantum 89
Cencius 113
Chalcedon, Council of 49, 74 n.49
Champagne 180
Chanson d'Antioche 23, 43, 120
chansons de geste 7, 8
Charlemagne 63 n.3, 85, 149, 157
Charleville Poet 19
  Historia Vie Hierosolimitanae 19
Chartres 130
Chenu, M.D. 26
Children of Israel 15, 17, 18, 20, 37, 47, 91, 96, 115, 169, 189
Chosroes 106
Chosen People 40, 42
Christ 3, 16, 17, 18, 21, 22, 23, 36, 38, 39, 40, 43, 44, 45, 46, 50, 54, 55, 81, 85, 90, 91, 94, 97, 99, 103, 104, 105, 108, 110, 119, 120, 135, 137, 144, 146, 147, 148, 154, 155, 162, 163, 169, 170, 176, 178, 189
  Advent 96
  as feudal lord 45, 192
  as Messiah 42
  Ascension 3, 5, 32, 78, 104, 145
  body of 4, 94
  burial 5, 17, 18, 20, 38, 58, 63, 66, 70, 71
  Church of 95
  cleansing the Temple 92, 106
  Cross of 3, 22, 43, 79, 81, 120, 132, 144, 147, 162, 179
  Crown of Thorns 65
  Crucifixion 5, 23, 32, 43, 69, 72, 83, 86, 89, 91, 97, 98, 104, 113, 162, 176
  curse on Jerusalem 92, 93, 114, 138
  death 3, 5, 14, 73
  Deposition 104
  desire to suffer with 66
  family of 44
  humanity of 66, 67, 76, 90, 91
  Incarnation 32, 73, 176
  Last Supper 81, 88, 102, 104
  life 1, 2, 5, 14, 17, 18, 20, 76
  Passion 1, 2, 3, 4, 5, 14, 17, 18, 20, 23, 32, 38, 40, 56, 69, 76, 83, 85, 86, 87, 90, 162, 167, 185, 193
  portrayal on Cross 67
  prison of 85–6, 193
  prophecy of destruction of Jerusalem 43, 92, 93, 114
  Redemption of mankind 22, 57, 71, 176
  relationship with the faithful 45
  Repose of 86
  Resurrection 2, 3, 5, 23, 32, 38, 40, 56, 58, 63, 66, 70, 71, 73, 78, 82, 91, 147, 167, 169
  sites commemorating episodes in life 75
  sufferings 66–7
  Tomb of 64
  traditions of the life 80
Christendom 22, 36, 40, 45, 49, 57, 67, 91, 93, 114, 152, 155, 157, 162, 163, 171, 173, 176, 180, 186, 187, 189, 190, 192, 193
Christi hereditas 37
Christian Knight(s) 7, 69, 106, 116, 130
Christianity 1, 2, 19, 23, 25, 35, 37, 42, 46, 53, 106, 107, 109, 123, 139, 141, 142, 145, 149, 170
Christians 35, 43, 44, 47, 51, 57, 71, 91, 105, 108, 109, 123, 138, 151, 157, 162, 163, 169, 170, 172, 173, 175, 176, 182, 186, 190, 192, 193
chroniclers of crusades 13, 14, 18, 22, 37, 39, 40, 119, 124, 143, 154, 162, 166, 169, 180, 186
Church(es) 4, 21, 23, 25, 27, 32, 94, 105, 106, 107, 116, 117, 135, 138, 145, 146, 157
  dedication rites 112
  Eastern 9, 36, 176
  Holy 4, 109
  institutional structure 49–55
  prayers of 23
  primitive 51
  Roman 60

Triumphant 94
Universal 52, 151
Church Fathers 4, 21, 71, 93, 105, 109, 110, 146
Cistercians 68, 84, 128, 131, 133, 162, 185
Cîteaux, abbot of 131
Citez de Jherusalem 72
Civitatis litterarum 61
Clairvaux 129, 138, 164
Clement III 186
Clermont, Council of 9, 10, 14, 15, 16, 17, 18, 37, 41, 72, 100, 101, 119, 15
cloister 135–6
Cole, Penny 38
Colonia Aelia Capitolina 1
Conon of Béthune 45, 124, 179
Conrad III of Germany 122, 154
Conrad of Montferrat 69, 168, 172
Conrad, bishop 63
Constable, Giles 126
Constantine I 1, 4, 47, 70, 91, 93, 103, 106, 107, 110, 142, 173
Constantinople 35, 49, 52, 55, 84, 90, 118, 127
Constantius 156
coronation ceremony 101
Cosmas 68
cowardice 172
Creation 22, 32
cross-staurotheca 84
crucifixes 83
crusade / crusading 117, 183
armies of the poor 149–50
as act of God 123
as pilgrimage 118
cross 66, 69, 78
for laity 129
for remission of sins 119
histories 16, 76
idea of 7, 116, 187
ideology 69, 116
movement 128
oath 65–6
opposition to 127
participation of abbots 131
prohibition of clergy 137
crusader
charters 10, 11, 25, 78
movement 7
queen(s) ix, 88
crusader(s) 21, 22, 24, 28, 29, 38, 39, 41, 42, 47, 56, 65, 69, 72, 74, 79, 80, 84, 87, 90, 96, 98, 99, 104, 118, 119, 120, 124, 127, 132, 149, 150, 159, 174, 179, 190, 192
Crusades
European 117; Jerusalem 117, 190:
First 6, 8, 9–20, 21, 25, 26, 31, 35, 36, 37, 38, 39, 40, 43, 49, 58, 63, 65, 66, 72, 76, 77, 79, 85, 87, 90, 96, 100, 101, 115, 116, 117, 118, 119, 120, 121, 126, 127, 136, 137, 138, 139, 140, 143, 149, 152, 153, 157, 161, 173, 174, 176, 181, 185, 186, 187, 189, 192;
Second 42, 44, 45, 57, 60, 79, 121, 122, 128, 129, 153, 154, 170, 176;
Third 45, 46, 48, 77, 78, 79, 116, 122, 123, 124, 134, 137, 138, 153, 155, 156, 169, 176, 178, 179, 181, 182, 183, 185, 186, 187;
Fourth 77, 118, 174;
Fifth 46, 185
Cumaean Sibyl 148
cures 64
Curia Ihesu Christi (Court of Christ) 116, 155, 176, 180
Custodia Terrae Sanctae 86, 193
'cycle marial' 89
Cyril of Jerusalem 141

Daimbert of Jerusalem 26, 41, 150
Damascus 36
Daniel 104
Daniel of Chernigov 68, 100, 144
David, King 1, 26, 91, 94, 96–7, 103, 104, 105 n.67, 108, 115, 142, 150, 169, 179, 180, 183, 189
David I of Scotland 85
De Civitate Dei 59
Denmark 82
descriptiones Terrae Sanctae 76
devil 129, 133
Devotio moderna 193
devotion to
Calvary 83
humanity of Christ 66, 74, 83, 85, 90

the Cross 67, 69, 83
the Holy Lance 67, 83
the Holy Sepulchre 113
the Five Wounds of Christ 67, 83
Virgin Mary 88, 89
Diego Gelmirez of Santiago of Compostela 117
dignity 52, 53
Dionysius of Alexandria 146
Divine Grace 51
divine judgement 164, 172, 173, 174, 176, 192
divine power(s) 41, 64
divine presence 145
divine providence 41
divine punishment 171, 175, 183, 184, 186, 192
divine right 38
divine will 190, 192
divine wrath 170, 172, 174, 175
Doomsday 151

Ebstorf map 143
Ecclesia(e) 4, 42, 91
ecclesiastical hierarchy, grades 50
ecclesiastical offices 53
Edessa 38
Egeria 40, 91, 93
Egypt 107, 169
Eigil 63
Ekkehard of Aura 22, 24–5, 27, 37, 100, 149, 151
Hierosolymita 100
Elias 104
Elijah 154
Elizabeth of Schönau 134
Elvingus of Corbie 135 n.138
Emiraldus 125
encyclicals 76–8, 116, 121, 160
England 123, 143, 184
Enoch 154
Eretz Israel 141
eschatology 141, 145, 146, 152, 192
Eucharist 67
Eudocia, Empress 89
Eugenius III 24, 39, 44, 56, 57, 68, 77, 128
Quantum predecessores 24, 39, 57
eulogiae 16, 83 n.97, 85
Eusebius 42, 56 n.26, 70, 73, 92, 110
Vita Constantini 73, 92
Eustace of Boulogne 27
Everard of Barres 80
exegesis, scriptural 25, 27, 32

Feast of the Assumption of the Virgin 87, 88
Feast of the Exaltation of the Holy Cross 48, 101–2, 181
Feast of the Liberation (Conquest) of Jerusalem 27–31, 37, 43, 116
Feast of the Presentation of the Lord in the Temple 88, 101, 102
Feast of the Purification of the Blessed Virgin 88, 101, 102
Felix Faber 193
feudalization 191
fideles Sancti Petri 36
fief(s) 44, 45, 47, 69, 190, 192
papal 117
flagella 79
Flanders 10, 180
flogging (flagellation) 67, 68, 113
Florence 63
Fourth Lateran Council 48
France 10, 64, 80, 138, 154, 176, 182, 185
Franciscans 86, 157, 193
Franks 91, 96, 101, 139, 154, 172, 174, 175, 186, 189, 193
Frederick I Barbarossa 59, 116, 154, 161, 180
Frederico Visconti 83
Fulcher of Chartres 14, 21, 23, 29, 30, 38, 41, 97, 99
Historia Hierosolymitana 41
Fulk III 'Nerra' of Anjou 13, 64, 113
Fulk of Anjou 84, 97
Fulk of Chartres 28, 101
funerary chapels 64

Gabriel 99, 104
Galilee 71
Garnier of Rochefort 164
Gaucherius of Aureil 136
Gelduin of Le Puiset 132
Genoa 160
Genoese letter 173
Gentiles 17, 18, 35, 38, 39, 117, 151, 156, 183, 184

Geoffrey of Chalard 132
Geoffrey of Vendôme 127, 133
Gerald of Cahors 11
Gerald of La Sauve-Majeure of Corbie 125
Gerald of Wales 163, 176
Gerberga 148
Gerhoh of Reichersberg 31, 60, 71, 138
Inclamatio contra Jerusalem terrenam 138
Libellus de investigatione Antichristi 60, 138
German
knighthood 116
princes 78, 163, 176, 180
Germany 26, 100, 134, 138, 154, 176, 185
Gervin of Saint-Riquier 125
Gesta Francorum 14, 65, 100
Gethsemane 80, 81, 165
Gibelin of Jerusalem 51, 56
Gilo 23, 27
Gisors 180
Glossa Ordinaria 105
Godfrey of Bouillon 30, 65, 96, 97, 118, 150
Godric of Finchale 68
Golgotha 16, 81, 85, 86, 89, 103, 147
Goliath 26, 169
Gospels 2, 3, 52, 75, 92, 151
Grado 51
Graindor of Douai 43
Grandselve 84
Gratian 53, 127
Decretum 46, 52, 137
Greeks 36, 146, 154
Gregory I, the Great 94–5, 156
Gregory of Nyssa 56 n.26, 74, 111
Gregory VII 9, 26, 35, 36, 113
Gregory VIII 45, 57, 77, 123, 163, 164, 167, 169, 173, 176, 177, 178, 182, 186
Audita tremendi 45, 57, 123, 161, 163, 166, 167, 169, 170, 173, 175, 176, 177, 178
Nunquam melius superni 170–1, 173, 182
Guibert of Nogent 13, 14, 16, 17, 18, 24, 25, 38, 39, 70 n.37, 115, 151
Gesta Dei per Francos 13
Guido of La Grande Chartreuse 133
Gunthelm 133–4
Guy of Lusignan 155

Habakkuk 104
Hadrian, Caesar Aelius 1
Hagar 3, 21, 47
hair, as relic 82, 90
Halberstadt 135 n.138
Hamilton, Bernard 74
Hamo of Cerne 128, 131
Hanover 76
Hattin, battle of 44, 155, 159, 160, 164, 173, 176
Hatto of Trier 136
Haymo of Halbersdtadt 148 n.43
heathen 36, 57
Hebron 91
Helena 56 n.26, 73, 84, 86, 106, 107
Helius Adrianus 95
Henri Sdyck of Olmütz 75
Henry II of England 84, 160, 161, 162, 164, 180
Henry IV, Emperor 26, 35, 36, 79, 113, 148, 151, 154
Henry Jasomirgott of Austria 122
Henry of Albano 77, 78, 123, 138, 161, 162, 163, 170, 171, 176, 178, 181, 182, 183, 186, 187
De Peregrinante Civitate Dei 138, 176, 178
Henry of Champagne 79, 165
Henry of Strasbourg 77, 78, 178
Henry the Lion 76, 80, 84
Heraclius 80, 84, 102, 106, 107, 155, 159, 173
Hereford map 143
heretics 116
Herman of Metz 64
Hermenger, Hospitaller 172
hermits 69
Herod 22, 156
Herveus of the Church of the Holy Cross 113
Hezekiah 26
Hilary of Poitiers 4, 114
Hildegard of Bingen 134
Hildesheim 84
'History of the Crusade of Frederick I'

(*Historia de Expeditione Frederici I*) 124, 160
Historia peregrinorum 19
holiness, deterritorialization of concept 5
Holy City 1, 5, 7, 8, 10, 11, 14, 23, 27, 36, 38, 42, 44, 56, 58, 70, 76, 91, 103, 110, 134, 147, 155, 160, 165, 172, 174, 186, 187, 190
Holy Cross 68, 102, 172, 181, 182, 186
  Chapel of 82, 103
  Moslem capture 163
Holy Fire 40–42, 71, 82, 98
Holy Geography 66, 76, 85, 90, 91, 93, 94, 96, 98, 107, 108, 142, 190
Holy Land 15, 36, 40, 44, 45, 46, 47, 48, 76, 77, 78, 79, 82, 89, 91, 106, 116, 123, 126, 127, 128, 131, 143, 156, 169, 172, 173, 174, 175, 176, 177, 181, 183, 186, 187, 190, 192
Holy Places 15, 27, 36, 37, 39, 40, 43, 46, 47, 52, 56 n.26, 68, 76, 80, 85, 93, 107, 116, 117, 120, 133, 135, 160, 165, 165, 176, 181, 182, 190
holy rock 99–100
Holy Saturday 40, 41, 71, 179
Holy War 69
Honorius of Autun 114, 115
Horace 135
Hosea 104
hospices 63
Hospitallers 87, 165
Hubert of Salisbury 79
Hugh, abbot of monastery of St Mary in the Valley of Jehoshaphat 56
Hugh Catula of Châlons 131
Hugh of Flavigny 113
Hugh of St Fleury 22
Hugh of Fouilly 105, 135
  De Claustro Animae ('On the Cloister of the Soul') 105, 135
Hugh of St Victor 115, 153
Hugh of Cluny 79
Huguccio 46–7, 53
  Summa 46
Hungarian raids 26
hymns 112

iconography 67
ideology of poverty 150
Igor Bigod 90
Iherusalem mirabilis 118
Imitatio Christi (imitation of Christ) 7, 66, 67–70, 109
'In Festivitate Sancte Hierusalem' 30
  additional commemorative services 30
  sermons 30–31
India 47
inheritance 48
  celestial 124
  of Christ 37, 38, 39, 40, 43, 44, 47, 48, 78, 167, 176, 182, 185, 186, 190, 192
  of St Peter 36
  of the Lord 45, 124, 187
Innocent II 52, 56
Innocent III 45–6, 52, 56, 77, 117–8, 169, 185
  Quia maior 46, 185
Innominatus VIII 102
Invention of the Holy Cross 84
Investiture Contest 54, 148
Ireland 143
Irenaeus 110, 146
Isaac 104, 142
Isaiah 97, 104
Isenbold of St Paul 135 n.138
Isidore of Seville 50, 142
  Etymologiae 50, 142
Islam 19, 35, 37, 42, 106, 147, 155, 163, 170
Israel 1, 3, 40, 97, 162, 171, 192
Israelites, biblical 169, 171, 173, 175
Italy 63
Itinerarium peregrinorum 155
Itinerarium Peregrinorum et Gesta Regis Ricardi 164, 174, 175
iter Hierosolimitanum 127
Iter/via Sancti Sepulchri 65
Ivo of Chartres 25–6
ius gentium 46

Jacques of Vitry 77, 193
Jacob 99, 104
James, first bishop of Jerusalem 54, 55
James, the apostle 85
James (Jacob) the Just 50, 55, 168
Jeremiah 60, 166, 167, 169
Jericho 87, 88, 91

Jerome 1 n.2, 15, 17, 40, 55, 70, 92, 93–4, 111, 114, 141, 145, 146, 171
Jerusalem 25
al-Aqsa mosque 96, 102, 103
allegorical 25, 109, 111
Anastasis 63, 70, 82, 83, 84, 85, 93, 103, 110
as mother 136–7, 167
as sacrament 130, 133
attitude to 14, 27, 32, 58, 90, 95, 189
Biblical 61, 91
Catholicon 144
centre of the world (*medium mundi*) 2, 8, 14, 18, 41, 61, 70, 71, 103, 141–57, 192
Christian 30
Christian attitudes to 6, 139
Christian rule 18, 49, 70, 106, 189, 192
Churches
Chapel of St James 144
Chapel of St Mary 142
of Dormition 165
of St Anna 86, 88, 165
of St Mary in the Valley of Jehoshaphat 56, 87, 88, 89, 98, 132, 165
of Stephen Protomartyr 87
of the Ascension 165
of the Bath (Manger of the Lord) 81, 88
of the Holy Sepulchre 1, 10, 23, 30, 41, 51, 57, 70, 72, 84, 86, 88, 94, 97, 103, 104, 105, 108, 145, 189, 190
of the Invention of the Holy Cross 142
of the Resurrection 63, 72, 90, 110, 142, 189
St Mary Latina 89, 90
St Mary 'La Grande' 69
city of Christ 17, 20, 167, 183, 190
city of Christ's resurrection 70, 189
City of the Holy Sepulchre 63, 90, 94, 107, 185
city of (King) David 21, 91, 97, 115, 189, 190
city of the Jews 109, 114, 115
City of the Life (Humanity) of Christ 66, 73, 77, 78, 80, 97, 185, 189, 192
City of Peace 59
city of the New Testament 91, 105, 114, 167, 189, 193
city of the Old Testament 91, 105, 114, 167, 189
Civitas Crucifixi 182
Civitatis Dei 2
Civitatis Perfida 2
conquest as divine act 21–2, 29, 32, 35, 151
copies 83
crusader conquest 37, 47, 50, 54, 56, 60, 63, 65, 85, 90, 96, 98, 107, 114, 118, 138, 139, 149, 150, 151, 154, 157, 162, 169, 185, 190, 192
Damascus Gate 87
daughters of 61
Dome of the Rock 24, 28, 30, 103, 106
Earthly 1, 2, 3, 4, 5, 8, 26, 27, 30, 32, 59, 60, 83, 106, 109–40, 144, 150, 152, 153, 155, 156, 157, 178, 179, 187, 190, 191, 192
exegesis of term 114–5
fall of 15, 57, 58, 112, 124, 138, 145, 155, 156, 157, 159–87, 192
as God's will 176, 185
explanations for 170–7, 186
flagellation pillar 67, 85, 86
Future 2, 4
Gate of David (Jaffa Gate) 100, 103
gateway to salvation/ the Heavenly Kingdom 123, 124, 130, 135, 139, 157, 178, 186, 187, 190, 192
Golden Gate 88, 102, 103
Heavenly 1, 2, 3, 4, 5, 8, 27, 30, 59, 60, 63 n.3, 64, 83, 92, 94, 95, 105, 109–40, 143, 144, 150, 152, 156, 157, 178, 179, 187, 192
historical 25
Holy Sepulchre 5, 13, 16, 19, 20, 24, 28, 29, 30, 32, 35, 36, 38, 39, 41, 43, 52, 56–8, 63–90, 93, 96, 97, 98, 99, 100, 101, 102, 103, 117, 125, 127, 132, 133, 142, 144, 147, 148, 149, 156, 157, 160, 162, 165, 168, 174, 179, 181, 182, 186, 189, 193
aedicula 63, 104
copies/imitations of 63–4, 82, 114
devotion to 63, 82

inscriptions 72
liberation 9, 10, 11, 12, 16, 18, 20, 65, 117, 138, 149
models of
omphalos 103
purification 16, 18, 20
rotunda 63, 70, 82, 83, 103, 104
sanctity 71
significance 66
Hospital of St John 68
House of Pilate 86
Jewish (biblical) past 42, 95, 96, 107, 189
justifications of conquest 8
Kingdom of 6, 39, 41, 44, 45, 52, 56, 65, 70, 97, 130, 139
King(s) 51, 57, 65, 84, 88, 96, 97, 101, 189
Latin occupation 7
liberation 18, 19, 20, 21, 22, 23, 24, 29, 42, 58, 138, 177
Lithostrotos, chapel of 67
liturgy 29–31, 33
loss of 8, 152, 175, 176
Mary Magdalen, convent of 69
Moslem conquest 94, 111, 139, 175, 192
Moslem presence 15, 38, 132
Moslem rule 16, 37, 43, 46, 50, 84, 157, 175
New 59, 60, 92, 109, 110, 112, 142, 156
Palace of Solomon 96, 102
papal concept 76
patriarch(s) of 50, 52, 68, 71, 84, 88, 89, 98, 101, 102, 131, 142, 164, 173, 189
patriarchate 49, 52–3
personal possession of Christ 37, 38, 44
place of salvation 57
place of the Last Days 152
place where Christ's feet had trodden 56, 58, 66, 73, 74, 77, 78, 131, 167, 174, 184, 185, 190
praetorium 86, 87
prevalent image 11
rebuilding by crusaders 72, 74, 76
reconquest 36, 130, 157
recovery of 45, 156, 157, 161, 175, 179, 180, 181, 183, 187, 192
relationship between Earthly and Heavenly 27, 32, 178, 190
role as capital of Christian Crusading Kingdom 7
role in the Last Days 141, 145, 185, 192
role in salvation 124, 139, 140, 167, 19
sanctity 12, 14, 16, 17, 20, 47, 49, 52, 55, 56, 57, 63, 66, 71, 74, 87, 91, 96, 98, 103, 133, 138, 183, 189, 190
Sheep Pool 86
siege of 65, 87, 119, 136
Sorrowful Gate 86
spiritual and invisible 129, 156
spiritualization 5
status 47, 49, 52, 53, 139, 190
St Stephen's Gate 87, 193
symbol of the Church 5
Temple Esplanade 30, 81, 86, 101
Temple Mount 30, 68, 91, 93, 94, 96, 98, 100, 101, 102, 105, 107, 141, 154, 165, 168
Temple of the Lord 23, 27, 28, 29, 30, 38, 41, 71, 75, 80, 81, 88, 91, 96, 97, 99, 100, 101, 102, 103, 104, 105, 106, 108, 144, 165
Temple of Solomon 2, 16, 23, 24, 68, 94, 95, 96, 97, 100, 101, 102, 103, 105, 106
Templum Domini 38, 75, 81, 100, 103, 104, 160, 165, 190
Tower of David 97, 103
true 129, 134
Valley of Jehoshaphat 56, 81, 98, 102, 103, 135, 145, 152, 153, 168
Jews 1, 2, 17, 31, 35, 42, 43, 47, 59, 80, 81, 93, 107, 149, 152, 154, 156, 164, 171
Jherusalem Laetare 29, 31, 115–6, 138
Joachim of Fiore 156, 157
John III 125
John XV 147
John Beleth 53, 115
John Chrysostom 111
John Fécamp 125, 135, 136
John Kamateros 55

John Liutius 125
John of Würzburg 31, 55, 72, 74, 79, 81, 86, 104, 106, 144
John the Baptist 15, 79, 85, 104
Jordan River 85, 135
Joseph of Arimathea 22, 104
Josephus Flavius 55, 95
Joshua 97
Josiah 26
Judaea 1 n.2, 2, 4, 25, 71, 96, 97, 102, 110, 115, 142, 154
  biblical kings 96, 103
Judah 111
Judaism 91, 96, 107, 141
Judas Iscariot 138, 172, 182
Judas Maccabaeus 26, 97
Just Judges 151
'Just War' 36, 37, 46, 47, 118, 190
justifications of Christian rights to Jerusalem 35–48, 190
Justinian 106, 107

Kappenberg 82
Kempe, Margery viii
Kidron Valley 98, 145
knighthood 122, 135, 139, 178, 191
knights 131, 179, 180
Knights Templar 102

L'Estoire d'Eracles 166
Lactantius 145
Lambert of Arras 21
Lambert of Saint-Omer 154
  Liber Floridus 154
Landulf of Milan 116
Last Days 1, 2, 17, 18, 93, 141, 146, 147, 148, 149, 150, 151, 152, 157, 170, 185, 192
  Angelic Pope (*pastor angelicus*) 157
  End of the World (End of Times) 8, 17, 35, 141, 145, 146
  Last Emperor 35, 107, 146–7, 148–9, 153, 154, 156
  Last Judgement (Day of Judgement) 2, 59, 107, 135, 140, 145, 146, 147, 148, 149, 150, 152, 153, 180, 182, 192
  Second Coming 2, 145, 147, 149, 150, 151
law (s)
  canon 52, 127
  divine 46, 54, 173
  feudal 44, 45, 52
  in medieval cartography 143
  Jewish 44
  of nations 46
laymen (laity) 129, 130, 140, 178, 187, 191
Lazarus 81, 104
Leonius of Saint-Bertin 125
lepers 6
Leopold V of Austria 160, 172
Lethbald of Burgundy 5
Leucata 111
Levant 7
Libellus de expugnatione Terrae Sanctae per Saladinum 160, 165, 168, 172
Liber de penitentia et tentationibus religiosorum (Book on the penitence and temptations of monks) 134
'Life of Constantine' 102
literary genre(s) 76, 161, 189
literary activity 161
liturgy 27, 30, 67, 75, 95, 116, 177, 181, 182–5
Livonia 48
loca sancta 15
Lombardy 45
London 183
Longinus 22
Louis of Outremer 148
Louis VII of France 79, 153
Louis VIII of France 85
Lubac, Henri de 109
luxury 174
Lyons 51, 146

Maccabees 26, 97, 101, 169, 189
Madaba, map 142
Magdeburg 51
Magnus of Reichersberg 124
Mainz 78, 116, 155, 176, 180
Maistre Renaut 45, 182
Mamre 91
Manasses of Rheims 21, 116, 118–9
Manegold of Lautenbach 26
mappae mundi (maps of known world) 143, 192

Mark 52, 53, 85
Martin of Pairis 77
martyrdom 31, 32, 59, 106, 120, 121, 179, 184
martyrs 46, 120, 121, 124
Mary Magdalen 89
Mary, the wife of Cleopas 89
Mass of the Resurrection 72
massacre of non-Christians 37
Maurice of Craon 85
medius terrae 141
Meinwerk of Padeborn 64
  Vita 64
Melisande of Jerusalem 88, 105 n.67
Melsemutus 156
Memphis 107
Mesopotamia 146
Messianic age 4
messianic expectations 149
Messina 156
Methodius of Patara 146
metropolitan(s) 50, 53
Midrashic interpretation of Ezekiel 141
Milan 82
miles Christi (knights of Christ) 7, 106, 117, 135
miles dei 26
Millennium 113, 145, 146, 147
miracles 41, 42, 64, 78, 82
Mohammed 156, 162, 164, 170
Moissac 43, 116
Mongols vii, viii
monasteries 63, 140
monasticism 125
monk(s) 125, 127, 130, 131, 132 ,134, 135, 140, 191
monk-pilgrims 133
Monomachus, Constantine IX, emperor 85
Monte Cassino 22, 125
Moors 116
mosaics, frescoes and bas-reliefs 75, 103–4, 142, 144
Moses 135
Moslems 1, 10, 23, 24, 25, 27, 35, 36, 38, 42, 48, 117, 119, 147, 149, 164, 165, 169, 175, 184, 192
Mother of the believers 136
Mother of the faith 52, 193
Mother of the faithful 52, 193
Mount Calvary 72, 81
Mount Gerizim 91, 94
Mount Moriah 91, 93, 97, 98, 101, 107, 138, 142, 144, 145
Mount Nebo 91
Mount of Olives 17, 35, 68, 74, 80, 81, 93, 107, 135, 145, 148, 153, 165
Mount of Transfiguration 82
Mount Sinai 3, 91, 129
Mount Zion 67, 75, 80, 81, 86, 87, 89, 98, 102, 115, 154, 160, 165, 169, 172

Nablus 159
Nachman of Braslav viii
Narbonne 51
navel
  of Jerusalem 144
  of the world 141, 142, 143, 144
Nazareth 71, 82
Nero 156
Neuvy St Sepulchre 64
New Israel 40, 106, 171
New Israelites 97
New Testament 3, 8, 24, 26, 44, 80, 87, 90, 92, 93, 94, 96, 98, 99, 102, 103, 104, 108, 145, 190, 192
Nibley, High 98
Nicholas of Sauve Majeure 125
Nicodemus 104
Nivelo the castellan of Freteval 25, 118
Noah 115
nobility
  European 139, 140, 179, 191
  evangelization 113
Norbert of Xanten or Magdeburg 132, 152
Norman Anonymous 54
Normandy 125
Northampton 82
nuns 61, 69, 88, 134

obedience 126, 127, 140, 190
Odenheim 134
Odilo of Cluny 126
Odo of Marmoutier 133
Odo of Deuil 31
Old Testament 2, 3, 8, 24, 26, 30, 44, 93, 95, 96, 98, 99, 100, 102, 103, 104, 108, 141, 142, 145, 146, 166, 167, 169, 186, 190, 192

Ombra, monastery of the Holy Cross in 67
Order of Grandmont 126
Order of the Temple 121, 160
Orderic Vitalis 13, 23, 96, 136, 152
ordo episcoporum 50, 53
ordo peregrinationis 96, 99, 102, 105
ordo processionis 99, 105
ordo Templum Domini 96
Origin 93, 110
Orleans 113
Ornan the Jebusite 115
Osmund of Salisbury 128
Osterlar 82
Otto I of Germany 148
Otto II of Germany 147
Otto of Freising 43, 59, 152–3
  Chronica sive Historia de Duabus Civitatibus 59
  The Two Cities 44
Otto the Great 26
Outremer 37, 172, 173, 174, 175, 176

Paderborn 114
pagan Rome 59
pagans 21, 23, 39, 46, 87, 106, 117, 132, 176, 187
Palestine 12, 27, 40, 70, 74 n.49, 111, 141, 142, 146
Palm Sunday 85, 101, 102, 103
papacy 7, 8, 36, 117, 138, 156, 162, 186
  and Jerusalem 49–61, 160
Paradise 112, 119, 122, 123, 136, 180
Paris 25, 46, 61, 143, 153, 164
Paschal II 21, 51, 56, 57, 59, 117
  Dictatus Papae 51
patriarch 49, 50, 53, 193
patriarchate(s) 49–61
patrimony of Christ (*patrimonium*) 44, 167, 185, 186, 192
Paul, apostle 2, 54, 58, 59, 85, 92, 110, 136, 145, 151
  epistles 2, 3, 67, 92
Paulinus of Nola 56 n.26, 74, 94
paupers 149
Peace of God (*Pax Dei*) movement 118
peccatis exigentibus hominum 170, 171, 172, 186, 192
Pegau 124
penance 113, 121, 123, 134, 148, 159, 162, 177, 179, 183
peregrinatio religiosa 113
peregrinatores S. Sepulcri 65
peregrini milites 22
Persecutio Saladini 174
Persians 102, 192
Peter Abelard 61, 67, 143
  'Cambridge Commentary' 67
Peter Comestor 46, 153
  Historia Scholastica 153
Peter of Aulps 12
Peter of Blois 78, 79, 123–4, 160, 161, 166, 169, 176, 181, 186
  De Hierosolymitana Peregrinatione Acceleranda 78
  Passio Reginaldi (The Passion of Reynald of Châtillon) 124
Peter of Celle 135
Peter the Deacon 19
Peter the Hermit 16, 37, 39, 119
Peter the Venerable of Cluny 39, 42, 44, 71, 77, 78, 83, 128, 130, 131, 132, 133, 136, 143, 144, 178, 191
  Adversus Judeorum 42
  In Laudem Sepulchri Domini 71, 131, 144
  On Miracels (*De Miraculis*) 131
Peter Tudebode 14, 38
Petrine theory 50
Philip II Augustus 85, 153, 180, 182
Philip, clerk to Alexander of Lincoln 128
Philip of Harvengt 61
Philip, apostle 85
Phrygia 109
pictures in churches 75, 134
Pilate 22, 86
pilgrimage(s) 1, 2, 5, 6, 7, 12, 13, 56, 58, 67, 69, 73, 74, 76, 79, 80, 115, 120, 121, 124, 178, 191
  abbots 125, 131, 132
  as act of Devil 134
  clerical 137
  for European nobility 125
  German 11, 113–4, 147
  group 113
  laity 125, 127, 132
  monastic 125 126, 128, 129, 131, 133, 135

opposition to 124–40, 190, 191
penitential 13, 113
physical 139, 178
prohibition of 126, 131, 133
spiritual, inner 126, 139, 178
to Jerusalem 12, 55, 68, 79, 90, 111, 112, 113, 122, 140, 147, 148, 190
to the Holy Sepulchre 11, 63
to tombs of saints 12
vow 127
pilgrim(s) 5, 12, 14, 24, 27, 37, 55, 56, 58, 65, 67, 75, 76, 79, 80, 81, 82, 84, 85, 88, 90, 96, 102, 103, 105, 106, 113, 114, 115, 143, 144, 147, 152, 174, 190, 193
crosses 81
guide 72
Icelandic 42
-knights 22
roads 63
traffic 63
Pisa 82, 83
poem(s) 45, 61, 77, 78, 79, 122–3, 134, 161, 179–80, 189
poet(s) 45, 69, 124, 143, 162, 165, 166, 169, 178, 179–80, 182, 186
Poitiers 12
Poitou 10, 180
pollution 15, 23, 37, 38, 47, 92, 110, 163, 164, 165, 175, 181, 187, 190
Pontius of Cluny 131, 132
Poppo of Stavelot 125
Portugal 67
Prawer, Joshua vii, viii, 139
Premonstratensian Order 132, 152
Prepositinus of Cremona 164
priesthood, orders of 94
primae sedes 50
primate(s) 49, 50, 51, 52, 53
Promised Land 39, 40, 43, 45, 47, 80, 126, 133, 144, 160, 167, 169, 173, 174, 192
propaganda, crusade 6, 7, 39, 58, 77, 122, 129, 139, 161, 179, 183, 186
prophecy/prophecies 15, 17, 24–5, 26, 32, 92, 96, 104, 106, 109, 145, 151, 155, 156, 157, 166, 169, 170, 185, 192
prophets 104, 105, 108, 114, 130, 132, 138, 150, 156, 166, 190, 193
Provence 30
provincia–regnum 49, 50, 51, 52
Psalter Map 143
Pseudo-Isidore 49, 50, 51
Pseudo-Methodius 146, 154
Revelation of 146, 147, 148
Pulcheria 89
purification 16, 20, 23, 37, 38, 39, 165
purgation 39
Puy, cathedral 11

Rabanus Maurus 63, 95
Radulf Glaber 113
Raedts, Peter 27, 129
Rahewin 59
Raimond of Aguilers 16, 21, 23, 24, 28, 30, 31, 38, 40, 65, 100, 102, 150
Raimond of Antioch 164
Raimond of St Gilles 11, 150
Ralph Niger 138, 139, 174, 175, 178, 182
De re militari 139
Ralph of Caen 120
Ralph Glaber 147
Ranger of Lucca 26
Ranulf Higden 143
Raoul of Mont-Saint-Michel 125
Raynald of Châtillon 169
recluse 68
recovery 36, 37, 45
redemption, personal 67
Reform papacy 1
reformers 117
Reiner of Liège 68
relics 16, 82, 84–5, 181
from the Holy Sepulchre 64, 131
of Christ 90
of Christ's passion 67
of John the Baptist 85
of St Stephen 132
of the apostles 85
of the True Cross 84, 159
of the Virgin Mary 85, 89–90, 131
reliquaries 85, 90
Renna, Thomas 105
Rheims 51
Synod of 147
Rhineland 149
Richard I of England 156, 164, 180, 182, 183

Richard of Devizes 79
Richard of Poitiers 79
Richard of St Vanne 113, 125
Richard of St Victor 105
Ripoll 30, 77 n.64
Robert I of Flanders 117, 149
Robert I of Normandy 13
Robert II of Flanders 11, 85
Robert the Monk 14, 16, 18, 22, 23, 31, 36–7, 38, 119, 143
  Historia Iherosolimitana 143
Roger of Hoveden 156, 160, 166, 169
Roman Empire 35, 36, 46, 107
Romans 1, 43, 59, 114, 171, 173, 177, 192
  King of 147
Rome 15, 35, 49, 50, 52, 53, 54, 55, 56, 58, 59, 60, 68, 148, 162, 193
Rorgo Fretellus 80, 86, 97, 102
Rostangus of Cluny 174
Rouen 54
Rupert of Deutz 27

S. Giovani del Sepolcro, Brindisi 82
Sabas and Theodosius 74 n.49
Saewulf 80, 89 n.120, 99
Saint-Denis 85
  Crusading Window 31, 32
Saint-Florent of Saumur 10
Saint-Laud in Angers 84
Saint-Pères of Chartres 25, 118
saints 55, 87
Saladin 23, 58, 72, 79, 81, 107, 155, 156, 157, 159, 163, 165, 166, 167, 171, 172, 173, 175, 192
Salerno 51
Salimbene of Parma 157
Salonica 51
salvation 64, 67, 70, 73, 82, 112, 114, 118, 119, 121, 122, 128, 129, 130, 132, 133, 135, 139, 140, 142, 143, 176, 178, 179, 187, 190, 193
Samson of Bury St Edmunds 162
San Sepolcro de Torres el Rio, Navarre 82
Sanctuarium Dei 47
Santiago de Compostella 68, 82
Santo Stefano, Bologna 82
Saracens 23, 37, 42, 43, 46, 47, 51, 57, 100, 132, 148, 150, 157, 160, 164, 169, 173, 174, 181, 187, 193
Saul, King 97, 183
Saxony 76, 84, 124
Schein, Sylvia vii–ix
schismatics 116
Schönau 68
Scripture(s) 2, 4, 25, 26, 32, 47, 54, 55, 75, 109, 111, 136, 146, 153, 173
  Acts of the Apostles 47, 52, 69, 81, 99, 125
  Chronicles, first book of 136
  Chronicles, second book of 41
  Corinthians, first letter to 44, 67, 73, 92
  Daniel 17, 146, 150 n.56, 153
  Deuteronomy 47, 60
  Exodus 106, 69
  Ezekiel 141, 142, 157
  Galatians, letter to 3, 25, 26, 54, 109, 110, 121, 129, 136, 167
  Genesis 99, 106, 125
  Hebrews, Epistle to the 3, 125
  Isaiah 2, 3–4, 17, 21, 24, 25, 28, 29, 31, 33, 53, 54, 56, 93, 104, 115, 121, 130, 138, 145, 148, 151, 157, 166, 177
  Jeremiah 155, 170, 177
  Job 67, 73
  Joel 98, 145, 153
  John, gospel of 3, 81, 85, 92, 98, 99, 106, 135
  Kings, first book of 41, 44
  Lamentations 60, 106, 165, 166, 167, 169, 179, 192
  Leviticus 22, 47
  Luke, gospel of 3, 18, 52, 69, 75, 81, 92, 93, 95, 99, 101, 103, 104, 114, 125, 151, 179
  Maccabees 101
  Mark, gospel of 3, 52, 86, 179
  Matthew, gospel of 3, 21, 31, 43, 46, 47, 52, 69, 75, 80, 81, 86, 92, 93, 95, 99, 101, 103, 104, 106, 114, 121, 125, 138, 163, 179
  Numbers 173
  Peter, First Epistle 58
  Peter, Second Epistle 28, 53
  Proverbs 53
  Psalms 3, 15, 21, 25, 28, 29, 30, 31, 33, 38, 44, 45, 46, 54, 56, 58, 60, 66, 71, 72, 74, 77, 78, 98, 106, 115,

124, 132, 138, 141, 142, 155, 162, 163, 166, 167, 170, 172, 175, 183–4, 192
Revelation, book of 2, 3, 4, 24, 28, 30, 54, 58, 59, 92, 95, 109, 112, 134, 135, 145, 146
Samuel, second book of 100
Thessalonians, second letter to 145, 148 n.43, 151
Tobit 2, 4
Zechariah 4, 24, 29, 44, 56, 58, 145, 167
seal of the kings of Jerusalem 97
Second Temple period 145
self-sanctification 116
Seljuks 9
Sergius IV 10, 43, 116
Seven Deadly Sins 155
shekhina 145
Shem 115
shrine(s) 75, 99, 102, 103, 108, 190
Byzantine 74
Christian 75
crusader 74
Marian 89
miracle 82
pilgrims' 74
St James of Compostella 82
St Peter's 82
Sibylla, Queen of Jerusalem 159
Sicard of Cremona 53
Mitrale seu De Officiis Ecclesiasticis Summa 53
Sicily 111, 117, 185
Sigebert of Gembloux 59, 152
signs 151–2
Sigur of Norway 84
Siloam 102
Simeon 99, 101, 104
Simeon the Just 81, 88
simoniac 117
Slavs 116, 117
Smail, Otto vii
Societas Christiana 7
Solomon 1, 26, 41, 93, 97, 99, 104, 142
song(s)
crusade 77
marching 69, 118
popular 77
popular German 65
Sophronius of Jerusalem 70 n.35, 94, 142
souvenirs 85
Spain 116, 117
Spanish *Reconquista* 117
spiritual worship 110
spirituality
Benedictine 14
Christian 1, 6, 14, 141, 192
crusading 124
European 76
lay 83
monastic 83
of late eleventh and twelfth century 66
of twelfth century 91
Western 189
St Anne ?(Anna) 88, 89
St Choloman 114
St David's 163
St Gerard of Csanud 125
St Gertrude of Liège 68
St Hildegund 68
St Hubert in Ardenne 64
St James 50, 52, 54, 56, 189
St Jean at Le Liget 82
St Joachim 88
St John 50, 52
St John the Evangelist 88, 104
St John of Parma 125
St Joseph, sepulchre of, in Nazareth 82
St Maixent 10
St Martin 135
St Martin of Tours 12
St Mary's in Nazareth 80, 97
St Maurice at Constance 64
St Meginwerd 114
St Michael at Fulda 63, 64
St Paul's 183
St Pelagia 68
St Peter 36, 49, 50, 52, 53, 54, 55, 58, 59, 60, 85, 104, 135, 149
St Peter of Moissac 10
St Simeon 85
St Simeon of Trier 125
St Stephen, proto-martyr 15, 54
St Teotenicus 80
stability 126, 128, 137, 140, 190
Stephen of Blois 84
Stephen of Muret 126
Stephen of Neublens 79

Stephen of Tournai 53, 80
Stephen of Valence 87
stylite 68
Suger of St Denis 31, 77, 78, 79, 132
Sulpicius Severus 74
Sultan, Egyptian (Fatimid) 16
Sybilline prophecies 153
Synagoga 42, 91
Syria 47, 174

Tafurs 150
Tancred 23
Templars 68, 73, 80, 86, 87, 102, 106, 121, 128
Temple, Jewish 1, 4, 14, 70, 91–108, 115, 141, 142, 145, 146, 154, 172, 183, 184
  as a metaphor 94, 105
  as a symbol 92, 94, 105, 106
  as a *typus* 94, 95, 105, 107
  spiritualization of term 92
Teotonicus 67–8
terra sancta 15
Terricus, Templar 160, 172, 181
Tertullian 110, 145
Theobald of St Colomba 131
Theodorich 55, 67, 76, 79, 81, 86, 107, 144, 152
  Libellus de Locis Sanctis 80
theologians 137, 142, 143
theophany of resurrection 70
Thierry of St Evroul 125
Tiber 111
Titus 23, 42, 43, 47, 167, 177
Toledo 51
tomb(s)
  of apostles Peter and Paul 58, 59
  of Christ 64, 82, 179
  of Isaiah 103
  of King David 102
  of King Jehoshaphat 103
  of King Solomon 102
  of St Pelagia 68
  of the kings of Judaea 102
Toulouse 11
Tours 13
Troa 111
troubadour(s) 45, 124, 178, 180
Troyes 46
True Cross 64, 83, 84
  cult of 83, 84
  fragment(s) of 84
  nail from 85
True Israel 43, 98
Turks 43, 51, 57, 160, 165
Turold of Bec 136
Tyre 49, 52, 69, 99

ubi steterunt 80
Ulm 193
Ulrich of Cluny 125
umbilicus terrae 141
Universal Chronicle 76
Urban II 9–20, 37, 38, 41, 72, 87, 100, 101, 119, 120, 126, 127, 133, 137, 151
Urban III 162, 173
Urbild 5, 139
Urbs Beata Hierusalem 30, 112

Valenciennes 143
Vallombrosa 11, 126
vassals 45, 192
veneration *see* devotion
Venjance Nostre Seigneur
Verdun 113, 125
Vespasian 23, 42, 43, 47
Vézelay 131, 136
Via Dolorosa 86, 193
Victor IV 59, 60
Victorinus of Pettau 141, 145
Vienne 51
Vigilantius 111
violence of crusaders 38
Virgil 111
Virgin Mary 7, 14, 18, 55, 71, 79, 81, 82, 85, 87, 88, 89, 99, 103, 104, 105, 108, 133, 167, 182, 189, 190
  Assumption of 89
  blood of 90
  cult of 87, 89, 90
  Dormition 88, 89, 102, 168
  hair of 90
  sepulchre 168
  traditions 88, 89, 90
virgins 61
Vision of Peace (*visio pacis*) 4, 59, 109, 112, 127, 134, 135, 139, 178

vision(s) 87, 120, 133, 157
  apocalyptic 8, 107, 143, 153, 156
Visitatio Sepulchri 72–3 n.44
visual media 75
Vita Apostolica 7
Vorbild 61
Vulgate 56

Walafrid Strabo 95, 112, 114
  '*On the Destruction of Jerusalem*' 95
Walter Map 163, 167
  'Courtiers' Trifles' (*De nugis curialum*) 167
war-pilgrimage 116
warrior, status of 7
Way of the Cross 86, 87, 193
Welf of Bavaria 122
Wends 117
Westminster 183
William Hugh 38
William I of Jerusalem 52, 84, 131
William II of Sicily 162
William IV of Montpellier 84
William of Malmesbury 19, 37, 90, 120
William of Maraval 68
William of Newburgh 47, 170, 173
William of Tyre 15, 37, 44, 72, 76, 97, 101, 136, 155, 156, 173
women, adulterous 61
Wyno of Helmershausen 64

York 51

Zechariah the priest 99
Zion 25, 44, 53, 54, 60, 98, 112, 121, 166
  Celestial 114, 148
  daughters of 61
  Earthly 3
Zwettl 162
Zwiefalten reliquaries 85